Chemical Principles
in the Laboratory

Chemical Principles in the Laboratory

Tenth Edition

Emil J. Slowinski
Macalester College
St. Paul, Minnesota

Wayne C. Wolsey
Macalester College
St. Paul, Minnesota

Robert C. Rossi
Macalester College
St. Paul, Minnesota

BROOKS/COLE
CENGAGE Learning

Australia • Brazil • Japan • Korea • Mexico • Singapore • Spain • United Kingdom • United States

BROOKS/COLE
CENGAGE Learning™

**Chemical Principles in the Laboratory,
Tenth Edition**
Emil J. Slowinski, Wayne C. Wolsey,
and Robert C. Rossi

Acquisitions Editor: Christopher Simpson

Production Manager: Matt Ballantyne

Editorial Assistant: Laura Bowen

Media Editor: Stephanie VanCamp

Technology Project Manager: Lisa Weber

Marketing Manager: Nicole Hamm

Marketing Assistant: Julie Stefani

Marketing Communications Manager: Linda Yip

Content Project Management: PreMediaGlobal

Creative Director: Rob Hugel

Art Director: John Walker

Print Buyer: Rebecca Cross

Permissions Editor: Bob Kauser

Production Service: PreMediaGlobal

Cover Designer: Riezebos Holzbaur/Brie Hattey

Cover Image Credit: Bogdan Chesaru

For product information and technology assistance, contact us at
Cengage Learning Customer & Sales Support, 1-800-354-9706

For permission to use material from this text or product,
submit all requests online at **www.cengage.com/permissions**
Further permissions questions can be emailed to
permissionrequest@cengage.com

Library of Congress Control Number: 2010942983

ISBN-13: 978-0-8400-4834-9

ISBN-10: 0-8400-4834-3

Brooks/Cole
10 Davis Drive
Belmont, CA 94002-3098
USA

Cengage Learning is a leading provider of customized learning solutions with office locations around the globe, including Singapore, the United Kingdom, Australia, Mexico, Brazil, and Japan. Locate your local office at **www.cengage.com/global.**

Cengage Learning products are represented in Canada by Nelson Education, Ltd

To learn more about Brooks/Cole, visit **www.cengage.com/brookscole**

Purchase any of our products at your local college store or at our preferred online store **www.cengagebrain.com.**

Printed in the United States of America
2 3 4 5 6 7 15 14 13 12

Preface

A Few Words to the Students

In spite of its many successful theories, chemistry remains, and probably always will remain, an experimental science. As the world of the computer has developed, there are those who feel that one can learn chemistry by working with software and observing reactions on the computer screen rather than in the laboratory. We do not agree with this approach. We believe that there is really no substitute for hands-on laboratory experience, if one is to learn what chemistry is all about.

It is not easy to do good experimental work. It requires experience, thought, and care. As beginning students, you have not had much opportunity to do experiments. We feel that the effort you put into your laboratory sessions can pay off in many ways. You can gain a better understanding of how the chemical world works, manual dexterity in manipulating apparatus, an ability to apply mathematics to chemical systems, and, perhaps most importantly, a way of thinking that allows you to better analyze many problems in and out of science. Who knows, perhaps you will find you enjoy doing chemistry and go on to a career as a chemist, as many of our students have.

In writing this manual, we attempted to illustrate many established principles of chemistry with experiments that are as interesting and challenging as possible. These principles are basic to the science, but are usually not intuitively obvious. With each experiment we introduce the theory involved, state in detail the procedures that are used, describe how to draw conclusions from your observations, and, in an Advance Study Assignment, ask you to answer questions similar to those you will encounter in the experiment. Before coming to lab, you should read over the experiment for that week, and do the Advance Study Assignment. If you prepare for lab as you should, you will get more out of it. To give an experiment a bit of a challenge, we occasionally ask you to work with chemical unknowns.

In this edition, we have included a new appendix entitled "Statistical Treatment of Laboratory Data." When professional scientists collect experimental data, especially in doing chemical analyses to determine compositions, they usually summarize their results by means of an average (the arithmetic mean) and a measure of how consistent the data are (through a statistical parameter called the standard deviation). In several experiments of this type you will be asked to calculate these quantities from your results. Your instructor may decide to make these calculations an optional part of the experiments, but these values are easily obtained on many standard calculators or via any spreadsheet (such as Excel).

A Few More Words, This Time to the Instructors

In this, the tenth edition of our manual, we bring a new, younger co-author on board: Dr. Robert Rossi, Laboratory Supervisor for the Department of Chemistry at Macalester College. Professor Emeritus Emil Slowinski, who was deeply involved in the many earlier editions of this manual—since the inception of *Chemical Principles in the Laboratory* in 1969—decided not to participate in the preparation of this edition.

We continue with the same format as that used in the ninth edition, but have made some important changes, some of which were suggested by users. We have changed Experiment 7, "Analysis of an Unknown Chloride," from the familiar Mohr method (which uses potassium chromate as the indicator) to the Fajans method, which uses dichlorofluorescein as an adsorption indicator. The new method is commonly used in the industrial sector and is the official method for chloride of the Committee on Analytical Reagents of the American Chemical Society. By making this change, the experiment can be said to be "greener" in two respects—no chromate disposal problem and a lower, less costly silver nitrate concentration of 0.02 molar. See the Instructor's Manual for some important experimental details before using this experiment.

We have introduced an optional part within Experiment 14, "Heat Effects and Calorimetry," to facilitate the experimental validation of Hess's Law. In Experiment 29, "Synthesis and Analysis of a Coordination Compound," we have improved the alternative procedure, **A.2**, for the preparation of the violet complex.

In the interest of safety, throughout the manual we have substituted heptane for hexane, as heptane offers a reduced neurotoxicity hazard. As with all chemical educators, we are concerned about the use of mercury and its compounds in the general chemistry laboratory. Bismuth has now been substituted for mercury(II) in the Paper Chromatography experiment (#2) and we hope to have finalized a mercury-free means of preserving starch solutions by the time of publication of this edition. Mercury compounds continue to be used in the qualitative analysis experiments.

As you will note in the student introductory section, we have increased the use of statistical analysis in several experiments. While your institution always has the option of not requiring the students to calculate means and standard deviations, we would encourage you to give your students this easy introduction to the important role of statistics in the real world. Chemical educators have frequently overlooked statistical analysis in the lower level courses and depended upon their analytical chemistry colleagues to provide this introduction. Certainly the analytical chemistry or quantitative analysis courses are where the more theoretical basis of statistics will continue to be housed, but there is no reason that the operational approach used in our Appendix VIII cannot be used by introductory chemistry students.

The order of experiments in this manual makes it compatible with the order of topics in the text, *Chemistry: Principles and Reactions*, Seventh Edition, by William Masterton, Cecile N. Hurley, and Edward Neth. We believe, however, that the overall set of experiments should be appropriate for most standard texts in general chemistry.

If this is the first time that you are using this manual, we have done what we could to make the transition to a new manual as easy as possible. The **Instructor's Manual,** available in hard copy or online, contains, for each experiment, a list of required equipment and chemicals, directions for preparing solutions, and suggestions for dealing with the disposal of chemical waste. We also note the time it will take to do the experiments and an approximate cost per student. In the second part of the Instructor's Manual, directed to the lab supervisor, we offer comments and suggestions for each experiment that may be helpful, sample data and calculations, and answers to the Advance Study questions.

As with any endeavor, there are many people who contribute to the effort. We gratefully acknowledge the assistance given by Amy Rice of our department, who helped develop the changes in several experiments and aided with the preparation of the manuscript for this edition. We acknowledge the assistance of Barbara Ekeberg in the preparation of the several editions of the Instructor's Manual. We also acknowledge the input offered by those responding to a survey conducted by Joshua Duncan and the editorial offices of our publisher, Brooks-Cole.

It has been a great experience being involved with this laboratory manual over its many editions. We have appreciated the support of our users. We encourage any comments, questions, or suggestions you may have. Please send them to wolsey@macalester.edu.

Finally, we acknowledge the patient support of our families in the preparation of this edition.

Emil J. Slowinski
Wayne C. Wolsey
Robert C. Rossi

Safety in the Laboratory

Read this section before performing any of the experiments in this manual

A chemistry laboratory can be, and should be, a safe place in which to work. Yet each year in academic and industrial laboratories accidents occur that in some cases injure seriously, or even kill, chemists. Most of these accidents could have been foreseen and prevented, had the chemists involved used the proper judgment and taken proper precautions.

The experiments you will be performing have been selected at least in part because they can be done safely. Instructions in the procedures should be followed carefully and in the order given. Sometimes even a change in concentration of one reagent is sufficient to change the conditions of a chemical reaction so as to make it occur in a different way, perhaps at a highly accelerated rate. So, do not deviate from the procedure given in the manual when performing experiments unless specifically told to do so by your instructor.

Eye Protection: One of the simplest, and most important, things you can do in the laboratory to avoid injury is to protect your eyes by routinely wearing safety glasses. Your instructor will tell you what eye protection to use, and you should use it. Glasses worn up on the hair may look hip, but they will not protect your eyes. If you use contact lenses, it is even more critical that you wear safety glasses as well.

Chemical Reagents: Chemicals in general are toxic materials. This means that they can act as poisons or carcinogens (causes of cancer) if they get into your digestive or respiratory system. Never taste a chemical substance, and avoid getting any chemical on your skin. If contact should occur, wash off the affected area promptly with plenty of water. Also, wash your face and hands when you are through working in the laboratory. Never pipet by mouth; when pipetting, use a rubber bulb or other device to suck up the liquid. Avoid breathing vapors given off by reagents or reactions. If directed to smell a vapor, do so cautiously. Use a fume hood when the directions call for it.

Some reagents, such as concentrated acids or bases, or bromine, are caustic, which means that they can cause chemical burns on your skin and eat through your clothing. Where such reagents are being used, we note the potential danger with a **CAUTION:** sign at that point in the procedure. Be particularly careful when carrying out that step. Always read the label on a reagent bottle before using it; there is a lot of difference between the properties of 1 M H_2SO_4 and those of concentrated (18 M) H_2SO_4.

A few of the chemicals we use are flammable. These include heptane, ethanol, and acetone. Keep your Bunsen burner well away from any open beakers containing such chemicals, and be careful not to spill them on the laboratory bench where they might be easily ignited.

When disposing of the chemical products from an experiment, use good judgment. Some dilute, nontoxic solutions can be poured down the sink, flushing with plenty of water. Insoluble or toxic materials should be put in the waste crocks provided for that purpose. Your lab instructor may give you instructions for treatment and disposal of the products from specific experiments.

Safety Equipment: In the laboratory there are various pieces of safety equipment, which may include a safety shower, an emergency eye wash, a fire extinguisher, and a fire blanket. Learn where these items are, so that you will not have to look all over if you ever need them in a hurry.

Laboratory Attire: Come to the laboratory in sensible clothing. Long, flowing robes are out, as are bare feet. Sandals and open-toed shoes offer less protection than regular shoes. Keep long hair tied back, out of the way of flames and reagents. Tight-fitting jewelry (especially rings) can exacerbate the harm caused by chemical exposure, and should not be worn in lab.

If an Accident Occurs: During the laboratory course a few accidents will probably occur. For the most part these will not be serious, and might involve a spilled reagent, a beaker of hot water that gets tipped over, a dropped test tube, or a small fire.

A common response in such a situation is panic. A student may respond to an otherwise minor accident by doing something irrational, like running from the laboratory when the remedy for the accident is close at hand. If a mishap befalls a fellow student, watch for signs of panic and tell the student what to do; if it seems necessary, help him or her do it. Call the instructor for assistance.

Most chemical spills are best handled by quickly absorbing wet material with a paper towel and then washing the area with water from the nearest sink. Use the eye wash fountain if you get something in your eye. In case of a severe chemical spill on your clothing or shoes, use the emergency shower and take off the affected clothing. In case of a fire in a beaker, on a bench, or on your clothing or that of another student, do not panic and run. Smother the fire with an extinguisher, with a blanket, or with water, as seems most appropriate at the time. If the fire is in a piece of equipment or on the lab bench and does not appear to require instant action, have your instructor put the fire out. If you cut yourself on a piece of broken glass, tell your instructor, who will assist you in treating it.

A Message to Foreign Students: Many students from foreign countries take courses in chemistry before they are completely fluent in English. If you are such a student, it may be that in some experiments you will be given directions that you do not completely understand. If that happens, do not try to do that part of the experiment by simply doing what the student next to you seems to be doing. Ask that student, or the instructor, what the confusing word or phrase means, and when you understand what you should do, go ahead. You will soon learn the language well enough, but until you feel comfortable with it, do not hesitate to ask others to help you with unfamiliar phrases and expressions.

Although we have spent considerable time here describing some of the things you should be concerned with in the laboratory from a safety point of view, this does not mean you should work in the laboratory in fear and trepidation. Chemistry is not a dangerous activity when practiced properly. Chemists as a group live longer than other professionals, in spite of their exposure to potentially dangerous chemicals. In this manual we have attempted to describe safe procedures and to employ chemicals that are safe when used properly. Many thousands of students have performed these experiments without having accidents, so you can too. However, we authors cannot be in the laboratory when you carry out the experiments to be sure that you observe the necessary precautions. You and your laboratory supervisor must, therefore, see to it that the experiments are done properly and assume responsibility for any accidents or injuries that may occur.

Contents

Experiment 1

The Densities of Liquids and Solids

Given a sample of a pure liquid, we can measure many of its characteristics. Its temperature, mass, color, and volume are among the many properties we can determine. We find that, if we measure the mass and volume of different samples of the liquid, the mass and volume of each sample are related in a simple way; if we divide the mass by the volume, the result we obtain is the same for each sample, independent of its mass. That is, for samples A, B, and C, of the liquid at constant temperature and pressure,

$$\text{Mass}_A/\text{Volume}_A = \text{Mass}_B/\text{Volume}_B = \text{Mass}_C/\text{Volume}_C = \text{a constant}$$

That constant, which is clearly independent of the size of the sample, is called its **density**, and is one of the fundamental properties of the liquid. The density of water is exactly 1.00000 g/cm^3 at 4°C, and is slightly less than one at room temperature (0.9970 g/cm^3 at 25°C). Densities of liquids and solids range from values that are less than that of water to values that are much greater. Osmium metal has a density of 22.5 g/cm^3 and is probably the densest material known at ordinary pressures.

In any density determination, two quantities must be determined—the mass and the volume of a given quantity of matter. The mass can easily be determined by weighing a sample of the substance on a balance. The quantity we usually think of as "weight" is really the mass of a substance. In the process of "weighing" we find the mass, taken from a standard set of masses, that experiences the same gravitational force as that experienced by the given quantity of matter we are weighing. The mass of a sample of liquid in a container can be found by taking the difference between the mass of the container plus the liquid and the mass of the empty container.

The volume of a liquid can easily be determined by means of a calibrated container. In the laboratory a graduated cylinder is often used for routine measurements of volume. Accurate measurement of liquid volume is made using a pycnometer, which is simply a container having a precisely definable volume. The volume of a solid can be determined by direct measurement if the solid has a regular geometric shape. Such is not usually the case, however, with ordinary solid samples. A convenient way to determine the volume of a solid is to measure accurately the volume of liquid displaced when an amount of the solid is immersed in the liquid. The volume of the solid will equal the volume of liquid which it displaces.

In this experiment we will determine the density of a liquid and a solid by the procedure we have outlined. First we weigh an empty flask and its stopper. We then fill the flask completely with water, measuring the mass of the filled stoppered flask. From the difference in these two masses we find the mass of water and then, from the known density of water, we determine the volume of the flask. We empty and dry the flask, fill it with an unknown liquid, and weigh again. From the mass of the liquid and the volume of the flask we find the density of the liquid. To determine the density of an unknown solid metal, we add the metal to the dry empty flask and weigh. This allows us to find the mass of the metal. We then fill the flask with water, leaving the metal in the flask, and weigh again. The increase in mass is that of the added water; from that increase, and the density of water, we calculate the volume of water we added. The volume of the metal must equal the volume of the flask minus the volume of water. From the mass and volume of the metal we calculate its density. The calculations involved are outlined in detail in the Advance Study Assignment.

Experimental Procedure

A. Mass of a Coin

After you have been shown how to operate the analytical balances in your laboratory, read the section on balances in Appendix IV. Take a coin and measure its mass to 0.0001 g. Record the mass on the Data page. If your balance has a TARE bar, use it to re-zero the balance. Take another coin and weigh it, recording its mass. Remove both coins, zero the balance, and weigh both coins together, recording the total mass. If you have no TARE bar on your balance, add the second coin and measure and record the mass of the two coins. Then remove both coins and find the mass of the second one by itself. When you are satisfied that your results are those you would expect, go to the stockroom and obtain a glass-stoppered flask, which will serve as a pycnometer, and samples of an unknown liquid and an unknown metal.

B. Density of a Liquid

If your flask is not clean and dry, clean it with detergent solution and water, rinse it with a few cubic centimeters of acetone, and dry it by letting it stand for a few minutes in the air or by *gently* blowing compressed air into it for a few moments.

Weigh the dry flask with its stopper on the analytical balance, or the top-loading balance if so directed, to the nearest milligram. Fill the flask with distilled water until the liquid level is nearly to the *top* of the ground surface in the neck. Put the stopper in the flask in order to drive out *all* the air and any excess water. Work the stopper gently into the flask, so that it is firmly seated in position. Wipe any water from the outside of the flask with a towel and soak up all excess water from around the top of the stopper.

Again weigh the flask, which should be completely dry on the outside and full of water, to the nearest milligram. Given the density of water at the temperature of the laboratory and the mass of water in the flask, you should be able to determine the volume of the flask very precisely. Empty the flask, dry it, and fill it with your unknown liquid. Stopper and dry the flask as you did when working with the water, and then weigh the stoppered flask full of the unknown liquid, making sure its surface is dry. This measurement, used in conjunction with those you made previously, will allow you to accurately determine the density of your unknown liquid.

C. Density of a Solid

Pour your sample of liquid from the flask into its container. Rinse the flask with a small amount of acetone and dry it thoroughly. Add small chunks of the metal sample to the flask until the flask is at least half full. Weigh the flask, with its stopper and the metal, to the nearest milligram. You should have at least 50 g of metal in the flask.

Leaving the metal in the flask, fill the flask with water and then replace the stopper. Roll the metal around in the flask to make sure that no air remains between the metal pieces. Refill the flask if necessary, and then weigh the dry, stoppered flask full of water plus the metal sample. Properly done, the measurements you have made in this experiment will allow a calculation of the density of your metal sample that will be accurate to about 0.1%.

DISPOSAL OF REACTION PRODUCTS. Pour the water from the flask. Put the metal in its container. Dry the flask and return it with its stopper and your metal sample, along with the sample of unknown liquid, to the stockroom.

Experiment 1

Data and Calculations: The Densities of Liquids and Solids

A. **Mass of coin 1** _____ g   **Mass of coin 2** _____ g

    Mass of coins 1 and 2 weighed together _____ g

    What general law is illustrated by the results of this experiment?

B. **Density of unknown liquid**

    Mass of empty flask plus stopper   _____ g

    Mass of stoppered flask plus water   _____ g

    Mass of stoppered flask plus liquid   _____ g

    Mass of water   _____ g

    Temperature in the laboratory   _____ °C

    Volume of flask (obtain density of water from Appendix I)   _____ cm^3

    Mass of liquid   _____ g

    Density of liquid   _____ g/cm^3

    To how many significant figures can the liquid density be
    properly reported? (See Appendix V.)   _____

C. **Density of unknown metal**

    Mass of stoppered flask plus metal   _____ g

    Mass of stoppered flask plus metal plus water   _____ g

    Mass of metal   _____ g

    Mass of water   _____ g

    Volume of water   _____ cm^3

(continued on following page)

Volume of metal _____ cm³

Density of metal _____ g/cm³

To how many significant figures can the density of the metal be properly reported? _____

Explain why the value obtained for the density of the metal is likely to have a larger percentage error than that found for the liquid.

Unknown liquid no. _____   Unknown solid no. _____

Experiment 1

Advance Study Assignment: Densities of Solids and Liquids

The advance study assignments in this laboratory manual are designed to assist you in making the calculations required in the experiment you will be doing. We do this by furnishing you with sample data and showing in some detail how that data can be used to obtain the desired results. In the advance study assignments we will often include the guiding principles as well as the specific relationships to be employed. If you work through the steps in each calculation by yourself, you should have no difficulty when you are called upon to make the necessary calculations on the basis of the data you obtain in the laboratory.

1. **Finding the volume of a flask.**

   A student obtained a clean, dry glass-stoppered flask. She weighed the flask and stopper on an analytical balance and found the total mass to be 33.695 g. She then filled the flask with water and obtained a mass for the full stoppered flask of 64.356 g. From these data, and the fact that at the temperature of the laboratory the density of water was 0.9978 g/cm³, find the volume of the stoppered flask.

   a. First we need to obtain the mass of the water in the flask. This is found by recognizing that the mass of a sample is equal to the sum of the masses of its parts. For the filled stoppered flask:

   Mass of filled stoppered flask = mass of empty stoppered flask + mass of water,
   so mass of water = mass of filled flask − mass of empty flask

   Mass of water = _____ g − _____ g = _____ g

   Many mass and volume measurements in chemistry are made by the method used in 1a. This method is called measuring by difference, and is a very useful one.

   b. The density of a pure substance is equal to its mass divided by its volume:

   $$\text{Density} = \frac{\text{mass}}{\text{volume}} \quad \text{or} \quad \text{volume} = \frac{\text{mass}}{\text{density}}$$

   The volume of the flask is equal to the volume of the water it contains. Since we know the mass and density of the water, we can find its volume and that of the flask. Make the necessary calculation.

   Volume of water = volume of flask = _____ cm³

2. **Finding the density of an unknown liquid.**

   Having obtained the volume of the flask, the student emptied the flask, dried it, and filled it with an unknown whose density she wished to determine. The mass of the stoppered flask when completely filled with liquid was 58.540 g. Find the density of the liquid.

   a. First we need to find the mass of the liquid by measuring by difference:

   Mass of liquid = _____ g − _____ g = _____ g

(continued on following page)

b. Since the volume of the liquid equals that of the flask, we know both the mass and volume of the liquid and can easily find its density using the equation in 1b. Make the calculation.

Density of liquid = _____ g/cm³

3. Finding the density of a solid.

The student then emptied the flask and dried it once again. To the empty flask she added pieces of a metal until the flask was about three-fourths full. She weighed the stoppered flask and its metal contents and found that the mass was 144.076 g. She then filled the flask with water, stoppered it, and obtained a total mass of 162.292 g for the flask, stopper, metal, and water. Find the density of the metal.

a. To find the density of the metal we need to know its mass and volume. We can easily obtain its mass by the method of differences:

Mass of metal = _____ g – _____ g = _____ g

b. To determine the volume of metal, we note that the volume of the flask must equal the volume of the metal plus the volume of water in the filled flask containing both metal and water. If we can find the volume of water, we can obtain the volume of metal by the method of differences. To obtain the volume of the water we first calculate its mass:

Mass of water = mass of (flask + stopper + metal + water) – mass of (flask + stopper + metal)

Mass of water = _____ g – _____ g = _____ g

The volume of water is found from its density, as in 1b. Make the calculation.

Volume of water = _____ cm³

c. From the volume of the water, we calculate the volume of metal:

Volume of metal = volume of flask – volume of water

Volume of metal = _____ cm³ – _____ cm³ = _____ cm³

From the mass of and volume of metal, we find the density, using the equation in 1b. Make the calculation.

Density of metal = _____ g/cm³

Now go back to Question 1 and check to see that you have reported the proper number of significant figures in each of the results you calculated in this assignment. Use the rules on significant figures as given in your chemistry text or Appendix V.

Experiment 2

Resolution of Matter into Pure Substances, I. Paper Chromatography

The fact that different substances have different solubilities in a given solvent can be used in several ways to effect a separation of substances from mixtures in which they are present. We will see in an upcoming experiment how fractional crystallization allows us to obtain pure substances by relatively simple procedures based on solubility properties. Another widely used resolution technique, which also depends on solubility differences, is chromatography.

In the chromatographic experiment a mixture is deposited on some solid adsorbing substance, which might consist of a strip of filter paper, a thin layer of silica gel on a piece of glass, some finely divided charcoal packed loosely in a glass tube, or even a resin coating microscopic glass beads which are then packed into a length of copper tubing.

The components of a mixture are adsorbed on the solid to varying degrees, depending on the nature of the component, the nature of the adsorbent, and the temperature. A solvent is then made to flow through the adsorbent solid under applied or gravitational pressure or by the capillary effect. As the solvent passes the deposited sample, the various components tend, to varying extents, to be dissolved and swept along the solid. The rate at which a component will move along the solid depends on its relative tendency to be dissolved in the solvent and adsorbed on the solid. The net effect is that, as the solvent passes slowly through the solid, the components separate from each other and move along as rather diffuse zones. With the proper choice of solvent and adsorbent, it is possible to resolve many complex mixtures by this procedure. If necessary, we can usually recover a given component by identifying the position of the zone containing the component, removing that part of the solid from the system, and eluting the desired component with a suitable (good) solvent.

The name given to a particular kind of chromatography depends upon the manner in which the experiment is conducted. Thus, we have column, thin-layer, paper, and gas chromatography, all in very common use (Fig. 2.1). Chromatography in its many possible variations offers the chemist one of the best methods, if not the best method, for resolving a mixture into pure substances, regardless of whether that mixture consists of a gas, a volatile liquid, or a group of nonvolatile, relatively unstable, complex organic compounds.

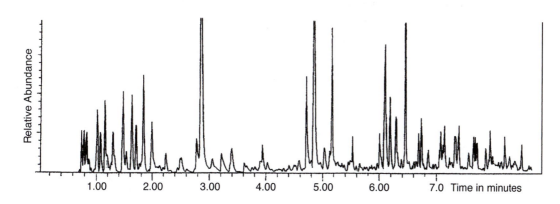

Figure 2.1 This is a gas chromatogram of a sample of unleaded gasoline. Each peak corresponds to a different molecule, so gasoline has many components, at least 50, each of which can be identified. The molar masses vary from about 50 to about 150, with the largest peak, at about 3 min, being due to toluene, $C_6H_5CH_3$. Sample size for the chromatogram was less than 10^{-6} grams, <0.001 mg! Gas chromatography offers the best method for resolution of complex volatile mixtures. Chromatogram courtesy of Prof. Becky Hoye at Macalester College.

In this experiment we will use paper chromatography to separate a mixture of metallic ions in solution. A sample containing a few micrograms of ions is applied as a spot near one edge of a piece of filter paper. That edge is immersed in a solvent, with the paper held vertically. As the solvent rises up the paper by capillary action, it will carry the metallic ions along with it to a degree that depends upon the relative tendency of each ion to dissolve in the solvent and adsorb on the paper. Because the ions differ in their properties, they move at different rates and become separated on the paper. The position of each ion during the experiment can be recognized if the ion is colored, as some of them are. At the end of the experiment their positions are established more clearly by treating the paper with a staining reagent which reacts with each ion to produce a colored product. By observing the position and color of the spot produced by each ion, and the positions of the spots produced by an unknown containing some of those ions, you can readily determine the ions present in the unknown.

It is possible to describe the position of spots such as those you will be observing in terms of a quantity called the R_f ("Retention factor") value. In the experiment the solvent rises a certain distance, say L centimeters. At the same time a given component will usually rise a smaller distance, say D centimeters. The ratio of D/L is called the R_f value for that component:

$$R_f = \frac{D}{L} = \frac{\text{distance component moves}}{\text{distance solvent moves}} \tag{1}$$

The R_f value is a characteristic property of a given component in a chromatography experiment conducted under particular conditions. Within some limits, it does not depend upon concentration or upon the other components present. Hence it can be reported in the literature and used by other researchers doing similar analyses. In the experiment you will be doing, you will be asked to calculate the R_f values for each of the cations studied.

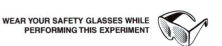

WEAR YOUR SAFETY GLASSES WHILE PERFORMING THIS EXPERIMENT

Experimental Procedure

From the stockroom obtain an unknown and a piece of filter paper about 19 cm long and 11 cm wide. Along the 19-cm edge, draw a pencil line about 1 cm from that edge. Starting 1.5 cm from the end of the line, mark the line at 2-cm intervals. Label the segments of the line as shown in Figure 2.2, with the formulas of the ions to be studied and the known and unknown mixtures.

Put two or three drops of 0.1 M solutions of the following compounds in small micro test tubes, one solution to a tube:

$$AgNO_3 \quad Co(NO_3)_2 \quad Cu(NO_3)_2 \quad Fe(NO_3)_3 \quad Bi(NO_3)_3$$

In solution these substances exist as ions. The metallic cations are Ag^+, Co^{2+}, Cu^{2+}, Fe^{3+}, and Bi^{3+}, respectively. One drop of each solution contains about 50 micrograms of cation. Into a sixth micro test tube put two

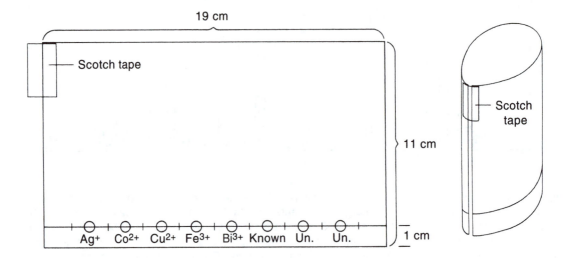

Figure 2.2

drops of each of the five solutions; swirl until the solutions are well mixed. This mixture will be our known, since we know it contains all of the cations.

Your instructor will furnish you with a fine capillary tube, which will serve as an applicator. Test the application procedure by dipping the applicator into one of the colored solutions and touching it momentarily to a round piece of filter paper. The liquid from the applicator should form a spot no larger than 8 mm in diameter. Practice making spots until you can reproduce the spot size each time.

Clean the applicator by dipping it about 1 cm into distilled water and then touching the round filter paper to remove the liquid. Continue contact until all the liquid in the tube is gone. Repeat the cleaning procedure one more time. Dip the applicator into one of the cation solutions and put a spot on the line on the rectangular filter paper in the region labeled for that cation. Clean the applicator twice, and repeat the procedure with another solution. Continue this approach until you have put a spot for each of the five cations and the known and unknown on the paper, cleaning the applicator between solutions. Dry the paper by moving it in the air or holding it briefly in front of a hair dryer or heat lamp (low setting). Apply the known and unknown three more times to the same spots; the known and unknown are less concentrated than the cation solutions, so this procedure will increase the amount of each ion present in the spots. Be sure to dry the spots between applications, since otherwise they will get larger. Don't heat the paper more than necessary, just enough to dry the spots.

Draw about 15 mL of eluting solution from the supply on the reagent shelf. This solution is made by mixing a solution of hydrochloric acid (HCl), with ethanol and butanol, which are organic solvents. Pour the eluting solution into a 600-mL beaker and cover with a watch glass.

Check to make sure that the spots on the filter paper are all dry. Place a 4- to 5-cm length of Scotch tape along the upper end of the left edge of the paper, as shown in Figure 2.2, so that about half of the tape is on the paper. Form the paper into a cylinder by attaching the tape to the other edge, in such a way that the edges are parallel but do not overlap. When you are finished, the pencil line at the bottom of the cylinder should form a circle (approximately, anyway) and the two edges of the paper should not quite touch. Stand the cylinder up on the lab bench to check that such is the case and readjust the tape if necessary. *Do not* tape the lower edges of the paper together.

Place the cylinder in the eluting solution in the 600-mL beaker, with the sample spots down near the liquid surface. The paper should not touch the wall of the beaker. Cover the beaker with the watch glass. The solvent will gradually rise by capillary action up the filter paper, carrying along the cations at different rates. After the process has gone on for a few minutes, you should be able to see colored spots on the paper, showing the positions of some of the cations.

While the experiment is proceeding, you can test the effect of the staining reagent on the different cations. Put an 8-mm spot of each of the cation solutions on a clean piece of round filter paper, labeling each spot and cleaning the applicator between solutions. Dry the spots as before. Some of them will have a little color; record those colors on the Data page. Put the filter paper on a paper towel, and, using the spray bottle on the lab bench, spray the paper evenly with the staining reagent, getting the paper moist but not really wet. The staining reagent is a solution containing potassium ferrocyanide and potassium iodide. This reagent forms colored precipitates or reaction products with many cations, including all of those used in this experiment. Note the colors obtained with each of the cations. Considering that each spot contains less than 50 micrograms of cation, the tests are quite definitive.

When the eluting solution has risen to within about 2 cm of the top of the filter paper (it will take about 75 minutes), remove the cylinder from the beaker and take off the tape. Draw a pencil line along the solvent front. Dry the paper with gentle heat until it is quite dry. Note any cations that must be in your unknown by virtue of your being able to see their colors. Then, with the filter paper set out flat on a paper towel, spray it as before with the staining reagent. Any cations you identified in your unknown before staining should be observed, as well as any that require staining for detection.

Measure the distance from the straight line on which you applied the spots to the solvent front, which is distance L in Equation 1. Then measure the distance from the pencil line to the center of the spot made by each of the cations, when pure and in the known; this is distance D. Calculate the R_f value for each cation. Then calculate R_f values for the cations in the unknown. How do the R_f values compare?

DISPOSAL OF REACTION PRODUCTS.   When you are finished with the experiment, pour the eluting solution into the waste crock, not down the sink. Wash your hands before leaving the laboratory.

Experiment 2

Data and Calculations: Resolution of Matter into Pure Substances,
I. Paper Chromatography

	Ag^+	Co^{2+}	Cu^{2+}	Fe^{3+}	Bi^{3+}
Colors (if observed)					
Dry	_____	_____	_____	_____	_____
After staining	_____	_____	_____	_____	_____
Pure Cations					
Distance solvent moved (L)	_____	_____	_____	_____	_____
Distance cation moved (D)	_____	_____	_____	_____	_____
R_f	_____	_____	_____	_____	_____
Known Mixture					
Distance solvent moved	_____	_____	_____	_____	_____
Distance cation moved	_____	_____	_____	_____	_____
R_f	_____	_____	_____	_____	_____
Unknown Mixture					
Cations identified					
Dry	_____	_____	_____	_____	_____
After staining	_____	_____	_____	_____	_____
Distance solvent moved	_____	_____	_____	_____	_____
Distance cation moved	_____	_____	_____	_____	_____
R_f	_____	_____	_____	_____	_____
Composition of unknown	_____	_____	_____	_____	_____

Unknown no. _____

Name _____ **Section** _____

Experiment 2

Advance Study Assignment: Resolution of Matter into Pure Substances,
I. Paper Chromatography

1. A student chromatographs a mixture, and after developing the spots with a suitable reagent he observes the following:

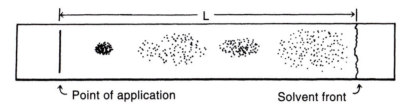

|← ———————————— L ———————————— →|

↳ Point of application   Solvent front ↗

What are the R_f values?

2. Explain, in your own words, why samples can often be separated into their components by chromatography.

3. The solvent moves 3 cm in about 10 minutes. Why shouldn't the experiment be stopped at that time instead of waiting 75 minutes for the solvent to move 10 cm?

4. In this experiment it takes about 10 microliters of solution to produce a spot 1 cm in diameter. If the $Co(NO_3)_2$ solution contains about 6 g Co^{2+} per liter, how many micrograms of Co^{2+} ion are there in one spot?

_____ micrograms

Experiment 3

Resolution of Matter into Pure Substances, II. Fractional Crystallization

One of the important problems faced by chemists is that of determining the nature and state of purity of the substances with which they work. In order to perform meaningful experiments, chemists must ordinarily use essentially pure substances, which are often prepared by separation from complex mixtures.

In principle the separation of a mixture into its component substances can be accomplished by carrying the mixture through one or more physical changes, experimental operations in which the nature of the components remains unchanged. Because the physical properties of various pure substances are different, physical changes frequently allow an enrichment of one or more substances in one of the fractions that is obtained during the change. Many physical changes can be used to accomplish the resolution of a mixture, but in this experiment we will restrict our attention to one of the simpler ones in common use, namely, fractional crystallization.

The solubilities of solid substances in different kinds of liquid solvents vary widely. Some substances are essentially insoluble in all known solvents; the materials we classify as macromolecular are typical examples. Most materials are noticeably soluble in one or more solvents. Those substances that we call salts often have very appreciable solubility in water but relatively little solubility in any other liquids. Organic compounds, whose molecules contain carbon and hydrogen atoms as their main constituents, are often soluble in organic liquids such as benzene or carbon tetrachloride.

We also often find that the solubility of a given substance in a liquid is sharply dependent on temperature. Most substances are more soluble in a given solvent at high temperatures than at low temperatures, although there are some materials whose solubility is practically temperature-independent and a few others that become less soluble as temperature increases.

By taking advantage of the differences in solubility of different substances we often find it possible to separate the components of a mixture in essentially pure form.

In this experiment you will be given a sample containing silicon carbide, potassium nitrate, and copper sulfate. Your goal will be to separate two pure components from the mixture, using water as the solvent. Silicon carbide, SiC, is a black, very hard material; it is the classic abrasive, and completely insoluble in water. Potassium nitrate, KNO_3, and copper sulfate, $CuSO_4 \cdot 5\ H_2O$, are water-soluble ionic substances, with different solubilities at different temperatures, as indicated in Figure 3.1. The copper sulfate we will use is blue as a crystalline hydrate, and in solution. Its solubility increases fairly rapidly with temperature. Potassium nitrate is a white solid, clear and colorless in solution. Its solubility in water increases about 20-fold between 0°C and 100°C.

Given a mixture containing roughly equal amounts of SiC and KNO_3 and a small amount of $CuSO_4 \cdot 5\ H_2O$, we separate out the silicon carbide first. This is done by simply stirring the mixture with water, which dissolves all of the potassium nitrate and copper sulfate in the mixture. The insoluble silicon carbide remains behind and is filtered off.

The solution obtained after filtration contains KNO_3 and $CuSO_4$ in a rather large amount of water. Some of the water is removed by boiling, and then the solution is cooled to 0°C. At that point the KNO_3 is not very soluble, and most of it crystallizes from solution. Since $CuSO_4$ is not present in large amount, its solubility is not exceeded and it remains in solution. The solid KNO_3 is separated from the solution by filtration. This procedure, by which a substance can be separated from an impurity, is called fractional crystallization.

The solid potassium nitrate one recovers is contaminated by a small amount of copper sulfate. The purity of the solid can be markedly increased by stirring it with a small amount of water and then filtering off the dissolved $CuSO_4$. The purity can be established by the intensity of the color produced by the copper impurity when treated with ammonia, NH_3.

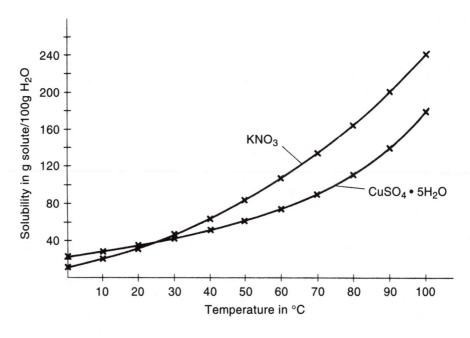

Figure 3.1

Experimental Procedure

Obtain from the stockroom a Buchner funnel, a suction flask, and a sample (about 20 grams) of your unknown solid mixture.

Weigh a 150-mL beaker, a 50-mL beaker, and a piece of filter paper, one item at a time, on a top-loading balance. These, and all other weighings in this experiment, should be made to ± 0.1 g. (Do not use an analytical balance for any weighings.) Use the masses you obtain as needed in your calculations.

Add your sample to the 150-mL beaker and weigh again. Then add about 40 mL of distilled water, which will be enough to dissolve the soluble solids. Light your Bunsen burner and adjust the flame so it is blue, quiet, and of moderate size.

Separation of SiC

Support the beaker with its contents on a piece of wire gauze on an iron ring. Warm gently to about 50°C, while stirring the mixture. When the blue and white solids are all dissolved, pour the contents of the beaker into a Buchner funnel while gentle suction is being applied (see Fig. 3.2 and Appendix IV). Transfer as much of the black solid carbide as you can to the funnel, using your rubber policeman. Transfer the blue filtrate to the (cleaned) 150-mL beaker and add 15 drops of 6 M nitric acid, HNO_3, which will help ensure that the copper sulfate remains in solution in later steps. Re-assemble the funnel, apply suction, and wash the SiC on the filter paper with distilled water. Continue the suction for a few minutes to dry the SiC. Turn off the suction, and, using your spatula, lift the filter paper and the SiC crystals from the funnel, and put the paper on the lab bench so that the crystals may dry in the air. When you are finished with the rest of the experiment, weigh the dry SiC on its piece of filter paper.

Separation of KNO₃

Heat the blue filtrate in the beaker to the boiling point, and then boil gently until the white crystals of KNO_3 are visible in the liquid.

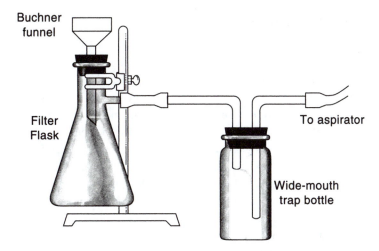

Figure 3.2   To operate the Buchner funnel, put a piece of circular filter paper in the funnel. Turn on suction and spray filter paper with distilled water from wash bottle. Keep suction on while filtering sample.

CAUTION: *The hot liquid may have a tendency to bump, so do not heat it too strongly.*
While the filtrate is being heated, prepare some ice-cold distilled water by putting your wash bottle in an ice-water bath.

When white crystals are clearly apparent in the boiling liquid (the solution may appear cloudy at that point) stop heating and add 12 mL distilled water to the solution. Stir the mixture with a glass stirring rod to dissolve the solids, including any on the wall; if necessary, warm the solution but do not boil it.

Cool the solution to room temperature in a water bath, and then to about 0°C in an ice bath. White crystals of KNO_3 will come out of solution. Stir the cold slurry of crystals for several minutes. Check the temperature of the slurry with your thermometer. It should be no more than 3°C. Continue stirring until your mixture gets to that temperature or even a bit lower.

Assemble the Buchner Funnel. Chill it by adding about 100 mL of ice-cold distilled water from your wash bottle, and, after about a minute, by drawing the water through with suction. Filter the KNO3 slurry through the cold Buchner funnel. Your rubber policeman will be helpful when you transfer the last of the crystals. Press the crystals dry with a clean piece of filter paper, and continue to apply suction for another 30 seconds. Turn off the suction. Lift the filter paper and the crystals from the funnel, and put the paper and crystals on the lab bench.

DISPOSAL OF REACTION PRODUCTS.   By this procedure you have separated most of the KNO_3 in your sample from the $CuSO_4$, which is present in the solution in the suction flask. This solution may now be discarded, so dispose of it as directed by your instructor.

Clean and dry your 150-mL beaker and to it add the KNO_3 crystals from the filter paper. Weigh the beaker and its contents.

Analysis of the Purity of the KNO₃

The KNO_3 crystals you have prepared contain a small amount of $CuSO_4$ as an impurity. To find the amount of $CuSO_4$ present, weigh out 0.5 g of the crystals into your weighed 50-mL beaker. Dissolve the crystals in 3 mL distilled water, and then add 3 mL 6 M ammonia (NH_3). The copper impurity will form a blue solution in the NH_3. Pour the solution into a small test tube. Compare the intensity of the blue color with that in a series of standard solutions prepared by your instructor. Estimate the relative concentration of $CuSO_4 \cdot 5\ H_2O$ in your product.

Recrystallization of the KNO₃

Given the mass of the KNO_3 you recovered (less 0.5 g), use Figure 3.1 to estimate the amount of water needed to dissolve the solid at 100°C. Add three times that amount of distilled water to the KNO_3. Stir the crystals for a minute or two to ensure that all of the $CuSO_4$ impurity goes into solution.

Cool the mixture in an ice bath. This will recrystallize most of the KNO_3 that dissolved. After stirring for several minutes, check the temperature of the mixture with your thermometer. Continue cooling until the temperature is below 3°C. Then filter the slurry through an ice-cold Buchner funnel, transferring as much of the solid as possible to the funnel. Press the crystals dry with a piece of filter paper and continue to apply suction for a minute or so. Lift the filter paper from the funnel and put it with its batch of crystals on the lab bench.

Transfer the purified crystals to a piece of dry filter paper. Weigh the paper and the crystals.

Determine the amount of $CuSO_4$ impurity in your recrystallized sample as you did with the first batch, and record that value. The recrystallization should have very significantly increased the purity of your KNO_3.

Weigh your dry SiC on its piece of filter paper. Show your samples of SiC and KNO_3 to your instructor for evaluation. Then turn the samples in to the stockroom.

Experiment 3

Data and Calculations: Resolution of Matter into Pure Substances, II. Fractional Crystallization

Mass of 150-mL beaker _____ g   Mass of 50-mL beaker _____ g

Mass of filter paper   _____ g

Separation of SiC

Mass of sample plus 150-mL beaker   _____ g

Mass of sample   _____ g

Mass of SiC plus filter paper   _____ g

Mass of SiC   _____ g

Percentage of SiC in sample, by mass   _____ %

Separation of KNO_3

Mass of 150-mL beaker plus KNO_3 _____ g

Mass of KNO_3 recovered   _____ g

Percentage (by mass) of sample recovered as KNO_3 _____ %

Analysis of Purity of KNO_3

Mass of 50-mL beaker plus KNO_3 _____ g

Mass of KNO_3 used in analysis   _____ g

Percentage (by mass) of $CuSO_4 \cdot 5 H_2O$ present in KNO_3 _____ %

(continued on following page)

Recrystallization of KNO_3

Mass of filter paper plus purified KNO_3 _____ g

Mass of purified KNO_3 obtained _____ g

Percentage (by mass) of sample recovered as pure KNO_3 _____ %

Mass of 50-mL beaker plus purified KNO_3 _____ g

Mass of purified KNO_3 used in analysis _____ g

Percentage (by mass) of $CuSO_4 \cdot 5 H_2O$ in purified KNO_3 _____ %

Experiment 3

Advance Study Assignment: Resolution of Matter into Pure Substances,
II. Fractional Crystallization

1. Using Figure 3.1, determine

    a. the number of grams of $CuSO_4 \cdot 5 H_2O$ that will dissolve in 100 g of H_2O at 100°C.

    _____ g $CuSO_4 \cdot 5 H_2O$

    b. the number of grams of water required to dissolve 3.5 g of $CuSO_4 \cdot 5 H_2O$ at 100°C. (Hint: your
        answer to Part a gives you the needed conversion factor for g $CuSO_4 \cdot 5 H_2O$ to g H_2O.)

    _____ g H_2O

    c. the number of grams of water required to dissolve 15 g KNO_3 at 100°C.

    _____ g H_2O

    d. the number of grams of water required to dissolve a mixture containing 15 g KNO_3 and 3.5 g
        $CuSO_4 \cdot 5 H_2O$, assuming that the solubility of one substance is not affected by the presence of
        another.

    _____ g H_2O

2. To the solution in Problem 1(d) at 100°C, 10 g of water are added, and the solution is cooled to 0°C.

    a. How much KNO_3 remains in solution?

    _____ g KNO_3

    b. How much KNO_3 crystallizes out?

    _____ g KNO_3

    c. How much $CuSO_4 \cdot 5 H_2O$ crystallizes out?

    _____ g $CuSO_4 \cdot 5 H_2O$

    d. What percent of the KNO_3 in the sample is recovered?

    _____ %

Experiment 4

Determination of a Chemical Formula

When atoms of one element combine with those of another, the combining ratio is typically an integer or a simple fraction; 1:2, 1:1, 2:1, and 2:3 are ratios one might encounter. The simplest formula of a compound expresses that atom ratio. Some substances with the ratios we listed include $CaCl_2$, KBr, Ag_2O, and Fe_2O_3. When more than two elements are present in a compound, the formula still indicates the atom ratio. Thus the substance with the formula Na_2SO_4 indicates that the sodium, sulfur, and oxygen atoms occur in that compound in the ratio 2:1:4. Many compounds have more complex formulas than those we have noted, but the same principles apply.

To find the formula of a compound we need to find the mass of each of the elements in a weighed sample of that compound. For example, if we resolved a sample of the compound NaOH weighing 40 grams into its elements, we would find that we obtained just about 23 grams of sodium, 16 grams of oxygen, and 1 gram of hydrogen. Since the atomic mass scale tells us that sodium atoms have a relative mass of 23, oxygen atoms a relative mass of 16, and hydrogen atoms a relative mass of just about 1, we would conclude that the sample of NaOH contained equal numbers of Na, O, and H atoms. Since that is the case, the atom ratio Na:O:H is 1:1:1, and so the simplest formula is NaOH. In terms of moles, we can say that that one mole of NaOH, 40 grams, contains one mole of Na, 23 grams, one mole of O, 16 grams, and one mole of H, 1 gram, where we define the molar mass to be that mass in grams equal numerically to the sum of the atomic masses in an element or a compound. From this kind of argument we can conclude that the atom ratio in a compound is equal to the mole ratio. We get the mole ratio from chemical analysis, and from that the formula of the compound.

In this experiment we will use these principles to find the formula of the compound with the general formula $Cu_xCl_y \cdot zH_2O$, where the x, y, and z are integers which, when known, establish the formula of the compound. (In expressing the formula of a compound like this one, where water molecules remain intact within the compound, we retain the formula of H_2O in the formula of the compound.)

The compound we will study, which is called copper chloride hydrate, turns out to be ideal for one's first venture into formula determination. It is stable, can be obtained in pure form, has a characteristic blue-green color which changes as the compound is changed chemically, and is relatively easy to decompose into the elements and water. In the experiment we will first drive out the water, which is called the water of hydration, from an accurately weighed sample of the compound. This occurs if we gently heat the sample to a little over 100°C. As the water is driven out, the color of the sample changes from blue-green to a tan-brown color similar to that of tobacco. The compound formed is anhydrous ("no water") copper chloride. If we subtract its mass from that of the hydrate, we can determine the mass of the water that was driven off, and, using the molar mass of water, find the number of moles of H_2O that were in the sample.

In the next step we need to find either the mass of copper or the mass of chlorine in the anhydrous sample we have prepared. It turns out to be much easier to determine the mass of the copper, and find the mass of chlorine by difference. We do this by dissolving the anhydrous sample in water, which gives us a green solution containing copper and chloride ions. To that solution we add some aluminum metal wire. Aluminum is what we call an active metal; in contact with a solution containing copper ions, the aluminum metal will react chemically with those ions, converting them to copper metal. The aluminum is said to reduce the copper ions to the metal, and is itself oxidized. The copper metal appears on the wire as the reaction proceeds, and has its typical red-orange color. When the reaction is complete, we remove the excess Al, separate the copper from the solution, and weigh the dried metal. From its mass we can calculate the number of moles of copper in the sample. We find the mass of chlorine by subtracting the mass of copper from that of the anhydrous copper chloride, and from that value determine the number of moles of chlorine. The mole ratio for Cu:Cl:H_2O gives us the formula of the compound.

Experimental Procedure

Weigh a clean, dry crucible, without a cover, accurately on an analytical balance. Place about 1 gram of the unknown hydrated copper chloride in the crucible. With your spatula, break up any sizeable crystal particles by pressing them against the wall of the crucible. Then weigh the crucible and its contents accurately. Enter your results on the Data page.

Place the uncovered crucible on a clay triangle supported by an iron ring. Light your Bunsen burner away from the crucible, and adjust the burner so that you have a small flame. Holding the burner in your hand, gently heat the crucible as you move the burner back and forth. Do not overheat the sample. As the sample warms, you will see that the green crystals begin to change to brown around the edges. Continue gentle heating, slowly converting all of the hydrated crystals to the anhydrous brown form. After all of the crystals appear to be brown, continue heating gently, moving the burner back and forth judiciously, for an additional two minutes. Remove the burner, cover the crucible to minimize rehydration, and let it cool for about 15 minutes. Remove the cover, and slowly roll the brown crystals around the crucible. If some green crystals remain, repeat the heating process. Finally, weigh the cool uncovered crucible and its contents accurately.

Transfer the brown crystals in the crucible to an empty 50-mL beaker. Rinse out the crucible with two 5- to 7-mL portions of distilled water, and add the rinsings to the beaker. Swirl the beaker gently to dissolve the brown solid. The color will change to green as the copper ions are rehydrated. Measure out about 20 cm of 20-gauge aluminum wire (~0.25 g) and form the wire into a loose spiral coil. Put the coil into the solution so that it is completely immersed. Within a few moments you will observe some evolution of H_2, hydrogen gas, and the formation of copper metal on the Al wire. As the copper ions are reduced, the color of the solution will fade. The Al metal wire will be slowly oxidized and enter the solution as aluminum ions. (The hydrogen gas is formed as the aluminum reduces water in the slightly acidic copper solution.)

When the reaction is complete, which will take about 30 minutes, the solution will be colorless, and most of the copper metal that was produced will be on the Al wire. Add 5 drops of 6 M HCl to dissolve any insoluble aluminum salts and clear up the solution. Use your glass stirring rod to remove the copper from the wire as completely as you can. Slide the unreacted aluminum wire up the wall of the beaker with your stirring rod, and, while the wire is hanging from the rod, rinse off any remaining Cu particles with water from your wash bottle. If necessary, complete the removal of the Cu with a drop or two of 6 M HCl added directly to the wire. Put the wire aside; it has done its duty.

In the beaker you now have the metallic copper produced in the reaction, in a solution containing an aluminum salt. In the next step we will use a Buchner funnel to separate the copper from the solution. Weigh accurately a dry piece of filter paper that will fit in the Buchner funnel, and record its mass. Put the paper in the funnel, and apply light suction as you add a few mL of water to ensure a good seal. With suction on, decant the solution into the funnel. Wash the copper metal thoroughly with distilled water, breaking up any copper particles with your stirring rod. Transfer the wash and the copper to the filter funnel. Wash any remaining copper into the funnel with water from your wash bottle. **All** of the copper must be transferred to the funnel. Rinse the copper on the paper once again with water. Turn off the suction. Add 10 mL of 95% ethanol to the funnel, and after a minute or so turn on the suction. Draw air through the funnel for about 5 minutes. With your spatula, lift the edge of the paper, and carefully lift the paper and the copper from the funnel. Dry the paper and copper under a heat lamp for 5 minutes. Allow it to cool to room temperature and then weigh it accurately.

DISPOSAL OF REACTION PRODUCTS.   Dispose of the liquid waste and copper produced in the experiment as directed by your instructor.

Name _____ **Section** _____

Experiment 4

Data and Calculations: Determination of a Chemical Formula

Atomic masses: Copper _____ Cl _____ H _____ O _____

Mass of crucible   _____ g

Mass of crucible and hydrated sample   _____ g

Mass of hydrated sample   _____ g

Mass of crucible and dehydrated sample   _____ g

Mass of dehydrated sample   _____ g

Mass of filter paper   _____ g

Mass of filter paper and copper   _____ g

Mass of copper   _____ g

No. moles of copper   _____ moles

Mass of water evolved   _____ g

No. moles of water   _____ moles

Mass of chlorine in sample (by difference)   _____ g

No. moles of chlorine   _____ moles

Mole ratio, chlorine:copper in sample   _____ :1

Mole ratio, water:copper in hydrated sample   _____ :1

Formula of dehydrated sample (round to nearest integer)   _____

Formula of hydrated sample   _____

Name _____ **Section** _____

Experiment 4

Advance Study Assignment: Determination of a Chemical Formula

1. To find the mass of a mole of an element, one looks up the atomic mass of the element in a table of atomic masses (see Appendix III or the Periodic Table). The molar mass of an element is simply the mass in grams of that element that is numerically equal to its atomic mass. For a compound substance, the molar mass is equal to the mass in grams that is numerically equal to the sum of the atomic masses in the formula of the substance. Find the molar mass of

   Cu _____ g   Cl _____ g   H _____ g   O _____ g H_2O _____ g

2. If one can find the ratio of the number of moles of the elements in a compound to one another, one can find the formula of the compound. In a certain compound of copper and oxygen, Cu_xO_y, we find that a sample weighing 0.6349 g contains 0.5639 g Cu.

   a. How many moles of Cu are there in the sample?

   $$\left(\text{No. moles} = \frac{\text{mass Cu}}{\text{molar mass Cu}} \right)$$

   _____ moles

   b. How many grams of O are there in the sample? (The mass of the sample equals the mass of Cu plus the mass of O.)

   _____ g

   c. How many moles of O are there in the sample?

   _____ moles

   d. What is the mole ratio (no. moles Cu/no. moles O) in the sample?

   _____ : 1

   e. What is the formula of the oxide? (The atom ratio equals the mole ratio, and is expressed using the smallest integers possible.)

   _____

   f. What is the molar mass of the copper oxide?

   _____ g/mol

Experiment 5

Identification of a Compound by Mass Relationships

In the previous experiment we showed how we can find the formula of a compound by analysis for the elements it contains. When chemical reactions occur, there is a relationship between the masses of the reactants and products that follows directly from the balanced equation for the reaction and the molar masses of the species that are involved. In this experiment we will use this relationship to identify an unknown substance.

Your unknown will be one of the following compounds, all of which are salts:

$$NaHCO_3 \quad Na_2CO_3 \quad KHCO_3 \quad K_2CO_3$$

In the first part of the experiment you will be heating a weighed sample of your compound in a crucible. If your sample is a carbonate, there will be no chemical reaction that occurs, but any small amount of adsorbed water will be driven off. If your sample is a hydrogen carbonate, it will decompose by the following reaction, using $NaHCO_3$ as the example:

$$2\ NaHCO_3(s) \rightarrow Na_2CO_3(s) + H_2O(g) + CO_2(g) \tag{1}$$

In this case there will be an appreciable decrease in mass, since some of the products will be driven off as gases. If such a mass decrease occurs, you can be sure that your sample is a hydrogen carbonate.

In the second part of the experiment, we will treat the solid carbonate in the crucible with HCl, hydrochloric acid. There will be considerable effervescence as CO_2 gas is evolved; the reaction that occurs is, using Na_2CO_3 as our example:

$$Na_2CO_3(s) + 2\ H^+(aq) + 2\ Cl^-(aq) \rightarrow 2\ NaCl(s) + H_2O(l) + CO_2(g) \tag{2}$$

(Since HCl in solution exists as ions, we write the equation in terms of ions.) We then heat the crucible strongly to drive off any excess HCl and any water that is present, obtaining pure, dry, solid NaCl as our product.

To identify your unknown, you will need to find the molar masses of the possible reactants and final products. For each of the possible unknowns there will be a different relationship between the mass of the original sample and the mass of the chloride salt that is produced in Reaction 2. If you know your sample is a carbonate, you need only be concerned with the mass relationships in Reaction 2, and should use as the original mass of your unknown the mass of the carbonate after it has been heated. If you have a hydrogen carbonate, the overall reaction your sample undergoes will be the sum of Reactions 1 and 2.

From your experimental data you will be able to calculate the change in mass that occurred when you formed the chloride from the hydrogen carbonate or the carbonate that you started with. That difference divided by the mass of the original salt will be different for each of the possible starting compounds and will not vary with the mass of the sample. Let's call that quantity Q.

$$Q = (\text{mass of chloride} - \text{original mass})/\text{original mass} \tag{3}$$

Your calculation of the theoretical values of Q for each of the possible compounds should allow you to determine the identity of your unknown. Since you will already know whether your compound is a carbonate or a hydrogen carbonate, you should then need only work out which of the two possible compounds of that type yours may be.

Experimental Procedure

Obtain an unknown from the stockroom.

Clean your crucible and its cover by rinsing them with distilled water and then drying them with a towel. Place the crucible with its cover slightly ajar on a clay triangle. Heat gently with your Bunsen burner flame for a minute or two and then strongly for two more minutes. Allow the crucible and cover to cool to room temperature (it will take about 10 minutes; it must cool completely, or your mass measurement will be artificially lowered by convection currents).

Weigh the crucible and cover accurately on an analytical balance. Record the mass on the Data page. With a spatula, transfer about 0.5 g of the unknown to the crucible. Weigh the crucible, cover, and the sample of unknown on the balance. Record the mass.

Put the crucible on the clay triangle, with the cover ajar. Heat the crucible, gently and intermittently, for a few minutes. Gradually increase the flame intensity, to the point where the bottom of the crucible is at red heat. Heat for 10 minutes. Allow the crucible to cool for 10 minutes, and then weigh it, with its cover and contents on the analytical balance, recording the mass.

At this point the sample in the crucible is a dry carbonate, since the heating process will convert any hydrogen carbonate to carbonate.

Put the crucible on the clay triangle, leaving the cover off. Add about 25 drops of 6 M HCl, a drop at a time, to the sample. As you add each drop, you will probably observe effervescence as CO_2 is produced. Let the action subside before adding the next drop, to keep the effervescence confined to the lower part of the crucible. We do not want the product to foam up over the edge! When you have added all of the HCl, the effervescence should have ceased, and the solid should be completely dissolved. Heat the crucible gently for brief periods to complete the solution process. If all of the solid is not dissolved, add 6 more drops of 6 M HCl and warm gently.

Place the cover on the crucible in an off-center position, to allow water to escape during the next heating operation. Heat the crucible, gently and intermittently, for about 10 minutes, to slowly evaporate the water and excess HCl. If you heat too strongly, spattering will occur and you may lose some sample. When the sample is dry, gradually increase the flame intensity, finally getting the bottom of the crucible to red heat. Heat at full flame strength for 10 minutes.

Allow the crucible to cool for 10 minutes. Weigh it, with its cover and contents, on the analytical balance, recording the mass.

DISPOSAL OF REACTION PRODUCTS. Dispose of the sample by dissolving it in water and pouring it down the drain.

Name _____ **Section** _____

Experiment 5

Data and Calculations: Identification of a Compound
by Mass Relationships

Atomic masses: Na _____ K_____ H_____ C_____ O _____ Cl _____

Molar masses: $NaHCO_3$ _____ g Na_2CO_3 _____ g   NaCl _____ g

KHCO_3 _____ g K_2CO_3 _____ g   KCl _____ g

Mass of crucible and cover plus unknown   _____ g

Mass of crucible and cover   _____ g

Mass of unknown   _____ g

Mass of crucible, cover, and unknown after heating   _____ g

Loss of mass of sample   _____ g

Unknown is a: carbonate   hydrogen carbonate   (circle your choice)

Mass of crucible, cover, and solid chloride   _____ g

Mass of solid chloride   _____ g

Change in mass when original compound was converted to a chloride   _____ g

Q = change in mass/original mass = _____ =   _____

Theoretical values of Q, as obtained by Equation 3

Possible sodium compound _____	Possible potassium compound _____
1 mole of compound → _____ moles NaCl	1 mole of compound → _____ moles KCl
_____ g of compound → _____ g NaCl	_____ g of compound → _____ g KCl
Change in mass _____ g Q = _____	Change in mass _____ g Q = _____

Identity of unknown _____   Unknown No. _____

Experiment 5

Advance Study Assignment: Identification of a Compound
by Mass Relationships

1. A student attempted to identify an unknown compound by the method used in this experiment. She found that when she heated a sample weighing 0.4862 g, the mass barely changed, dropping to 0.4855 g. When the product was converted to a chloride, the mass went up, to 0.5247 g.

    a. Is the sample a carbonate?   Yes / no (Circle one)

    Please provide your reasoning below.

    b. What are the two compounds that might be in the unknown?

                                                          _____ or _____

    c. Write the balanced chemical equation for the overall reaction that occurs when each of these two original compounds is converted to a chloride. If the compound is a hydrogen carbonate, use the sum of Reactions 1 and 2. If the sample is a carbonate, use Reaction 2. Write the equation for a sodium salt and then for a potassium salt.

    d. How many moles of the chloride salt would be produced from one mole of original compound?

                                                                          _____

    e. How many grams of the chloride salt would be produced from one molar mass of original compound?

    Molar masses: $NaHCO_3$ _____ g Na_2CO_3 _____ g   NaCl _____ g

                   $KHCO_3$ _____ g K_2CO_3 _____ g   KCl _____ g

            If a sodium salt, _____ g original compound → _____ g chloride

            If a potassium salt, _____ g original compound → _____ g chloride

    f. What is the theoretical value of Q, as found by Equation 3,

        if she has the Na salt? _____ if she has the K salt? _____

    g. What was the observed value of Q?   _____

    h. Which compound did she have as an unknown?   _____

Experiment 6

Properties of Hydrates

Most solid chemical compounds will contain some water if they have been exposed to the atmosphere for any length of time. In most cases the water is present in very small amounts, and is merely adsorbed on the surface of the crystals. Other solid compounds contain larger amounts of water that is chemically bound in the crystal. These compounds are usually ionic salts. The water that is present in these salts is called water of hydration and is usually bound to the cations in the salt.

The water molecules in a hydrate are removed relatively easily. In many cases, simply heating a hydrate to a temperature somewhat above the boiling point of water will drive off the water of hydration. Hydrated barium chloride is typical in this regard; it is converted to anhydrous $BaCl_2$ if heated to about 115°C:

$$BaCl_2 \cdot 2\ H_2O(s) \rightarrow BaCl_2(s) + 2\ H_2O(g)\ \text{at}\ t \geq 115°C$$

In the dehydration reaction the crystal structure of the solid will change, and the color of the salt may also change. In Experiment 4, when copper chloride hydrate was gently heated, it was converted to the brownish anhydride.

Some hydrates lose water to the atmosphere upon standing in air. This process is called efflorescence. The amount of water lost depends on the amount of water in the air, as measured by its relative humidity, and the temperature. In moist warm air, $CoCl_2$ is fully hydrated and exists as $CoCl_2 \cdot 6\ H_2O$, which is red. In cold dry air the salt loses most of its water of hydration and is found as anhydrous $CoCl_2$, which is blue. At intermediate humidities and 25°C, we find the stable form is the violet dihydrate, $CoCl_2 \cdot 2\ H_2O$. In the old days one could obtain inexpensive hygrometers that indicated the humidity by the color of the cobalt chloride they contained.

Some anhydrous ionic compounds will tend to absorb water from the air or other sources so strongly that they can be used to dry liquids or gases. These substances are called desiccants, and are said to be hygroscopic. A few ionic compounds can take up so much water from the air that they dissolve in the water they absorb; sodium hydroxide, NaOH, will do this. This process is called deliquescence.

Some compounds evolve water on being heated but are not true hydrates. The water is produced by decomposition of the compound rather than by loss of water of hydration. Organic compounds, particularly carbohydrates, behave this way. Decompositions of this sort are not reversible; adding water to the product will not regenerate the original compound. True hydrates typically undergo reversible dehydration. Adding water to anhydrous $BaCl_2$ will cause formation of $BaCl_2 \cdot 2\ H_2O$, or if enough water is added you will get a solution containing Ba^{2+} and Cl^- ions. Many ionic hydrates are soluble in water, and are usually prepared by crystallization from water solution. If you order barium chloride from a chemical supply house, you will probably get crystals of $BaCl_2 \cdot 2\ H_2O$, which is a stable, stoichiometrically pure, compound. The amount of bound water may depend on the way the hydrate is prepared, but in general the number of moles of water per mole of compound is either an integer or a multiple of ½.

In this experiment you will study some of the properties of hydrates. You will identify the hydrates in a group of compounds, observe the reversibility of the hydration reaction, and test some substances for efflorescence or deliquescence. Finally you will be asked to determine the amount of water lost by a sample of unknown hydrate on heating. From this amount, if given the formula or the molar mass of the anhydrous sample, you will be able to calculate the formula of the hydrate itself.

Experimental Procedure

A. Identification of Hydrates

Place about 0.5 g of each of the compounds listed below in small, dry test tubes, one compound to a tube. Observe carefully the behavior of each compound when you heat it gently with a burner flame. If droplets of water condense on the cool upper walls of the test tube, this is evidence that the compound may be a hydrate. Note the nature and color of the residue. Let the tube cool and try to dissolve the residue in a few cm^3 of water, warming very gently if necessary. A true hydrate will tend to dissolve in water, producing a solution with a color very similar to that of the original hydrate. If the compound is a carbohydrate, it will give off water on heating and will tend to char. The solution of the residue in water will often be caramel colored.

Nickel chloride	Sucrose
Potassium chloride	Calcium carbonate
Sodium tetraborate (borax)	Barium chloride

B. Reversibility of Hydration

Gently heat a few crystals, ~0.3 g, of hydrated cobalt(II) chloride, $CoCl_2 \cdot 6\ H_2O$, in an evaporating dish until the color change appears to be complete. Dissolve the residue in the evaporating dish in a few cm^3 of water from your wash bottle. Heat the resulting solution to boiling **CAUTION:** and carefully boil it to dryness. Note any color changes. Put the evaporating dish on the lab bench and let it cool.

C. Deliquescence and Efflorescence

Place a few crystals of each of the compounds listed below on separate watch glasses and put them next to the dish of $CoCl_2$ prepared in Part B. Depending on their composition and the relative humidity (amount of moisture in the air), the samples may gradually lose water of hydration to, or pick up water from, the air. They may also remain unaffected. To establish whether the samples gain or lose mass, weigh each of them on a top-loading balance to 0.01 g. Record their masses. Weigh them again after about an hour to detect any change in mass. Observe the samples occasionally during the laboratory period, noting any changes in color, crystal structure, or degree of wetness that may occur.

$Na_2CO_3 \cdot 10\ H_2O$ (washing soda)	$KAl(SO_4)_2 \cdot 12\ H_2O$ (alum)
$CaCl_2$	$CuSO_4$

D. Percent Water in a Hydrate

Clean a porcelain crucible and its cover with 6 M HNO_3. Any stains that are not removed by this treatment will not interfere with this experiment. Rinse the crucible and cover with distilled water. Put the crucible with its cover slightly ajar on a clay triangle and heat with a burner flame, gently at first and then to redness for about 2 minutes. Allow the crucible and cover to cool, and then weigh them to 0.001 g on an analytical balance. Handle the crucible with clean crucible tongs.

Obtain a sample of unknown hydrate from the stockroom and place about a gram of sample in the crucible. Weigh the crucible, cover, and sample on the balance. Put the crucible on the clay triangle, with the cover in an off-center position to allow the escape of water vapor. Heat again, gently at first and then strongly, keeping the bottom of the crucible at red heat for about 10 minutes. Center the cover on the crucible and let it cool to room temperature. Weigh the cooled crucible along with its cover and contents.

Examine the solid residue. Add water until the crucible is two thirds full and stir. Warm gently if the residue does not dissolve readily. Does the residue appear to be soluble in water?

DISPOSAL OF REACTION PRODUCTS.   Dispose of the residues in this experiment as directed by your instructor.

Experiment 6

Data and Calculations: Properties of Hydrates

A. Identification of Hydrates

	H_2O appears	Color of residue	Water soluble	Hydrate
Nickel chloride	_____	_____	_____	_____
Potassium chloride	_____	_____	_____	_____
Sodium tetraborate	_____	_____	_____	_____
Sucrose	_____	_____	_____	_____
Calcium carbonate	_____	_____	_____	_____
Barium chloride	_____	_____	_____	_____

B. Reversibility of Hydration

Summarize your observations on $CoCl_2 \cdot 6 H_2O$:

Is the dehydration and hydration of $CoCl_2$ reversible?

C. Deliquescence and Efflorescence

	Mass (sample + glass)		Observations	Conclusions
	Initial	Final		
$Na_2CO_3 \cdot 10 H_2O$	_____	_____	_____	_____
$KAl(SO_4)_2 \cdot 12 H_2O$	_____	_____	_____	_____
$CaCl_2$	_____	_____	_____	_____
$CuSO_4$	_____	_____	_____	_____
$CoCl_2$	_____	_____	_____	_____

(continued on following page)

D. Percent Water in a Hydrate

Mass of crucible and cover   _____ g

Mass of crucible, cover, and solid hydrate   _____ g

Mass of crucible, cover, and residue   _____ g

Calculations and Results

Mass of solid hydrate   _____ g

Mass of residue   _____ g

Mass of H_2O lost   _____ g

Percentage of H_2O in the unknown hydrate   _____ %

Formula mass of anhydrous salt (if furnished)   _____

Number of moles of water per mole of unknown hydrate   _____

Unknown no.   _____

Experiment 6

Advance Study Assignment: Properties of Hydrates

1. A student is given a sample of a blue nickel sulfate hydrate. She weighs the sample in a dry covered cru-
    cible and obtains a mass of 23.711 g for the crucible, cover, and sample. Earlier she had found that the
    crucible and cover weighed 21.594 g. She then heats the crucible to drive off the water of hydration, keep-
    ing the crucible at red heat for about 10 minutes with the cover slightly ajar. She then lets the crucible
    cool, and finds it has a lower mass; the crucible, cover and contents then weigh 22.840 g. In the process
    the sample was converted to greenish-yellow anhydrous $NiSO_4$.

    a. What was the mass of the hydrate sample?

    _____ g hydrate

    b. What is the mass of the anhydrous $NiSO_4$?

    _____ g $NiSO_4$

    c. How much water was driven off?

    _____ g H_2O

    d. What is the percent by mass of water in the hydrate?

    $$\% \text{ water} = \frac{\text{mass of water in sample}}{\text{mass of hydrate sample}} \times 100\%$$

    _____ $\%_{mass}$ H_2O

    e. How many grams of water would there be in 100.0 g hydrate? How many moles?

    _____ g H_2O; _____ moles H_2O

    f. How many grams of $NiSO_4$ are there in 100.0 g hydrate? How many moles? (What percentage of the
        hydrate is $NiSO_4$? Convert the mass of $NiSO_4$ to moles. The molar mass of $NiSO_4$ is 154.8 g.)

    _____ g $NiSO_4$; _____ moles $NiSO_4$

    g. How many moles of water are present per mole $NiSO_4$?

    _____

    h. What is the formula of the hydrate?

    _____

Experiment 7

Analysis of an Unknown Chloride

One of the important applications of precipitation reactions lies in the area of quantitative analysis. Many substances that can be precipitated from solution are so slightly soluble that the precipitation reaction by which they are formed can be considered to proceed to completion. Silver chloride is an example of such a substance. If a solution containing Ag^+ ion is slowly added to one containing Cl^- ion, the ions will react to form AgCl:

$$Ag^+(aq) + Cl^-(aq) \rightarrow AgCl(s) \tag{1}$$

Silver chloride is so insoluble that essentially all of the Ag^+ added will precipitate as AgCl until all of the Cl^- is used up. When the amount of Ag^+ added to the solution is equal to the amount of Cl^- initially present, the precipitation of Cl^- ion will be, for all practical purposes, complete.*

A convenient method for chloride analysis using AgCl has been devised. A solution of $AgNO_3$ is added to a chloride solution just to the point where the number of moles of Ag^+ added is equal to the number of moles of Cl^- initially present. We analyze for Cl^- by simply measuring how many moles of $AgNO_3$ are required. Surprisingly enough, this measurement is rather easily made by an experimental procedure called a titration.

In the titration a solution of $AgNO_3$ of known concentration (in moles $AgNO_3$ per liter of solution) is added from a calibrated buret to a solution containing a measured amount of unknown. The titration is stopped when a color change occurs in the solution, indicating that stoichiometrically equivalent amounts of Ag^+ and Cl^- are present. The color change is caused by a chemical reagent, called an indicator, that is added to the solution at the beginning of the titration.

The volume of $AgNO_3$ solution that has been added up to the time of the color change can be measured accurately with the buret, and the number of moles of Ag^+ added can be calculated from the known concentration of the solution.

In the Fajans method for the volumetric analysis of chloride, which we will employ in this experiment, the indicator used is dichlorofluorescein. This compound, in its anionic form, functions as an adsorption indicator, in that as silver nitrate is added to the chloride solution, silver ions from the first excess are adsorbed onto the surface of the colloidal silver chloride particles. The silver chloride now has a positive charge, and will attract dichlorofluorescein anions onto the surface. When this occurs, the silver chloride will take on a pink color in the presence of the yellowish, fluorescent dichlorofluorescein, indicating that the endpoint of the titration has been reached. This color change is best observed with indirect lighting. A small amount of dextrin (a degradation product of starch) is added to the titration to minimize the coagulation of the silver chloride colloidal particles.

In this experiment, you will be obtaining a 100 mL volumetric flask which contains an unknown chloride solution. You will dilute the solution to the 100 mL mark, and after mixing, pipet out 10.00 mL aliquots of the solution into separate flasks for titration with a standardized solution of $AgNO_3$. Using the respective volumes of titrant required to reach the endpoint in each titration, you will obtain three replicate measurements of the concentration of the diluted chloride solution in the volumetric flask. Given the molarity of the $AgNO_3$ titrant,

$$\text{no. of moles } Ag^+ = \text{no. of moles } AgNO_3 = M_{AgNO_3} \times V_{AgNO_3} \tag{2}$$

where the volume of $AgNO_3$ is expressed in liters and the molarity M_{AgNO_3} is in moles per liter of solution. At the end point of the titration,

$$\text{no. of moles } Ag^+ \text{ added} = \text{no. of moles } Cl^- \text{ present in unknown} \tag{3}$$

$$\text{no. of moles } Cl^- \text{ present in unknown} = M_{Cl} \times V_{Cl} \tag{4}$$

*See Experiment 26 for a discussion of the principles governing precipitations of this sort.

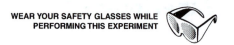

Experimental Procedure

From the stockroom, obtain a buret, a 100 mL volumetric flask containing your concentrated unknown chloride solution, and a 10 mL volumetric pipet. **USING CARE**, add distilled water slowly to the volumetric flask up to the mark on the narrow neck , squeezing your wash bottle gently as the bottom of the meniscus becomes just coincident with the ring—this may be best seen if the ring on the flask is at the same level as your eyes. Place the stopper in the flask (see Appendix IV), and holding the stopper in place with the index finger or palm of one hand, invert the flask 15 times to thoroughly mix the solution. Transfer via pipet (see Appendix IV) 10.00 mL of your diluted chloride solution to a 250 mL Erlenmeyer flask, and add about 50 mL of DI water. Add 0.1 g (+/– 0.02 g) of dextrin, and swirl to dissolve—it's not important that all of it dissolves. Add two drops of dichlorofluorescein indicator.

Using the graduated cylinder at the reagent shelf, measure out about 100 mL of the standardized $AgNO_3$ solution into a clean, *dry* 125-mL Erlenmeyer flask. This will be your total supply for the entire experiment, so do not waste it. Silver solutions will leave long-lasting black stains on skin and clothing, so take care not to spill your titrant. Clean your buret thoroughly with detergent solution, with the aid of a buret brush, and rinse it with distilled water. Pour 2 or 3 successive portions of the $AgNO_3$ titrant into the buret and tip it back and forth to rinse the inside walls. Allow the $AgNO_3$ to drain out of the buret tip each time. Fill the buret with $AgNO_3$ solution. Open the stopcock momentarily to flush any air bubbles out of the buret tip. Be sure that your stopcock fits snugly and that the buret does not leak. (See Appendix IV for procedures regarding titrations with a buret.)

Read the initial buret level to 0.01 mL. You may find it useful when making readings to put a white card, with a thick black stripe or black tape on it, behind the meniscus. If the black line is held just below the level to be read, its reflection in the surface of the meniscus will help you obtain an accurate reading. In indirect light, begin to add the $AgNO_3$ titrant solution to the chloride solution while swirling the flask. Colloidal AgCl will begin to form, giving a slight cloudiness to the yellowish solution. At the beginning of the titration, you can add the $AgNO_3$ fairly rapidly, a few milliliters at a time, swirling the flask to mix the solution. You will find that a pinkish color will form in the solution and disappear as you mix by swirling.

Gradually decrease the rate at which you add the $AgNO_3$ as the pink color persists longer. At some stage you will find it convenient to set your buret stopcock to deliver titrant slowly, drop by drop while you swirl the flask. The endpoint of the titration will be that point where the mixture takes on a permanent pink color while swirling. Just before the endpoint you may note that the solution takes on a peach color. If you are careful, you can hit the endpoint within 1 drop of $AgNO_3$. When you have reached the endpoint, stop the titration and record the buret level, again to ± 0.01 mL.

Pour the solution you have just titrated into a 250 mL beaker. Analyze two more 10.00 mL diluted chloride samples, as described above. Note the volume of $AgNO_3$ used in the first titration and refill the buret as necessary, to be certain that you will not go below the 50.00 mL mark at the bottom of the buret.

Calculate the concentration of chloride in the diluted solution within your 100 mL volumetric flask, as independently indicated by each of your three titrations. Also calculate the mean and the standard deviation of these three results. (Consult Appendix VIII.)

Optional Experiment: Find the mass percent Cl^- or NaCl in a bouillon cube, or canned chicken and rice, or French onion soup. Compare your results with those on the label.

> **DISPOSAL OF REACTION PRODUCTS.** Silver nitrate is very expensive. Pour all titrated solutions and any $AgNO_3$ remaining in your buret or flask into the waste bottles provided unless directed otherwise by your instructor.

Experiment 7

Data and Calculations: Analysis of an Unknown Chloride

Molarity of standard $AgNO_3$ solution _____ M

	I	**II**	**III**
Initial buret reading	_____ mL	_____ mL	_____ mL
Final buret reading	_____ mL	_____ mL	_____ mL
Volume of $AgNO_3$ used to titrate sample	_____ mL	_____ mL	_____ mL
No. of moles of $AgNO_3$ used to titrate sample	_____	_____	_____
No. of moles of Cl^- present in sample	_____	_____	_____
Volume of diluted chloride solution pipetted	_____ mL	_____ mL	_____ mL
Total moles of Cl^- in 100 mL vol flask	_____	_____	_____
Cl^- concentration in diluted solution in volumetric flask	_____ M	_____ M	_____ M
Mean chloride Concentration	_____ M		
Standard deviation	_____ M		
Unknown number	_____		

Experiment 7

Advance Study Assignment: Analysis of an Unknown Chloride

1. A student performed this experiment and obtained the following concentration values: 7.5360×10^{-4} M, 7.5477×10^{-4} M, and 7.5411×10^{-4} M.

    a. What is the mean concentration?

    _____ M

    b. What is the standard deviation of these results?

    _____ M

2. A 10.00 mL diluted chloride sample required 44.89 mL of 0.01982 M $AgNO_3$ to reach the Fajans end point.

    a. How many moles of Cl^- ion were present in the sample? (Use Eqs. 2 and 3.)

    _____ moles Cl^-

    b. What was the concentration of chloride in the diluted solution?

    _____ M

3. How would the following errors affect the mass percent Cl^- obtained in Question 2c? Give your reasoning in each case.

    a. The student read the molarity of $AgNO_3$ as 0.01982 M instead of 0.01892 M.

    b. The student was past the end point of the titration when he took the final buret reading.

Experiment 8

Verifying the Absolute Zero of Temperature— Determination of the Barometric Pressure

Gases differ from liquids and solids in that much of the ordinary physical behavior of gases can be described by simple laws that apply to all gases. Few such relations exist for liquids and solids, so information about their physical properties must be obtained by direct experimentation on the liquid or solid of interest.

Gases are easily compressed relative to liquids and solids. If one doubles the pressure on a gas, the volume decreases by a factor of two. Over rather wide ranges of pressure, the pressure-volume product, PV, remains nearly constant as long as the temperature remains constant. This relationship is called Boyle's Law, after Robert Boyle, who discovered it in 1660. It was one of the first natural laws that scientists recognized.

When a gas is heated at constant volume, the pressure goes up. In 1787, about 120 years after Boyle did his work, Charles found that the pressure increased linearly with Celsius temperature. Boyle was in no position to discover Charles' Law, since in 1660 the idea of temperature was not well developed. In 1884, Lord Kelvin, at the ripe old age of 24, recognized that one could set up an absolute temperature scale on which the pressure was actually proportional to temperature. This idea was relatively sophisticated and led to the observation that many natural laws, not just the gas laws, are most simply expressed on the absolute temperature scale, now called the Kelvin scale.

The gas we encounter most often is the air about us. Its physical behavior is like that of other gases, so under ordinary conditions it obeys Boyle's and Charles' Laws. The pressure of the air is called the barometric pressure. It varies a little from day to day, and at sea level it is equal to the pressure exerted by a column of mercury about 76 cm high, or a column of water 10 meters high; this is about 1×10^5 Pascals, or in common American units, about 15 psi (lbs/in^2). Ordinarily we are not aware of that rather substantial pressure because it is nicely balanced by our bodies, but if it decreases suddenly, as it does in a tornado, it will definitely get our attention, since it can, among other things, cause a house to fly apart. The barometric pressure can be found by various instruments, but probably the most accurate way is to measure the height to which a column of mercury will rise in an evacuated tube immersed in a pool of mercury. The mercury barometer is commonly found in most science laboratories.

In the first part of this experiment we will try to verify the absolute zero of temperature. We will simply heat a sample of air, keeping its volume constant while we measure the increase in pressure. We assume the validity of Charles' Law, and calculate the temperature at which the pressure would become zero. That is the absolute zero of temperature and the basis of the Kelvin temperature scale.

In the rest of the experiment we will measure the barometric pressure using a simple apparatus that allows us to find the change in volume when a sample of air is slightly compressed at constant temperature. The changes in volume and pressure are related to the barometric pressure in a remarkably simple way.

WEAR YOUR SAFETY GLASSES WHILE
PERFORMING THIS EXPERIMENT

Experimental Procedure

You may work in pairs in this experiment. Obtain from the stockroom a 1000-mL beaker, a 500-mL Erlenmeyer flask with stopper and glass tubing insert, a digital or sensitive mercury thermometer, a glass U-tube manometer, two lengths of rubber tubing, and a disposable Pasteur pipet.

A. Verifying the Absolute Zero of Temperature

In this part of the experiment we will examine how the pressure of a sample of air of fixed volume increases when we raise the temperature. Given that $PV = nRT$ for the air, if we hold the volume and amount of air constant, we can see that the Ideal Gas Law reduces to $P = AT$, so the pressure is proportional to the Kelvin temperature T. Our goal is to establish T.

Assemble the apparatus shown in Figure 8.1, placing a 1000-mL beaker on a magnetic stirrer if one is available. Put a magnetic stirring bar in the beaker, and clamp a **dry** 500-mL Erlenmeyer flask in the beaker as shown. (If the flask is not dry, add a few milliliters of acetone and shake well. Pour out the acetone and gently, gently, blow compressed air into the flask for a minute or two.) Fill the beaker with room temperature water which has been stored overnight in the lab. Start the stirrer. If you don't have a magnetic stirrer available, stirring with your stirring rod should be satisfactory, and indeed you should use it in addition to the magnetic stirrer. Connect the short piece of rubber tubing to the glass tubing in the stopper, and then moisten the rubber stopper and insert it firmly into the flask. Clamp the U-tube manometer into position. You can rest the bottom of the U-tube on the lab bench surface if that is convenient. Using your Pasteur pipet, add a few mL of water to the U-tube manometer, so that the level in the left arm is at about 4.0 cm above the lowest part of the U. Mark the level in that arm with masking tape. Connect the rubber tubing to the left arm of the manometer. If the level in the left arm changes, bring it back to the mark by adding a little water to the right arm, or by adding a little warm water to the water bath. Record the levels in the left and right arms to the nearest 0.1 cm. Record the temperature to the nearest 0.1°C. Record the barometric pressure in mm Hg.

If the system is tight, and the water in the beaker is at room temperature, with stirring the levels will remain steady for several minutes. If necessary, adjust the bath temperature until the levels hold steady, and then take new readings.

Using the long piece of rubber tubing and a pipet bulb, siphon about 100 mL of water from the bath into a beaker. With your Pasteur pipet add a few mL of water to the right arm of the manometer, raising the level in the **left** arm about 3 cm. Add some warm water (at about 40°C) from the tap to a beaker, and very slowly add the warm water to the bath. The level in the left arm will go down as the air in the flask expands. Add warm water as necessary until the level in the left arm is at the location you marked. If the level goes too low, add a little room temperature water to the right arm to bring it to the proper position. Wait a minute or so, with stirring, to see that levels hold steady and the temperature is not changing. Read the two levels and the temperature.

Repeat the experiment at two or three higher temperatures, being sure to wait long enough for the levels and temperature to become steady. At the higher temperatures, the water bath will tend to cool, so you may need to add warm water to make it hold its temperature. Your final level in the right arm should be near the top of the manometer. You should have at least four runs for which you have recorded water levels and temperature.

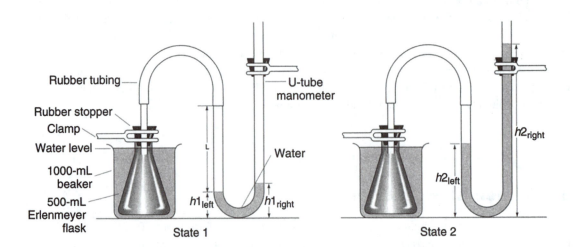

Figure 8.1

You can process the data you have obtained by making a graph of total pressure, P vs the Celsius temperature, t. The total pressure of the gas will equal the measured barometric pressure plus the difference, Δh, between the level in the right arm of the manometer and the level in the left arm. You have measured pressures in cm of water, whereas our barometric pressure is usually in mm Hg. For convenience we will use cm H_2O as our unit, and convert cm Hg to cm H_2O by a conversion factor:

$$P_{bar} \text{ in cm } H_2O = P_{bar} \text{ in } \mathbf{cm} \text{ Hg} \times \text{density of Hg/density of } H_2O$$
$$= 13.57 \times P_{bar} \text{ in } \mathbf{cm} \text{ Hg} \tag{1}$$

$$\text{In units of cm } H_2O, \quad P_{total} = P_{bar} + \Delta h \tag{2}$$

Using Equations 1 and 2, make a table of temperatures t in °C and the values of P_{total} in cm H_2O. Then using graph paper or a software program like Excel, make a graph of total pressure vs temperature (see Appendix V). The P vs t graph should be a straight line, with the general equation $y = mx + b$, or, with our variables, $P = mt + b$, where m is the slope of the line and b is a constant. Using your graph, find the equation of the line. Then find the temperature t_o in °C at which P equals zero. Kelvin realized that if he added minus that temperature to t, defining T on his absolute temperature scale, he would obtain the simple equation $P = mT$, rather than $P = mt + b$. The Advance Study Assignment demonstrates how this is accomplished.

B. Determining the Barometric Pressure

In this part of the experiment we will use the same apparatus as in the first part. Disconnect the rubber tubing from the manometer and pour out the water. Take the flask out of the beaker and pour out the water. Clamp the flask in the beaker and firmly insert the stopper. Fill the beaker with room temperature water. Check with your thermometer to see that the water is indeed at room temperature. In this experiment it is imperative that the temperature of the flask remains constant.

Using your Pasteur pipet, add a few milliliters of water to the manometer, so that the levels are about 4 cm above the bottom of the U. Then connect the rubber tubing to the left arm of the manometer. The apparatus at that point should look like the left-hand sketch in Figure 8.1.

You have now isolated the air in the flask in a fixed initial state, which we will call State 1. That state is defined by its volume, pressure, and temperature. The volume, V_1, of the air is essentially that of the flask, V_{flask} (we will ignore the small volume of the air in the tubing). The pressure, P_1, is equal to the barometric pressure, P_{bar}, plus the difference between the heights of the liquid levels in the manometer, $h1_{right} - h1_{left}$ (see Appendix IV).

$$P_1 = P_{bar} + h1_{right} - h1_{left} \tag{3}$$

As in the first part of the experiment, we will use cm H_2O as our unit for pressure.

Measure the heights $h1_{right}$ and $h1_{left}$ with a ruler or meter stick and record them to the nearest 0.1 cm. If the temperature is not changing, these levels should hold steady. Wait three minutes and check to see they have not moved. If they move, add a little warm or cold water to the bath to bring it to room temperature. Don't proceed until the levels hold steady for at least two minutes.

In the next step of the experiment we compress the air in the flask just a little, by adding water to the right arm of the manometer. Add the water slowly, using your Pasteur pipet. You will see that both levels go up, but the right-hand level goes up faster than the left, since the air in the flask, as its volume is decreased, is going to a higher pressure than before. Continue adding water until the level is near the top of the right arm. Measure the heights of the levels in the right and left arms of the manometer and record them. The levels should not change with time. The air sample is now in State 2 (see Fig. 8.1).

In State 2, the pressure P_2 of the air is equal to P_{bar} plus $h2_{right} - h2_{left}$. The volume V_2 is slightly less than V_1 by an amount equal to the volume of the air in the manometer that was displaced by water as the air was compressed. Let's call that change in volume ΔV, equal to $(h2_{left} - h1_{left})$ times the cross sectional area, A, of the manometer tube.

$$\Delta V = (h2_{left} - h1_{left}) \times A \tag{4}$$

Summarizing, for States 1 and 2:

$$P_1 = P_{bar} + h1_{right} - h1_{left} \qquad V_1 = V_{flask} \text{ (to be measured)} \tag{5}$$

$$P_2 = P_{bar} + h2_{right} - h2_{left} \qquad V_2 = V_{flask} - \Delta V \tag{6}$$

$$\Delta P = P_2 - P_1$$

Since Boyle's Law is valid for the change in state,

$$P_1 V_1 = P_2 V_2 \text{ (gas sample at constant temperature)}$$

$$\text{or,} \quad P_1 V_1 = (P_1 + \Delta P)(V_1 - \Delta V) \tag{7}$$

Solving this equation for P_1, we find that

$$P_1 = (V_1 \Delta P - \Delta V \, \Delta P)/\Delta V \tag{8}$$

We can use Equations 8 and 5 to find P_{bar}, but first we need to calibrate the apparatus by measuring V_{flask} and the cross sectional area A of the manometer.

Calibrating the System

The volume V_1 of the apparatus is essentially equal to the volume of the flask, V_{flask}, since it is much larger than the volume of the tubing or the manometer. To determine the volume of the flask, take it out of the beaker and fill it with water up to about 1 cm of the top. Pour the water into a tared beaker and record the mass of the water. We can assume that the density of water is 1g/mL, so the volume of the flask in mL is numerically equal to the mass of the water.

The cross sectional area of the manometer tube is found in a similar way. Fill the manometer with water, using your finger to stopper the tube as necessary. Pour the water into a tared small beaker and record its mass, and hence its volume. Then measure the lengths of the arms of the manometer, and the length of the U as $\pi \times r$, the radius. The length of the manometer is the sum of these lengths. Use the length and volume to find the cross sectional area A in cm^2.

Using the data you have obtained and the relevant equations, calculate P_1 and P_{bar}. The units of P_{bar} will be cm H$_2$O, which you can convert to cm Hg by dividing by 13.57. Report the value of P in both sets of units. Compare the value you obtain with that found with the barometer in the laboratory.

Complete all of your calculations before leaving the lab. When you are finished, turn off the digital thermometer if you have one and return all of the borrowed equipment to the stockroom.

Experiment 8

Data and Calculations: Verifying the Absolute Zero of Temperature
Determination of the Barometric Pressure

A. Verifying the Absolute Zero of Temperature

Manometer heights in cm H_2O		Temp, t °C	Total P, cm H_2O
$h1_{right}$ _____	$h1_{left}$ _____	t_1 _____	P_1 _____
$h2_{right}$ _____	$h2_{left}$ _____	t_2 _____	P_2 _____
$h3_{right}$ _____	$h3_{left}$ _____	t_3 _____	P_3 _____
$h4_{right}$ _____	$h4_{left}$ _____	t_4 _____	P_4 _____
$h5_{right}$ _____	$h5_{left}$ _____	t_5 _____	P_5 _____

P_{bar} in cm Hg _____ P_{bar} in cm H_2O _____

To convert P_{bar} from cm Hg to cm H_2O, multiply by 13.57.

$$\text{Total} \quad P = P_{bar} + h_{right} - h_{left} \ \text{(for P in cm H_2O)}$$

Using graph paper or Excel, make a graph of Total P vs t. The way to do this is the same as we used in our illustration in the Introduction to Excel in Appendix VII. If you have gone through through that, making the graph for this experiment should be no problem. If you haven't, this would be a good time to start becoming familiar with Excel. The graph should be a straight line, with the equation $P = mt + b$, where m is the slope and b is a constant. If you need to find the equation by hand, see Appendix V for the method. If you find the graph and equation using software, print out the graph and include it with this report. In any case, report your result below:

Equation for P vs t _____

Find the temperature t_0 in °C where P becomes zero.

$$t_0 = \text{_____} \ °C \qquad \text{Let } A = -t_0, \ \text{ and } \ T = t + A$$

On the Kelvin scale, $T = 0$ at absolute zero, t_0 °C.

In your equation, substitute $T - A$ for t, and show that, with your value of A, the equation reduces to $P = mT$.

(continued on following page)

B. Measuring the Barometric Pressure

Manometer Heights in cm

$h1_{right}$ _____ $h2_{right}$ _____ t _____ °C

$h1_{left}$ _____ $h2_{left}$ _____

Using Equations 5 and 6, evaluate P_1, P_2, and ΔP

$P_1 = P_{bar} +$ _____ cm H_2O $P_2 = P_{bar} +$ _____ cm H_2O $\Delta P =$ _____ cm H_2O

Evaluate V_{flask} as described in the Calibration section:

Mass of water in flask _____ $V_{flask} =$ _____ mL

Evaluate the cross section A of the manometer tube:

Length of left arm _____ cm   Length of right arm _____ cm

Length of U section _____ cm   Total length _____ cm

Mass of water in full manometer _____ $g = V_{man}$ in mL

Cross sectional area A

_____ cm^2

Change in volume, DV _____ mL   (Eq. 4)

Using Equation 8, evaluate P_1 and P_{bar}.

$P_1 =$ _____ cm H_2O $P_{bar} =$ _____ cm H_2O

Express P_{bar} in cm Hg and compare your value to that obtained on the laboratory barometer.

$P_{bar} =$ _____ cm Hg $P_{bar\ obs} =$ _____ cm Hg

Experiment 8

Advance Study Assignment: Verifying the Absolute Zero of Temperature—
Determination of the Barometric Pressure

1. The barometric pressure on a spring day in Minnesota was found to be 733 mm Hg.

    a. What would the pressure be in cm H_2O?

    _____ cm H_2O

    b. About how many meters of water would it take to exert that pressure?

    _____ m H_2O

2. In this experiment a student found that when she increased the temperature of a 544 mL sample of air from 23.5°C to 31.5°C, the pressure of the air went from 1013 cm H_2O up to 1039 cm H_2O. Since the air expands linearly with temperature, the equation relating P to t is of the form:

$$P = mt + b \tag{9}$$

    where m is the slope of the line and b is a constant.

    a. What is the slope of the line? (Find the change in P divided by the change in t.)

    $m =$ _____ cm H_2O/°C

    b. Find the value of b. (Substitute known values of P and t into Equation 9 and solve for b.)

    $b =$ _____ cm H_2O

    c. Express Equation 9 in terms of the values of m and b.

    $P =$

    d. At what temperature t will P become zero?

    $P = 0$ at $t =$ _____ °C $= t_o = -A$

    e. The temperature in Part d is the absolute zero of temperature. Lord Kelvin suggested that we set up a scale on which that temperature is 0 K. On that scale, $T = t + A$. Show that, on the Kelvin scale, your equation reduces to $P = mT$.

Experiment 8

For use as needed.

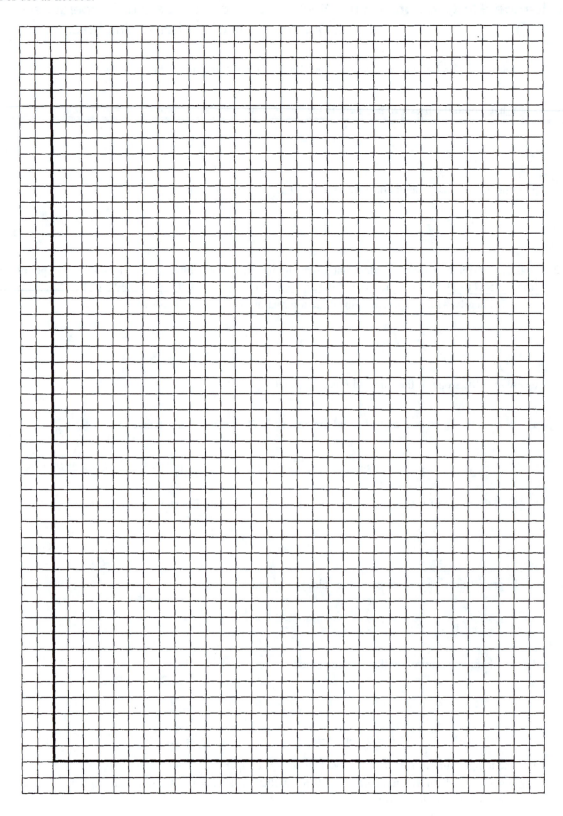

Experiment 9

Molar Mass of a Volatile Liquid

One of the important applications of the Ideal Gas Law is found in the experimental determination of the molar masses of gases and vapors. In order to measure the molar mass of a gas or vapor we need simply to determine the mass of a given sample of the gas under known conditions of temperature and pressure. If the gas obeys the Ideal Gas Law,

$$PV = nRT \tag{1}$$

If the pressure P is in atmospheres, the volume V in liters, the temperature T in K, and the amount n in moles, then the gas constant R is equal to 0.0821 L atm/(mole K).

From the measured values of P, V, and T for a sample of gas we can use Equation 1 to find the number of moles of gas in the sample. The molar mass in grams, MM, is equal to the mass g of the gas sample divided by the number of moles n.

$$n = \frac{PV}{RT} \qquad MM = \frac{g}{n} \tag{2}$$

This experiment involves measuring the molar mass of a volatile liquid using Equation 2. A small amount of the liquid is introduced into a weighed flask. The flask is then placed in boiling water, where the liquid will vaporize completely, driving out the air and filling the flask with vapor at barometric pressure and the temperature of the boiling water. If we cool the flask so that the vapor condenses, we can measure the mass of the vapor and calculate a value for MM.

Experimental Procedure*

WEAR YOUR SAFETY GLASSES WHILE
PERFORMING THIS EXPERIMENT

Obtain a special round-bottomed flask, a stopper and cap, and an unknown liquid from the storeroom. Support the flask on an evaporating dish or in a beaker at all times. If you should break or crack the flask, report it to your instructor immediately so that it can be repaired. With the stopper loosely inserted in the neck of the flask, weigh the empty, dry flask on the analytical balance. Use a copper loop, if necessary, to suspend the flask from the hook supporting the balance pan. With an automatic balance, use a cork ring to support the flask.

Pour about half your unknown liquid, about 5 mL, into the flask. Assemble the apparatus as shown in Figure 9.1. Place the cap on the neck of the flask. Add a few boiling chips to the water in the 600-mL beaker and heat the water to the boiling point. Watch the liquid level in your flask; the level should gradually drop as vapor escapes through the cap. After all the liquid has disappeared and no more vapor comes out of the cap, continue to boil the water gently for 5 to 8 minutes. Measure the temperature of the boiling water. Shut off the burner and wait until the water has stopped boiling (about $\frac{1}{2}$ minute) and then remove the cap and *immediately* insert the stopper used previously.

Loosen the clamp holding the flask in place. Remove the flask from the beaker of water, holding it by the neck, which will be cooler. Immerse the flask in a beaker of cool water to a depth of about 5 cm. After holding the flask in the water for about 2 minutes to allow it to cool, carefully remove the stopper *for not more than a second or two* to allow air to enter, and again insert the stopper. (As the flask cools the vapor inside condenses and the pressure drops, which explains why air rushes in when the stopper is removed.)

*See W. L. Masterton and T. R. Williams, J. Chem. Educ. *36*, 528 (1959). For an alternate apparatus, see instructor's manual.

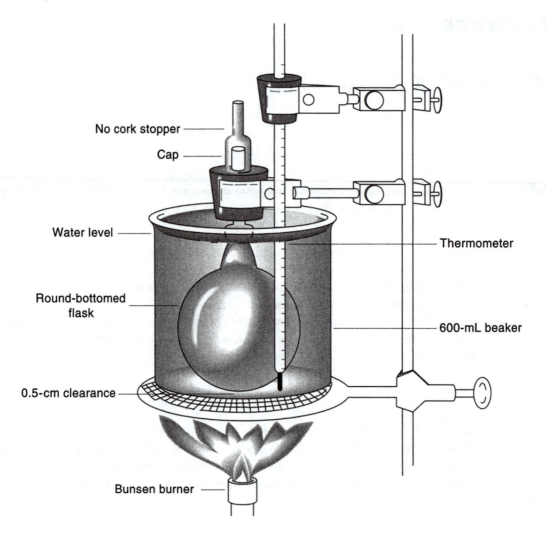

No cork stopper

Cap

Water level

Round-bottomed flask

0.5-cm clearance

Bunsen burner

Thermometer

600-mL beaker

Figure 9.1

Dry the flask with a towel to remove the surface water and let it cool to room temperature. Loosen the stopper again momentarily to equalize any pressure differences, and reweigh the flask. Read the atmospheric pressure from the barometer.

Repeat the procedure using another 5 mL of your liquid sample.

You may obtain the volume of the flask from your instructor. Alternatively, you may be directed to measure its volume by weighing the flask stoppered and full of water on a top-loading balance. *Do not* fill the flask with water unless specifically told to do so.

When you have completed the experiment, return the flask to the storeroom; do not attempt to wash or clean it in any way.

Experiment 9

Data and Calculations: Molar Mass of a Volatile Liquid

	Trial 1	**Trial 2**
Unknown no.	_____	
Mass of flask and stopper	_____ g	_____ g
Mass of flask, stopper, and condensed vapor	_____ g	_____ g
Mass of flask, stopper, and water (see directions)	_____ g	_____ g
Temperature of boiling water bath	_____ °C	_____ °C
Barometric pressure	_____ mm Hg	_____ mm Hg

Calculations and Results

	Trial 1	**Trial 2**
Pressure of vapor, P	_____ atm	_____ atm
Volume of flask (volume of vapor), V	_____ L	_____ L
Temperature of vapor, T	_____ K	_____ K
Mass of vapor, g	_____ g	_____ g
Number of moles of vapor, n	_____ M	_____ M
Molar mass of unknown, as found by substitution into Equation 2	_____ g	_____ g

Experiment 9

Advance Study Assignment: Molar Mass of a Volatile Liquid

1. A student weighs an empty flask and stopper and finds the mass to be 55.844 g. She then adds about 5 mL of an unknown liquid and heats the flask in a boiling water bath at 99.7°C. After all the liquid is vaporized, she removes the flask from the bath, stoppers it, and lets it cool. After it is cool, she momentarily removes the stopper, then replaces it and weighs the flask and condensed vapor, obtaining a mass of 56.101 g. The volume of the flask is known to be 248.1 mL. The barometric pressure in the laboratory that day is 752 mm Hg.

   a. What was the pressure of the vapor in the flask in atm?

   $P =$ _____ atm

   b. What was the temperature of the vapor in K? the volume of the flask in liters?

   $T =$ _____ K $V =$ _____ L

   c. What was the mass of vapor that was present in the flask?

   $g =$ _____ grams

   d. How many moles of vapor are present?

   $n =$ _____ moles

   e. What is the mass of one mole of vapor (Eq. 2)?

   $MM =$ _____ g/mole

2. How would each of the following procedural errors affect the results to be expected in this experiment? Give your reasoning in each case.

   a. Not all of the liquid was vaporized when the flask was removed from the water bath.

   b. The flask was not dried before the final weighing with the condensed vapor inside.

   c. The flask was left open to the atmosphere while it was being cooled, and the stopper was inserted just before the final weighing.

   d. The flask was stoppered and removed from the bath before the vapor had reached the temperature of the boiling water. All the liquid had vaporized.

Experiment 10

Analysis of an Aluminum-Zinc Alloy*

Some of the more active metals will react readily with solutions of strong acids, producing hydrogen gas and a solution of a salt of the metal. Small amounts of hydrogen are commonly prepared by the action of hydrochloric acid on metallic zinc:

$$Zn(s) + 2\ H^+(aq) \rightarrow H_2(g) + Zn^{2+}(aq) \tag{1}$$

From this equation it is clear that one mole of zinc produces one mole of hydrogen gas in this reaction. If the hydrogen were collected under known conditions, it would be possible to calculate the mass of zinc in a pure sample by measuring the amount of hydrogen it produced on reaction with acid.

Since aluminum reacts spontaneously with strong acids in a manner similar to that shown by zinc,

$$2\ Al(s) + 6\ H^+(aq) \rightarrow 2\ Al^{3+}(aq) + 3\ H_2(g) \tag{2}$$

we could find the amount of aluminum in a pure sample by measuring the amount of hydrogen produced by its reaction with an acid solution. In this case two moles of aluminum would produce three moles of hydrogen.

Since the amount of hydrogen produced by a gram of zinc is not the same as the amount produced by a gram of aluminum,

$$1\ \text{mole Zn} \rightarrow 1\ \text{mole H}_2, \quad 65.4\ \text{g Zn} \rightarrow 1\ \text{mole H}_2, \quad 1.00\ \text{g Zn} \rightarrow 0.0153\ \text{mole H}_2 \tag{3}$$

$$2\ \text{moles Al} \rightarrow 3\ \text{moles H}_2, \quad 54.0\ \text{g Al} \rightarrow 3\ \text{moles H}_2, \quad 1.00\ \text{g Al} \rightarrow 0.0556\ \text{mole H}_2 \tag{4}$$

it is possible to react an alloy of zinc and aluminum of known mass with acid, determine the amount of hydrogen gas evolved, and calculate the percentages of zinc and aluminum in the alloy, using Relations 3 and 4. The object of this experiment is to make such an analysis.

In this experiment you will react a weighed sample of an aluminum-zinc alloy with an excess of acid and collect the hydrogen gas evolved over water (Fig. 10.1). If you measure the volume, temperature, and total pressure of the gas and use the Ideal Gas Law, taking proper account of the pressure of water vapor in the system, you can calculate the number of moles of hydrogen produced by the sample:

$$P_{H_2}V = n_{H_2}RT, \quad n_{H_2} = \frac{P_{H_2}V}{RT} \tag{5}$$

The volume V and the temperature T of the hydrogen are easily obtained from the data. The pressure exerted by the dry hydrogen P_{H_2} requires more attention. The total pressure P of gas in the flask is, by Dalton's Law, equal to the partial pressure of the hydrogen P_{H_2} plus the partial pressure of the water vapor P_{H_2O}:

$$P = P_{H_2} + P_{H_2O} \tag{6}$$

The water vapor in the flask is present with liquid water, so the gas is saturated with water vapor; the pressure P_{H_2O} under these conditions is equal to the vapor pressure VP_{H_2O} of water at the temperature of the experiment. This value is constant at a given temperature, and is found in Appendix I at the end of this manual. The total gas pressure P in the flask is very nearly equal to the barometric pressure P_{bar}.

Substituting these values into (6) and solving for P_{H_2}, we obtain

$$P_{H_2} = P_{bar} - VP_{H_2O} \tag{7}$$

---

*W. L. Masterton, J. Chem. Educ. *38*, 558 (1961). For a source of alloy samples, see instructor's manual.

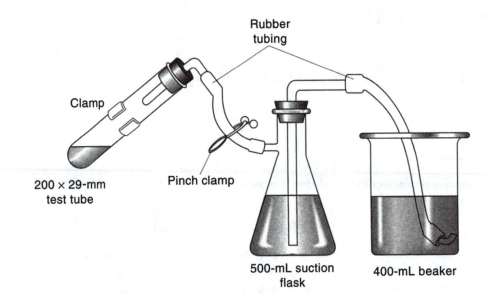

Figure 10.1

Using (5), you can now calculate n_{H_2}, the number of moles of hydrogen produced by your weighed sample. You can then calculate the percentages of Al and Zn in the sample by properly applying (3) and (4) to your results. For a sample containing g_{Al} grams Al and g_{Zn} grams Zn, it follows that

$$n_{H_2} = (g_{Al} \times 0.0556) + (g_{Zn} \times 0.0153) \tag{8}$$

For a one-gram sample, g_{Al} and g_{Zn} represent the mass fractions of Al and Zn, that is, % Al/100 and % Zn/100. Therefore

$$N_{H_2} = \left(\frac{\% \, Al}{100} \times 0.0556\right) + \left(\frac{\% \, Zn}{100} \times 0.0153\right) \tag{9}$$

where N_{H_2} = number of moles of H_2 produced *per gram* of sample.

    Since it is also true that

$$\% \, Zn = 100 - \% \, Al \tag{10}$$

Equation 9 can be written in the form

$$N_{H_2} = \left(\frac{\% \, Al}{100} \times 0.0556\right) + \left(\frac{100 - \% \, Al}{100} \times 0.0153\right) \tag{11}$$

We can solve Equation 11 directly for % Al if we know the number of moles of H_2 evolved per gram of sample. To save time in the laboratory and to avoid arithmetic errors, it is highly desirable to prepare in advance a graph giving N_{H_2} as a function of % Al. Then when N_{H_2} has been determined in the experiment, % Al in the sample can be read directly from the graph. Directions for preparing such a graph are given in Problem 1 in the Advance Study Assignment.

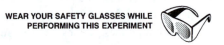

WEAR YOUR SAFETY GLASSES WHILE
PERFORMING THIS EXPERIMENT

Experimental Procedure

Obtain a suction flask, large test tube, stopper assemblies, and a sample of Al-Zn alloy from the stockroom. Assemble the apparatus as shown in Figure 10.1.

    Take a gelatin capsule from the supply on the lab bench and weigh it on the analytical balance to ± 0.0001g. Pour your alloy sample out on a piece of paper and add about half of it to the capsule. If necessary, break up the turnings into smaller pieces by simply tearing them. Cover the capsule and weigh it again. The mass of sample should be between 0.1500 and 0.2500 g. Use care in both weighings, since the sample is small and a small weighing error will produce a large experimental error. Put the remaining alloy back in its container.

Fill the suction flask and beaker about two thirds full of water. Moisten the stopper on the suction flask and insert it firmly into the flask. Open the pinch clamp and apply suction to the tubing attached to the side arm of the suction flask. Pull water into the flask from the beaker until the water level in the flask is 4 or 5 cm below the side arm. To apply suction, use a suction bulb or a short piece of rubber tubing attached temporarily to the tube that goes through the test tube stopper. Close the pinch clamp to prevent siphoning. The tubing from the beaker to the flask should be full of water, with no air bubbles.

Carefully remove the tubing from the beaker and put the end on the lab bench. As you do this, no water should leak out of the end of the tubing. Pour the water remaining in the beaker into another beaker, letting the 400-mL beaker drain for a second or two. Without drying it, weigh the empty beaker on a top-loading balance to ± 0.1 g. Put the tubing back in this beaker.

Pour 10. mL of 6 M HCl, hydrochloric acid, as measured in your graduated cylinder, into the large test tube. Drop the gelatin capsule into the HCl solution; if it sticks to the tube, poke it down into the acid with your stirring rod. Insert the stopper firmly into the test tube and open the pinch clamp. If a little water goes into the beaker at that point, pour that water out, letting the beaker drain for a second or two.

Within 3 or 4 minutes the acid will eat through the wall of the capsule and begin to react with the alloy. The hydrogen gas that is formed will go into the suction flask and displace water from the flask into the beaker. The volume of water that is displaced will equal the volume of gas that is produced. As the reaction proceeds you will probably observe a dark foam, which contains particles of unreacted alloy. The foam may carry some of the alloy up the tube. Wiggle the tube gently to make sure that all of the alloy gets into the acid solution. The reaction should be over within 5 to 10 minutes. At that time the liquid solution will again be clear, the foam will be essentially gone, the capsule will be all dissolved, and there should be no unreacted alloy. When the reaction is over, close the pinch clamp and take the tubing out of the beaker. Weigh the beaker and the displaced water to ± 0.1 g. Measure the temperature of the water and the barometric pressure.*

Optional Experiment: Using the procedure and apparatus in this experiment, find the % Al in an aluminum soda or beer can.

   DISPOSAL OF REACTION PRODUCTS.   Pour the acid solution into the waste crock. Reassemble the apparatus and repeat the experiment with the remaining sample of alloy.

---

*The pressure of the gas in the flask will be slightly different from the barometric pressure, because the water levels inside and outside the flask are not quite equal. The error arising from this effect is smaller than other experimental errors, so we shall ignore it.

Experiment 10

Data: Analysis of an Aluminum-Zinc Alloy

	Trial 1	Trial 2
Mass of gelatin capsule	_____ g	_____ g
Mass of alloy sample plus capsule	_____ g	_____ g
Mass of empty beaker	_____ g	_____ g
Mass of beaker plus displaced water	_____ g	_____ g
Barometric pressure	_____ mm Hg	
Temperature	_____ °C	

Calculations

	Trial 1	Trial 2
Mass of alloy sample	_____ g	_____ g
Mass of displaced water	_____ g	_____ g
Volume of displaced water (density of water, $\rho = 1.00$ g/mL)	_____ mL	_____ mL
Volume of H_2, V	_____ L	_____ L
Temperature of H_2, T	_____ K	
Vapor pressure of water at T, VP_{H_2O}, from Appendix I	_____ mm Hg	
Pressure of dry H_2, P_{H_2}	_____ mm Hg;	_____ atm
Moles H_2 from sample, n_{H_2}	_____ moles	_____ moles
Moles H_2 per gram of sample, N_{H_2}	_____ moles/g	_____ moles/g
% Al (read from graph)	_____ %	_____ %

Unknown no.   _____

Experiment 10

Advance Study Assignment: Analysis of an Aluminum-Zinc Alloy

1. On the following page, construct a graph of N_{H_2} vs. % Al. To do this, refer to Equation 11 and the discussion preceding it. Note that a plot of N_{H_2} vs. % Al should be a straight line (why?). To fix the position of a straight line it is necessary to locate only two points. The most obvious way to do this is to find N_{H_2} when % Al = 0 and when % Al = 100. If you wish you may calculate some intermediate points (for example, N_{H_2} when % Al = 50, or 20, or 70); all these points should lie on the same straight line. To use Excel, set up Equation 11 for different Al percentages. Graph N_{H_2} vs. % Al.

2. A student obtained the following data in this experiment. Fill in the blanks in the data and make the indicated calculations:

Mass of gelatin capsule   0.1284 g   Ambient temperature, t 22°C

Mass of capsule plus alloy
sample   0.3432 g   Ambient temperature, T _____ K

Mass of alloy sample, m _____ g   Barometric pressure   745 mm Hg

Mass of empty beaker   144.5 g   Vapor pressure of
                                             H_2O at t (Appendix I) _____ mm Hg

Mass of beaker plus displaced water   368.9 g

                                            Pressure of dry H_2,
Mass of displaced water _____ g P_{H_2} (Eq. 7)   _____ mm Hg

Volume of displaced water
(density = 1.00 g/mL) _____ mL   Pressure of dry H_2 _____ atm

Volume, V, of H_2 = Volume of displaced water _____ mL; _____ liters

Find the number of moles of H_2 evolved, n_{H_2} (Eq. 5; V in liters, P_{H_2} in atm, T in K, $R = 0.0821$ liter-atm/mole K).

                                                     _____ moles H_2

Find N_{H_2}, the number of moles of H_2 per gram of sample (n_{H_2}/m).

                                                     _____ moles H_2/g

Find the % Al in the sample from the graph prepared for Problem 1.   _____ % Al

Find the % Al in the sample by using Equation 11.

                                                     _____ % Al

Experiment 10

Analysis of an Aluminum-Zinc Alloy

(Advance Study Assignment)

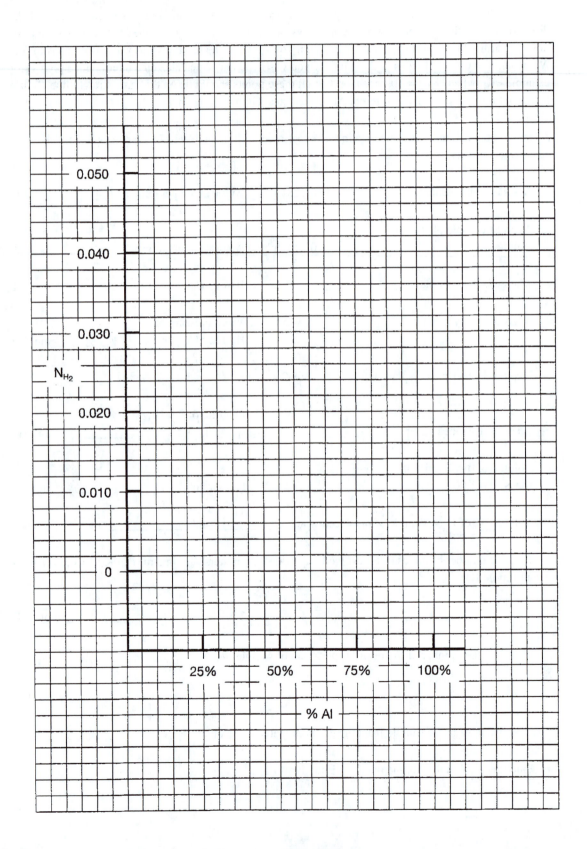

Experiment 11

The Atomic Spectrum of Hydrogen

When atoms are excited, either in an electric discharge or with heat, they tend to give off light. The light is emitted only at certain wavelengths that are characteristic of the atoms in the sample. These wavelengths constitute what is called the atomic spectrum of the excited element and reveal much of the detailed information we have regarding the electronic structure of atoms.

Atomic spectra are interpreted in terms of quantum theory. According to this theory, atoms can exist only in certain states, each of which has an associated fixed amount of energy. When an atom changes its state, it must absorb or emit an amount of energy that is just equal to the difference between the energies of the initial and final states. This energy may be absorbed or emitted in the form of light. The emission spectrum of an atom is obtained when excited atoms fall from higher to lower energy levels. Since there are many such levels, the atomic spectra of most elements are very complex.

Light is absorbed or emitted by atoms in the form of photons, each of which has a specific amount of energy, ϵ. This energy is related to the wavelength of light by the equation

$$\epsilon_{photon} = \frac{hc}{\lambda} \tag{1}$$

where h is Planck's constant, 6.62608×10^{-34} joule seconds, c is the speed of light, 2.997925×10^8 meters per second, and λ is the wavelength, in meters. The energy ϵ_{photon} is in joules and is the energy given off by one atom when it jumps from a higher to a lower energy level. Since total energy is conserved, the change in energy of the atom, $\Delta\epsilon_{atom}$, must equal the energy of the photon emitted:

$$\Delta\epsilon_{atom} = \epsilon_{photon} \tag{2}$$

where $\Delta\epsilon_{atom}$ is equal to the energy in the upper level minus the energy in the lower one. Combining Equations 1 and 2, we obtain the relation between the change in energy of the atom and the wavelength of light associated with that change:

$$\Delta\epsilon_{atom} = \epsilon_{upper} - \epsilon_{lower} = \epsilon_{photon} = \frac{hc}{\lambda} \tag{3}$$

The amount of energy in a photon given off when an atom makes a transition from one level to another is very small, on the order of 1×10^{-19} joules. This is not surprising since, after all, atoms are very small particles. To avoid such small numbers, we will work with 1 mole of atoms, much as we do in dealing with energies involved in chemical reactions. To do this we need only to multiply Equation 3 by Avogadro's number, N. Let

$$N\Delta\epsilon = \Delta E = N\epsilon_{upper} - N\epsilon_{lower} = E_{upper} - E_{lower} = \frac{Nhc}{\lambda}$$

Substituting the values for N, h, and c, and expressing the wavelength in nanometers rather than meters (1 meter = 1×10^9 nanometers), we obtain an equation relating energy change in kilojoules per mole of atoms to the wavelength of photons associated with such a change:

$$\Delta E = \frac{6.02214 \times 10^{23} \times 6.62608 \times 10^{-34}\,\text{J sec} \times 2.997925 \times 10^8\,\text{m/sec}}{\lambda\ (\text{in nm})} \times \frac{1 \times 10^9\,\text{nm}}{1\,\text{m}} \times \frac{1\,\text{kJ}}{1000\,\text{J}}$$

$$\Delta E = E_{upper} - E_{lower} = \frac{1.19627 \times 10^5\,\text{kJ/mole}}{\lambda\ (\text{in nm})} \quad \text{or} \quad \lambda\ (\text{in nm}) = \frac{1.19627 \times 10^5}{\Delta E\ (\text{in kJ/mole})} \tag{4}$$

Equation 4 is useful in the interpretation of atomic spectra. Say, for example, we study the atomic spectrum of sodium and find that the wavelength of the strong yellow line is 589.16 nm (see Fig. 11.1).

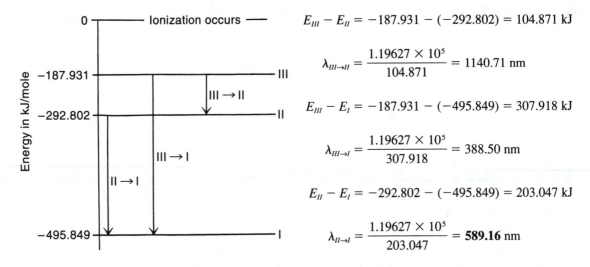

Figure 11.1 Calculation of wavelengths of spectral lines from energy levels of the sodium atom.

This line is known to result from a transition between two of the three lowest levels in the atom. The energies of these levels are shown in Figure 11.1. To determine which of the levels give rise to the 589.16 nm line, we note that there are three possible transitions, shown by downward arrows in the figure. We find the wavelengths associated with those transitions by first calculating ΔE ($E_{\text{upper}} - E_{\text{lower}}$) for each transition. Knowing ΔE we calculate λ by Equation 4. Clearly, the II $\rightarrow$ I transition is the source of the yellow line in the spectrum.

The simplest of all atomic spectra is that of the hydrogen atom. In 1886 Balmer showed that the lines in the spectrum of the hydrogen atom had wavelengths that could be expressed by a rather simple equation. Bohr, in 1913, explained the spectrum on a theoretical basis with his famous model of the hydrogen atom. According to Bohr's theory, the energies allowed to a hydrogen atom are given by the equation

$$\epsilon_n = \frac{-B}{n^2} \qquad (5)$$

where B is a constant predicted by the theory and n is an integer, 1, 2, 3, . . . , called a quantum number. It has been found that all the lines in the atomic spectrum of hydrogen can be associated with energy levels in the atom which are predicted with great accuracy by Bohr's equation. When we write Equation 5 in terms of a mole of H atoms, and substitute the numerical value for B, we obtain

$$E_n = \frac{-1312.04}{n^2} \text{ kilojoules per mole,} \quad n = 1, \ 2, \ 3, \dots \qquad (6)$$

Using Equation 6 you can calculate, very accurately indeed, the energy levels for hydrogen. Transitions between these levels give rise to the wavelengths in the atomic spectrum of hydrogen. These wavelengths are also known very accurately. Given both the energy levels and the wavelengths, it is possible to determine the actual levels associated with each wavelength. In this experiment your task will be to make determinations of this type for the observed wavelengths in the hydrogen atomic spectrum that are listed in Table 11.1.

Table 11.1

Some Wavelengths (in nm) in the Spectrum of the Hydrogen Atom as Measured in a Vacuum					
Wavelength	**Assignment** $n_{\text{hi}} \rightarrow n_{\text{lo}}$	**Wavelength**	**Assignment** $n_{\text{hi}} \rightarrow n_{\text{lo}}$	**Wavelength**	**Assignment** $n_{\text{hi}} \rightarrow n_{\text{lo}}$
97.25	_____	410.29	_____	1005.2	_____
102.57	_____	434.17	_____	1094.1	_____
121.57	_____	486.27	_____	1282.2	_____
389.02	_____	656.47	_____	1875.6	_____
397.12	_____	954.86	_____	4052.3	_____

Experimental Procedure

There are several ways we might analyze an atomic spectrum, given the energy levels of the atom involved. A simple and effective method is to calculate the wavelengths of some of the lines arising from transitions between some of the lower energy levels, and see if they match those that are observed. We shall use this method in our experiment. All the data are good to at least five significant figures, so by using your hand calculator you should be able to make very accurate determinations. An Excel spreadsheet can be used to good advantage in this experiment. Your instructor may give you some suggestions on how you might proceed.

A. Calculation of the Energy Levels of the Hydrogen Atom

Given the expression for E_n in Equation 6, it is possible to calculate the energy for each of the allowed levels of the H atom starting with $n = 1$. Using your calculator, calculate the energy in kJ/mole of each of the 10 lowest levels of the H atom. Note that the energies are all negative, so that the *lowest* energy will have the *largest* allowed negative value. Enter these values in the table of energy levels, Table 11.2. On the energy level diagram provided just before the ASA, plot along the y axis each of the six lowest energies, drawing a horizontal line at the allowed level and writing the value of the energy alongside the line near the y axis. Write the quantum number associated with the level to the right of the line.

B. Calculation of the Wavelengths of the Lines in the Hydrogen Spectrum

The lines in the hydrogen spectrum all arise from jumps made by the atom from one energy level to another. The wavelengths in nm of these lines can be calculated by Equation 4, where ΔE is the difference in energy in kJ/mole between any two allowed levels. For example, to find the wavelength of the spectral line associated with a transition from the $n = 2$ level to the $n = 1$ level, calculate the difference, ΔE, between the energies of those two levels. Then substitute ΔE into Equation 4 to obtain this wavelength in nanometers.

Using the procedure we have outlined, calculate the wavelengths in nm of all the lines we have indicated in Table 11.3. That is, calculate the wavelengths of all the lines that can arise from transitions between any two of the six lowest levels of the H atom. Enter these values in Table 11.3.

C. Assignment of Observed Lines in the Hydrogen Spectrum

Compare the wavelengths you have calculated with those listed in Table 11.1. If you have made your calculations properly, your wavelengths should match, within the error of your calculation, several of those that are observed. On the line opposite each wavelength in Table 11.1, write the quantum numbers of the upper and lower states for each line whose origin you can recognize by comparison of your calculated values with the observed values. On the energy level diagram, draw a vertical arrow pointing down (light is emitted, $\Delta E < 0$) between those pairs of levels that you associate with any of the observed wavelengths. By each arrow write the wavelength of the line originating from that transition.

There are a few wavelengths in Table 11.1 that have not yet been calculated. Enter those wavelengths in Table 11.4. By assignments already made and by an examination of the transitions you have marked on the diagram, deduce the quantum states that are likely to be associated with the as yet unassigned lines. This is perhaps most easily done by first calculating the value of ΔE, which is associated with a given wavelength. Then find two values of E_n whose difference is equal to ΔE. The quantum numbers for the two E_n states whose energy difference is ΔE will be the ones that are to be assigned to the given wavelength. When you have found n_{hi} and n_{lo} for a wavelength, write them in Table 11.1 and Table 11.4; continue until all the lines in the table have been assigned.

D. The Balmer Series

This is the most famous series in the atomic spectrum of hydrogen. The lines in this series are the only ones in the spectrum that occur in the visible region. Your instructor may have a hydrogen source tube and a spectroscope with which you may be able to observe some of the lines in the Balmer series. In the Data and Calculations section are some questions you should answer relating to this series.

Experiment 11

Data and Calculations: The Atomic Spectrum of Hydrogen

A. Calculation of the Energy Levels of the Hydrogen Atom

Energies are to be calculated from Equation 6 for the 10 lowest energy states.

Table 11.2

Quantum Number, n	Energy, E_n, in kJ/mole	Quantum Number, n	Energy, E_n, in kJ/mole
_____	_____	_____	_____
_____	_____	_____	_____
_____	_____	_____	_____
_____	_____	_____	_____
_____	_____	_____	_____

B. Calculation of Wavelengths in the Spectrum of the H Atom

In the upper half of each box write ΔE, the difference in energy in kJ/mole between $E_{n_{hi}}$ and $E_{n_{lo}}$. In the lower half of the box, write λ in nm associated with that value of ΔE.

Table 11.3

n_{hi}	6	5	4	3	2	1
n_{lo}						
1						
2						
3						
4						
5						

$$\Delta E = E_{n_{hi}} - E_{n_{lo}}$$

$$\lambda \text{ (nm)} = \frac{1.19627 \times 10^5}{\Delta E}$$

(continued on following page)

C. Assignment of Wavelengths

1. As directed in the procedure, assign n_{hi} and n_{lo} for each wavelength in Table 11.1 which corresponds to a wavelength calculated in Table 11.3.

2. List below any wavelengths you cannot yet assign.

Table 11.4

Wavelength λ Observed	ΔE Transition	Probable Transition $n_{hi} \rightarrow n_{lo}$	λ Calculated in nm (Eq. 4)
_____	_____	_____	_____
_____	_____	_____	_____
_____	_____	_____	_____
_____	_____	_____	_____

D. The Balmer Series

1. When Balmer found his famous series for hydrogen in 1886, he was limited experimentally to wavelengths in the visible and near ultraviolet regions from 250 nm to 700 nm, so all the lines in his series lie in that region. On the basis of the entries in Table 11.3 and the transitions on your energy level diagram, what common characteristic do the lines in the Balmer series have?

What would be the longest possible wavelength for a line in the Balmer series?

$\lambda = $ _____ nm

What would be the shortest possible wavelength that a line in the Balmer series could have? Hint: What is the largest possible value of ΔE that can be associated with a line in the Balmer series?

$\lambda = $ _____ nm

Fundamentally, why would any line in the hydrogen spectrum between 250 nm and 700 nm belong to the Balmer series? Hint: On the energy level diagram note the range of possible values of ΔE for transitions to the $n = 1$ level and to the $n = 3$ level. Could a spectral line involving a transition to the $n = 1$ level have a wavelength in the range indicated?

(continued on following page)

The Ionization Energy of Hydrogen

1. In the normal hydrogen atom the electron is in its lowest energy state, which is called the ground state of the atom. The maximum electronic energy that a hydrogen atom can have is 0 kJ/mole, at which point the electron would essentially be removed from the atom and it would become a H^+ ion. How much energy in kilojoules per mole does it take to ionize an H atom?

_____ kJ/mole

The ionization energy of hydrogen is often expressed in units other than kJ/mole. What would it be in joules per atom? In electron volts per atom? ($1 \text{ ev} = 1.602 \times 10^{-19}$ J)

_____ J/atom; _____ ev/atom

(The energy level diagram to be completed in Part A is on the following page.)

Experiment 11

The Atomic Spectrum of Hydrogen: Energy Level Diagram

(Data and Calculations)

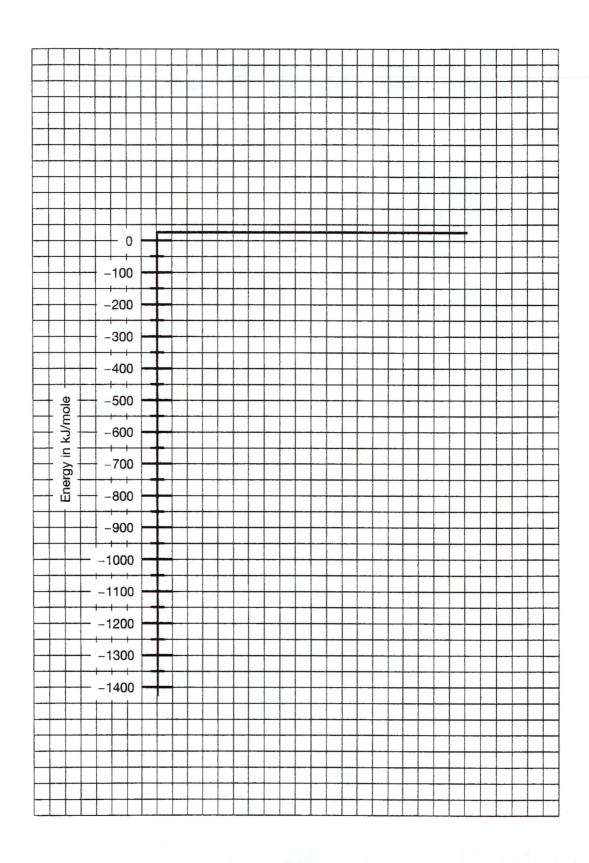

Experiment 11

Advance Study Assignment: The Atomic Spectrum of Hydrogen

1. The lithium dication, Li^{2+}, has an energy level formula analogous to that of the hydrogen atom, since both species have only one electron. The energy levels of the Li^{2+} ion are given by the equation

$$E_n = -\frac{11808}{n^2}\ kJ/mole \quad n = 1,\ 2,\ 3,\ldots$$

a. Calculate the energies in kJ/mole for the four lowest energy levels of the Li^{2+} ion.

$E_1 = $ _____ kJ/mole

$E_2 = $ _____ kJ/mole

$E_3 = $ _____ kJ/mole

$E_4 = $ _____ kJ/mole

b. One of the most important transitions for the Li^{2+} ion involves a jump from the $n = 2$ to the $n = 1$ level. ΔE for this transition equals $E_2 - E_1$, where these two energies are obtained as in Part a. Find the value of ΔE in kJ/mole. Find the wavelength in nm of the line emitted when this transition occurs; use Equation 4 to make the calculation.

$\Delta E = $ _____ kJ/mole; $\lambda = $ _____ nm

c. Three of the strongest lines in the Li^{2+} ion spectrum are observed at the following wavelengths: (1) 11.397 nm; (2) 54.030 nm; (3) 72.941 nm. Find the quantum numbers of the initial and final states for the transitions that give rise to these three lines. Do this by calculating, using Equation 4, the wavelengths of lines that can originate from transitions involving any two of the four lowest levels. You calculated one such wavelength in Part b. Make similar calculations with the other possible pairs of levels. When a calculated wavelength matches an observed one, write down n_{hi} and n_{lo} for that line. Continue until you have assigned all three of the lines.

(1) _____ $\rightarrow$ _____   (2) _____ $\rightarrow$ _____   (3) _____ $\rightarrow$ _____

Experiment 12

The Alkaline Earths and the Halogens—Two Families in the Periodic Table

The Periodic Table arranges the elements in order of increasing atomic number in horizontal rows of such length that elements with similar properties recur periodically; that is, they fall directly beneath each other in the Table. The elements in a given vertical column are referred to as a family or group. The physical and chemical properties of the elements in a given family change gradually as one goes from one element in the column to the next. By observing the trends in properties the elements can be arranged in the order in which they appear in the Periodic Table. In this experiment we will study the properties of the elements in two families in the Periodic Table, the alkaline earths (Group 2) and the halogens (Group 17).

The alkaline earths are all moderately reactive metals and include barium, beryllium, calcium, magnesium, radium, and strontium. (Since beryllium compounds are rarely encountered and often very poisonous, and radium compounds are highly radioactive, we will not include these two elements in this experiment.) All the alkaline earths exist in their compounds and in solution as M^{2+} cations (Mg^{2+}, Ca^{2+}, etc.). If a solution containing one of these cations is mixed with one containing an anion (CO_3^{2-}, SO_4^{2-}, IO_3^-, etc.), an alkaline earth salt will precipitate if the compound containing those two ions is insoluble.

For example:

$$M^{2+}(aq) + SO_4^{2-}(aq) \rightarrow MSO_4(s) \quad \text{if } MSO_4 \text{ is insoluble} \tag{1a}$$

$$M^{2+}(aq) + 2\ IO_3^-(aq) \rightarrow M(IO_3)_2(s) \quad \text{if } M(IO_3)_2 \text{ is insoluble} \tag{1b}$$

We would expect, and do indeed generally observe, that the salts composed of a given anion and the alkaline earth cations show a smooth trend in solubility, consistent with the order of the cations' corresponding elements in the Periodic Table. That is, as we go from one end of the alkaline earth family to the other, the solubilities of, say, the sulfate salts either gradually increase or decrease. Similar trends exist for the carbonates, oxalates, and iodates formed by those cations. By determining such trends in this experiment, you will be able to confirm the order of the alkaline earths in the Periodic Table.

The elemental halogens are also relatively reactive. They include astatine, bromine, chlorine, fluorine, and iodine. We will not study astatine or fluorine in this experiment, since the former is radioactive and the latter is too reactive to be safely employed in labs such as ours. Unlike the alkaline earths, the halogen tend to gain electrons, forming X^- anions (Cl^-, Br^-, etc.). Because of this property, the halogens are oxidizing agents, species that tend to oxidize (remove electrons from) other species. An interesting and simple example of the sort of reaction that may occur arises when a solution containing a halogen (Cl_2, Br_2, I_2) is mixed with a solution containing a (different) halide ion (Cl^-, Br^-, I^-). Taking X_2 to be the halogen, and Y^- to be a halide ion, the following reaction may occur, in which another halogen, Y_2, is formed:

$$X_2(aq) + 2\ Y^-(aq) \rightarrow 2\ X^-(aq) + Y_2(aq) \tag{2}$$

The reaction will occur if X_2 is a better oxidizing agent than Y_2, since then X_2 can produce Y_2 by removing electrons from the Y^- ions. If Y_2 is a better oxidizing agent than X_2, Reaction 2 will not proceed, but will instead be spontaneous in the opposite direction.

In this experiment we will mix solutions of halogens and halide ions to determine the relative oxidizing strengths of the halogens. These strengths show a smooth variation as one goes from one halogen to the next in the Periodic Table. We will be able to tell if a reaction occurs by the colors we observe. In water, and particularly in some organic solvents, the halogens have characteristic colors. The halide ions are colorless in aqueous solution and insoluble in organic solvents. Bromine (Br_2) in heptane, C_7H_{16} (HEP), is orange, while Cl_2 and I_2 in that solvent have quite different colors.

Say, for example, we shake an aqueous solution of Br_2 with a little heptane, which is lighter than and insoluble in water. The Br_2 is much more soluble in HEP than in water and goes into the HEP layer, giving it an orange color. To that mixture we add a solution containing a halide ion, say Cl^- ion, and mix well. If Br_2 is a better oxidizing agent than Cl_2, it will take electrons from the chloride ions and will be converted to bromide, Br^-, ions; the reaction would be

$$Br_2(aq) + 2\,Cl^-(aq) \rightarrow 2\,Br^-(aq) + Cl_2(aq) \tag{3}$$

If the reaction occurs, the color of the HEP layer will change, since Br_2 will be used up and Cl_2 will form. The color of the HEP layer will go from orange to that of a solution of Cl_2 in HEP. If the reaction does *not* occur, the color of the HEP layer will remain orange. By using this line of reasoning, and by working with the possible mixtures of halogens and halide ions, you should be able to arrange the halogens in order of increasing oxidizing power, which should correspond to their order in the Periodic Table.

One difficulty that you may have in this experiment involves terminology rather than actual chemistry. You must learn to distinguish the halogen *elements* from the halide *ions*, since the two kinds of species are not at all the same, even though their names are similar:

Elemental halogens	Halide ions
Bromine, Br_2	Bromide ion, Br^-
Chlorine, Cl_2	Chloride ion, Cl^-
Iodine, I_2	Iodide ion, I^-

The *halogens* are molecular substances and oxidizing agents, and all have odors. They are only slightly soluble in water and are much more soluble in HEP, where they have distinct colors. The *halide ions* exist in solution only in water, have no color or odor, and are *not* oxidizing agents. They do not dissolve in HEP.

Given the solubility properties of the alkaline earth cations, and the oxidizing power of the halogens, it is possible to develop a systematic procedure for determining the presence of any Group 2 cation and any Group 17 anion in a solution. In the last part of this experiment you will be asked to set up such a procedure and use it to establish the identity of an unknown solution containing a single alkaline earth halide.

WEAR YOUR SAFETY GLASSES WHILE
PERFORMING THIS EXPERIMENT

Experimental Procedure

I. Relative Solubilities of Some Salts of the Alkaline Earths

To each of four small test tubes add about 1 mL (approximately 12 drops) of 1 M H_2SO_4. Then add 1 mL of 0.1 M solutions of the nitrate salts of barium, calcium, magnesium, and strontium to those tubes, one solution to a tube. Stir each mixture with your glass stirring rod, rinsing the rod in a beaker of distilled water between stirs. Record your results on the solubilities of the sulfates of the alkaline earths in the Table, noting whether a precipitate forms, and any characteristics (such as color, amount, size of particles, and settling tendencies) that might distinguish it.

Rinse out the test tubes, and to each add 1 mL 1 M Na_2CO_3. Then add 1 mL of the solutions of the alkaline earth salts, one solution to a tube, as before. Record your observations on the solubility properties of the carbonates of the alkaline earth cations. Rinse out the tubes, and test for the solubilities of the oxalates of these cations, using 0.25 M $(NH_4)_2C_2O_4$ as the precipitating reagent. Finally, determine the relative solubilities of the iodates of the alkaline earths, using 1 mL 0.1 M KIO_3 as the test reagent.

II. Relative Oxidizing Powers of the Halogens

Place a few milliliters of bromine-saturated water in a small test tube, and add 1 mL of heptane. Stopper the test tube and shake until the bromine color is mostly in the HEP layer. **CAUTION:** *Avoid breathing the halogen vapors. Don't use your finger to stopper the tube, since a halogen solution can give you a bad chemical burn.* Repeat the experiment using chlorine water and iodine water with separate samples of HEP, noting any color changes as the bromine, chlorine, and iodine are extracted from the water layer into the HEP layer.

To each of three small test tubes add 1 mL bromine water and 1 mL HEP. Then add 1 mL 0.1 M NaCl to the first test tube, 1 mL 0.1 M NaBr to the second, and 1 mL 0.1 M NaI to the third. Stopper each tube and shake it. Note the color of the HEP phase above each solution. If the color is not that of Br_2 in HEP, a reaction indeed occurred, and Br_2 oxidized that anion, producing the halogen. In such a case, Br_2 is a stronger oxidizing agent than the halogen that was produced.

Rinse out the tubes, and this time add 1 mL chlorine water and 1 mL HEP to each tube. Then add 1 mL of the 0.1 M solutions of the sodium halide salts, one solution to a tube, as before. Stopper each tube and shake, noting the color of the HEP layer after shaking. Depending on whether the color is that of Cl_2 in HEP or not, decide whether Cl_2 is a better oxidizing agent than Br_2 or I_2. Again, rinse out the tubes, and add 1 mL iodine water and 1 mL HEP to each. Test each tube with 1 mL of a sodium halide salt solution, and determine whether I_2 is able to oxidize Cl^- or Br^- ions. Record all your observations in the Table.

III. Identification of an Alkaline Earth Halide

Your observations on the solubility properties of the alkaline earth cations should allow you to develop a method for determining which of those cations is present in a solution containing one Group 2 cation and no other cations. The method will involve testing samples of the solution with one or more of the reagents you used in Part I. Indicate on the Data page how you would proceed.

In a similar way you can determine which halide ion is present in a solution containing only one such anion and no others. There you will need to test a solution of an oxidizing halogen with your unknown to see how the halide ion is affected. From the behavior of the halogen-halide ion mixtures you studied in Part II you should be able to identify easily the particular halide that is present. Describe your method on the Data page, obtain an unknown solution of an alkaline earth halide, and then use your procedure to determine the cation and anion that it contains.

IV. Microscale Procedure for Determining Solubilities of Alkaline Earth Salts   Optional

Your instructor may have you carry out Part I of this experiment by a microscale approach. This method uses much smaller amounts of reagents. Plastic well plates are employed as containers, and reagents are measured out with small Beral pipettes.

Using Beral pipettes, add four drops 0.1 M $Ba(NO_3)_2$, barium nitrate, to wells A1–A4, four drops to each well. Similarly, add four drops 0.1 M $Ca(NO_3)_2$, calcium nitrate, to wells B1–B4; four drops of 0.1 M $Mg(NO_3)_2$, magnesium nitrate, to wells C1–C4, and four drops 0.1 M $Sr(NO_3)_2$, strontium nitrate, to wells D1–D4.

Then, with another Beral pipette, add four drops 1 M H_2SO_4, sulfuric acid, to wells A1–D1. In the Table, record your results on the solubilities of the sulfates of the alkaline earths. Note whether a precipitate formed, and any characteristics, such as amount, size of particles, and cloudiness, which might distinguish it.

With a different Beral pipette, add four drops of 1 M Na_2CO_3, sodium carbonate, to wells A2–D2. Record your observations on the solubilities of the carbonates of the alkaline earths. Then carry out the same sort of tests with 0.25 M $(NH_4)_2C_2O_4$, ammonium oxalate, in wells A3–D3, and finally with 0.1 M KIO_3, potassium iodate, in wells A4–D4. Note all of your observations in the Table.   ▪

Optional Experiment: Dissolve a sample of limestone in acid, and determine whether it contains Mg^{2+} as well as Ca^{2+} ions. Estimate the relative concentrations of the two ions.

DISPOSAL OF REACTION PRODUCTS: Dispose of the reaction products from this experiment as directed by your instructor.

Experiment 12

Observations and Conclusions: The Alkaline Earths and the Halogens

I. Solubilities of Salts of the Alkaline Earths

	1 M H_2SO_4	1 M Na_2CO_3	0.25 M $(NH_4)_2C_2O_4$	0.1 M KIO_3
$Ba(NO_3)_2$				
$Ca(NO_3)_2$				
$Mg(NO_3)_2$				
$Sr(NO_3)_2$				

Key: P = precipitate forms; S = no precipitate forms.

Note any distinguishing characteristics of precipitate, such as amount and degree of cloudiness.

Consider the relative solubilities of the Group 2 cations in the various precipitating reagents. On the basis of the trends you observed, list the four alkaline earths in the order in which they should appear in the Periodic Table. *Start with the one which forms the most soluble oxalate.*

most soluble _____ _____ _____ _____ least soluble

Why did you arrange the elements as you did? Is the order consistent with the properties of the cations in all of the participating reagents?

II. Relative Oxidizing Powers of the Halogens

   a. Color of the halogens in solution:

	Br_2	Cl_2	I_2
HEP	_____	_____	_____
Water	_____	_____	_____

(continued on following page)

b. Reactions between halogens and halides:

	Br$^-$	Cl$^-$	I$^-$
Br$_2$			
Cl$_2$			
I$_2$			

State your observations with each mixture, noting the initial and final colors of the HEP layer, and which halogen ends up in the HEP layer. Key: R = reaction occurs; NR = no reaction occurs.

Rank the halogens in order of their increasing oxidizing power.

weakest _____ _____ _____ strongest

Is this their order in the Periodic Table?

III. Identification of an Alkaline Earth Halide
Procedure for identifying the Group 2 cation:

Procedure for identifying the Group 17 anion:

Observations on unknown alkaline earth halide solution:

Cation present_____

Anion present_____

Unknown no._____

Experiment 12

Advance Study Assignment: The Alkaline Earths and the Halogens

1. As pure elements, all of the halogens are diatomic molecular species. Their melting points are: F_2, 85 K, Cl_2, 239 K, Br_2, 332 K, and I_2, 457 K. Using the Periodic Table, predict as best you can the molecular formula of elemental astatine, At, the only radioactive element in this family. Also predict whether it will be a solid, liquid, or gas at room temperature.

   Elemental formula: _____   Phase at room temperature: _____

2. Substances A, B, and C can all act as oxidizing agents. In solution, A is green, B is yellow, and C is red. In the reactions in which they participate, they are reduced to A^-, B^-, and C^- ions, all of which are colorless. When a solution of C is mixed with one containing B^- ions, the color changes from red to yellow.

   Which species is oxidized? _____

   Which is reduced? _____

   When a solution of C is mixed with one containing A^- ions, the color remains red.

   Is C a better oxidizing agent than A? _____

   Is C a better oxidizing agent than B? _____

   Arrange A, B, and C in order of increasing strength as an oxidizing agent.

3. You are given an unknown, colorless, solution that contains one of the following salts: NaA, NaB, NaC. In solution, each salt dissociates completely into the Na^+ ion and the anion A^-, B^-, or C^-, whose properties are given in Problem 2. The Na^+ ion is effectively inert. Given the availability of solutions of A, B, and C, develop a simple procedure for identifying the salt that is present in your unknown.

Experiment 13

The Geometrical Structure of Molecules— An Experiment Using Molecular Models

Many years ago it was observed that in many of its compounds the carbon atom formed four chemical linkages to other atoms. As early as 1870, graphic formulas of carbon compounds were drawn as shown:

$$\begin{array}{ccc}
 & H & \\
 & | & \\
H - & C & - H \\
 & | & \\
 & H &
\end{array}
\qquad\qquad
\begin{array}{cc}
H & H \\
| & | \\
C & = C \\
| & | \\
H & H
\end{array}$$

methane   ethylene

Although such drawings as these would imply that the atom-atom linkages, indicated by valence strokes, lie in a plane, chemical evidence, particularly the existence of only one substance with the graphic formula

$$\begin{array}{c}
Cl \\
| \\
H - C - Cl \\
| \\
H
\end{array}$$

requires that the linkages be directed toward the corners of a tetrahedron, at the center of which is the carbon atom.

The physical significance of the chemical linkages between atoms, expressed by the lines or valence strokes in molecular structure diagrams, became evident soon after the discovery of the electron. In a classic paper, G. N. Lewis suggested, on the basis of chemical evidence, that the single bonds in graphic formulas involve two electrons and that an atom tends to hold eight electrons in its outermost or valence shell.

Lewis' proposal that atoms generally have eight electrons in their outer shells proved to be extremely useful and has come to be known as the octet rule. It can be applied to many atoms, but is particularly important in the treatment of covalent compounds of atoms in the second row of the Periodic Table. For atoms such as carbon, oxygen, nitrogen, and fluorine, the eight valence electrons occur in pairs that occupy tetrahedral positions around the central atom core. Some of the electron pairs do not participate directly in chemical bonding and are called unshared or nonbonding pairs; however, the structures of compounds containing such unshared pairs reflect the tetrahedral arrangement of the four pairs of valence shell electrons. In the H_2O molecule, which obeys the octet rule, the four pairs of electrons around the central oxygen atom occupy essentially tetrahedral positions; there are two unshared nonbonding pairs and two bonding pairs that are shared by the O atom and the two H atoms. The $H-O-H$ bond angle is nearly but not exactly tetrahedral since the properties of shared and unshared pairs of electrons are not exactly alike.

$$\begin{array}{c}
\ddot{O} \\
\diagup \quad \diagdown \\
H \qquad H
\end{array}$$

Most molecules obey the octet rule. Essentially, all organic molecules obey the rule, and so do most inorganic molecules and ions. For species that obey the octet rule it is possible to draw electron-dot, or Lewis, structures. The previous drawing of the H_2O molecule is an example of a Lewis structure. Here are several others:

$$\begin{array}{ccccc}
\begin{array}{c}
:\ddot{Cl}: \\
| \\
H - C - H \\
| \\
:\ddot{Cl}:
\end{array}
&
\begin{array}{c}
\ddot{} \\
H - \ddot{N} - H \\
| \\
H
\end{array}
&
:\ddot{O} - H^{-}
&
\begin{array}{c}
H \quad H \\
| \quad | \\
H - C - C - H \\
| \quad | \\
H \quad H
\end{array}
&
\begin{array}{c}
H - C = C - H \\
| \quad | \\
H \quad H
\end{array}
\end{array}$$

In each of the previous structures there are eight electrons around each atom (except for H atoms, which always have two electrons). There are two electrons in each bond. When counting electrons in these structures, one considers the electrons in a bond between two atoms as belonging to the atom under consideration. In the CH_2Cl_2 molecule just shown, for example, the Cl atoms each have eight electrons, including the two in the single bond to the C atom. The C atom also has eight electrons, two from each of the four bonds to that atom. The bonding and nonbonding electrons in Lewis structures are all from the *outermost* shells of the atoms involved, and are the so-called valence electrons of those atoms. For the main group elements, the number of valence electrons in an atom is equal to the last digit in the group number of the element in the Periodic Table. Carbon, in Group 14, has four valence electrons in its atoms; hydrogen, in Group 1, has one; chlorine, in Group 17, has seven valence electrons. In an octet rule structure the valence electrons from all the atoms are arranged in such a way that each atom, except hydrogen, has eight electrons.

Often it is quite easy to construct an octet rule structure for a molecule. Given that an oxygen atom has six valence electrons (Group 16) and a hydrogen atom has one, it is clear that one O and two H atoms have a total of eight valence electrons; the octet rule structure for H_2O, which we discussed earlier, follows by inspection. Structures like that of H_2O, involving only single bonds and nonbonding electron pairs, are common. Sometimes, however, there is a "shortage" of electrons; that is, it is not possible to construct an octet rule structure in which all the electron pairs are either in single bonds or are nonbonding. C_2H_4 is a typical example of such a species. In such cases, octet rule structures can often be made in which two atoms are bonded by two pairs, rather than one pair, of electrons. The two pairs of electrons form a double bond. In the C_2H_4 molecule, shown above, the C atoms each get four of their electrons from the double bond. The assumption that electrons behave this way is supported by the fact that the $C=C$ double bond is both shorter and stronger than the $C-C$ single bond in the C_2H_6 molecule (see previous example). Double bonds, and triple bonds, occur in many molecules, usually between C, O, N, and/or S atoms.

Lewis structures can be used to predict molecular and ionic geometries. All that is needed is to assume that the four pairs of electrons around each atom are arranged tetrahedrally. We have seen how that assumption leads to the correct geometry for H_2O. Applying the same principle to the species whose Lewis structures we listed earlier, we would predict, correctly, that the CH_2Cl_2 molecule would be tetrahedral (roughly anyway), that NH_3 would be pyramidal (with the nonbonding electron pair sticking up from the pyramid made from the atoms), that the bond angles in C_2H_6 are all tetrahedral, and that the C_2H_4 molecule is planar (the two bonding pairs in the double bond being in a sort of banana bonding arrangement above and below the plane of the molecule). In describing molecular geometry we indicate the positions of the atomic nuclei, not the electrons. The NH_3 molecule is pyramidal, not tetrahedral.

It is also possible to predict polarity from Lewis structures. Polar molecules have their center of positive charge at a different point than their center of negative charge. This separation of charges produces a dipole moment in the molecule. Covalent bonds between different kinds of atoms are polar; all heteronuclear diatomic molecules are polar. In some molecules the polarity from one bond may be canceled by that from others. Carbon dioxide, CO_2, which is linear, is a nonpolar molecule. Methane, CH_4, which is tetrahedral, is also nonpolar. Among the species whose Lewis structures we have listed, we find that H_2O, CH_2Cl_2, NH_3, and OH^- are polar. C_2H_6 and C_2H_4 are nonpolar.

For some molecules with a given molecular formula, it is possible to satisfy the octet rule with different atomic arrangements. A simple example would be

$$
\begin{array}{cccccc}
& \text{H} & \text{H} & & \text{H} & \text{H} \\
& | & | & & | & | \\
\text{H}-&\text{C}-&\text{C}-\overset{..}{\underset{..}{\text{O}}}-\text{H} & \text{and} \quad \text{H}-&\text{C}-\overset{..}{\underset{..}{\text{O}}}-&\text{C}-\text{H} \\
& | & | & & | & | \\
& \text{H} & \text{H} & & \text{H} & \text{H}
\end{array}
$$

The two molecules are called isomers of each other, and the phenomenon is called isomerism. Although the molecular formulas of both substances are the same, C_2H_6O, their properties differ markedly because of their different atomic arrangements.

Isomerism is very common, particularly in organic chemistry, and when double bonds are present, isomerism can occur in very small molecules:

The first two isomers result from the fact that there is no rotation around a double bond, although such rotation can occur around single bonds. The third isomeric structure cannot be converted to either of the first two without breaking bonds.

With certain molecules, given a fixed atomic geometry, it is possible to satisfy the octet rule with more than one bonding arrangement. The classic example is benzene, whose molecular formula is C_6H_6:

These two structures are called resonance structures, and molecules such as benzene, which have two or more resonance structures, are said to exhibit resonance. The actual bonding in such molecules is thought to be an average of the bonding present in the resonance structures. The stability of molecules exhibiting resonance is found to be higher than that anticipated for any single resonance structure.

Although the conclusions we have drawn regarding molecular geometry and polarity can be obtained from Lewis structures, it is much easier to draw such conclusions from models of molecules and ions. The rules we have cited for octet rule structures transfer readily to models. In many ways the models are easier to construct than are the drawings of Lewis structures on paper. In addition, the models are three-dimensional and hence much more representative of the actual species. Using the models, it is relatively easy to see both geometry and polarity, as well as to deduce Lewis structures. In this experiment you will assemble models for a sizeable number of common chemical species and interpret them in the ways we have discussed.

Experimental Procedure

IN THIS EXPERIMENT YOUR INSTRUCTOR MAY ALLOW YOU TO WORK WITHOUT SAFETY GLASSES

In this experiment you may work in pairs during the first portion of the laboratory period.

The models you will use consist of drilled wooden balls, short sticks, and springs. The balls represent atomic nuclei surrounded by the inner electron shells. The sticks and springs represent electron pairs and fit in the holes in the wooden balls. The model (molecule or ion) consists of wooden balls (atoms) connected by sticks or springs (chemical bonds). Some sticks may be connected to only one atom (nonbonding pairs).

In this experiment we will deal with atoms that obey the octet rule; such atoms have four electron pairs around the central core and will be represented by balls with four tetrahedral holes in which there are four sticks or springs. The only exception will be hydrogen atoms, which share two electrons in covalent compounds, and which will be represented by balls with a single hole in which there is a single stick.

In assembling a molecular model of the kind we are considering, it is possible, indeed desirable, to proceed in a systematic manner. We will illustrate the recommended procedure by developing a model for a molecule with the formula CH_2O.

1. Determine the total number of valence electrons in the species. This is easily done once you realize that the number of valence electrons on an atom of a main group element is equal to the last digit of number of the group to which the atom belongs in the Periodic Table. For CH_2O,

C   Group 14   H   Group 1   O   Group 16

   Therefore each carbon atom in a molecule or ion contributes four electrons, each hydrogen atom one electron, and each oxygen atom six electrons. The total number of valence electrons equals the sum of the valence electrons on all of the atoms in the species being studied. For CH_2O this total would be $4 + (2 \times 1) + 6$, or 12 valence electrons. If we are working with an ion, we add one electron for each negative charge or subtract one for each positive charge on the ion.

2. Select wooden balls and sticks to represent the atoms and electron pairs in the molecule. You should use four-holed balls for the carbon atom and the oxygen atom, and one-holed balls to represent the hydrogen atoms. Since there are 12 valence electrons in the molecule and electrons occur in pairs, you will need six sticks to represent the six electron pairs. The sticks will serve both as bonds between atoms and as nonbonding electron pairs.

3. Connect the balls with some of the sticks. (Assemble a skeleton structure for the molecule, joining atoms by single bonds.) In some cases this can only be done in one way. Usually, however, there are various possibilities, some of which are more reasonable than others. In CH_2O the model can be assembled by connecting the two H atom balls to the C atom ball with two of the available sticks, and then using a third stick to connect the C atom and O atom balls.

4. The next step is to use the sticks that are left over in such a way as to fill all the remaining holes in the balls. (Distribute the electron pairs so as to give each atom eight electrons and so satisfy the octet rule.) In the model we have assembled, there is one unfilled hole in the C atom ball, three unfilled holes in the O atom ball, and three available sticks. An obvious way to meet the required condition is to use two sticks to fill two of the holes in the O atom ball, and then use two springs instead of two sticks to connect the C atom and O atom balls. The model as completed is shown in Figure 13.1.

5. Interpret the model in terms of the atoms and bonds represented. The sticks and spatial arrangement of the balls will closely correspond to the electronic and atomic arrangement in the molecule. Given our model, we would describe the CH_2O molecule as being planar with single bonds between carbon and hydrogen atoms and a double bond between the C and O atoms. The H—C—H angle is approximately tetrahedral. There are two nonbonding electron pairs on the O atom. Since all bonds are polar and the molecular symmetry

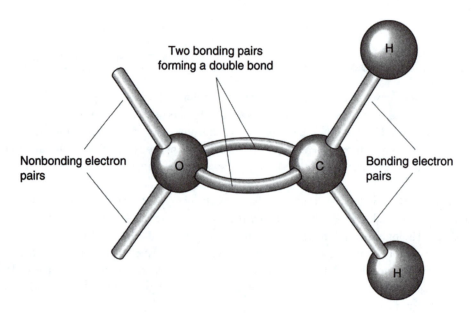

Figure 13.1

does not cancel the polarity in CH_2O, the molecule is polar. The Lewis structure of the molecule is given below:

$$\ddot{O}=C\begin{smallmatrix}H\\ \\H\end{smallmatrix}$$

(The compound having molecules with the formula CH_2O is well known and is called formaldehyde. The bonding and structure in CH_2O are as given by the model.)

6. Investigate the possibility of the existence of isomers or resonance structures. It turns out that in the case of CH_2O one can easily construct an isomeric form that obeys the octet rule, in which the central atom is oxygen rather than carbon. It is found that this isomeric form of CH_2O does not exist in nature. As a general rule, carbon atoms almost always form a total of four bonds; put another way, nonbonding electron pairs on carbon atoms are very rare. Another useful rule of a similar nature is that if a species contains several atoms of one kind and one of another, the atoms of the same kind will assume equivalent positions in the species. In SO_4^{2-}, for example, the four O atoms are all equivalent, and are bonded to the S atom and not to one another.

Resonance structures are reasonably common. For resonance to occur, however, the atomic arrangement must remain fixed for two or more possible electronic structures. (The sticks and springs can move, but the balls can not.) For CH_2O there are no resonance structures.

A. Using the procedure we have outlined, construct and report on models of the molecules and ions listed here and/or other species assigned by your instructor. Draw the complete Lewis structure for each molecule, showing nonbonding as well as bonding electrons. Given the structure, describe the geometry of the molecule or ion, and state whether the species is polar. Finally, draw the Lewis structures of any likely isomers or resonance forms.

CH_4	H_3O^+	N_2	C_2H_2	SCN^-
CH_2Cl_2	HF	P_4	SO_2	NO_3^-
CH_4O	NH_3	C_2H_4	SO_4^{2-}	HNO_3
H_2O	H_2O_2	$C_2H_2Br_2$	CO_2	$C_2H_4Cl_2$

B. Assuming that stability requires that each atom obey the octet rule, predict the stability of the following species:

$$PCl_3 \quad H_3O \quad CH_2 \quad CO$$

C. When you have completed parts A and B, see your laboratory instructor, who will check your results and assign you a set of unknown species. Working now by yourself, assemble models for each species as in the previous section, and report on the geometry and bonding in each of the unknown species on the basis of the model you construct. Also consider and report the polarity and the Lewis structures of any isomers and resonance forms for each species.

Name ______________________________ Section ______________________________

Experiment 13

Observations and Conclusions: The Geometrical Structure of Molecules

A. Species	Lewis structure	Molecular geometry	Polar?	Isomers or resonance structures
CH_4				
CH_2Cl_2				
CH_4O				
H_2O				
H_3O^+				
HF				
NH_3				
H_2O_2				
N_2				
P_4				
C_2H_4				

(continued on following page)

A.

Species	Lewis structure	Molecular geometry	Polar?	Isomers or resonance structures
$C_2H_2Br_2$				
C_2H_2				
SO_2				
SO_4^{2-}				
CO_2				
SCN^-				
NO_3^-				
HNO_3				
$C_2H_4Cl_2$				

B. Stability predicted for PCl_3 _____ H_3O _____ CH_2 _____ CO _____

C. Unknowns

_____

_____

_____

_____

Experiment 13

Advance Study Assignment: The Geometrical Structure of Molecules

You are asked by your instructor to construct a model of the NH_2Cl molecule. Being of a conservative nature, you proceed as directed in the section on Experimental Procedure.

1. First you need to find the number of valence electrons in NH_2Cl_2. For counting purposes with Lewis structures, the number of valence electrons in an atom of a main group element is equal to the last digit in the group number of that element in the Periodic Table.

    N is in Group _____   H is in Group _____   Cl is in Group _____

    In NH_2Cl there is a total of _____ valence electrons.

2. The model consists of balls and sticks. How many holes should be in the ball you select for the N atom? _____ the H atoms? _____ the Cl atom? _____ The electrons in the molecule are paired, and each stick represents a valence electron pair. How many sticks do you need? _____

3. Assemble a skeleton structure for the molecule, connecting the balls and sticks to make one unit. Use the rule that N atoms form three bonds, whereas Cl and H atoms usually form only one. Draw a sketch of the skeleton below:

4. How many sticks did you need to make the skeleton structure? _____ How many sticks are left over? _____ If your model is to obey the octet rule, each ball must have four sticks in it (except for hydrogen atom balls, which need and can only have one). (Each atom in an octet rule species is surrounded by four pairs of electrons.) How many holes remain to be filled? _____ Fill them with the remaining sticks, which represent nonbonding electron pairs. Draw the complete Lewis structure for NH_2Cl using lines for bonds and pairs of dots for nonbonding electrons.

5. Describe the geometry of the model, which is that of NH_2Cl. _____ Is the NH_2Cl molecule polar? _____ Why?

    Would you expect NH_2Cl to have any isomeric forms? _____ Explain your reasoning.

6. Would NH_2Cl have any resonance structures? _____ If so, draw them below.

Experiment 14

Heat Effects and Calorimetry

Heat is a form of energy, sometimes called thermal energy, that can pass spontaneously from an object at a high temperature to an object at a lower temperature. If the two objects are in contact, they will, given sufficient time, both reach the same temperature.

Heat flow is ordinarily measured in a device called a calorimeter. A calorimeter is simply a container with insulating walls, made so that essentially no heat is exchanged between the contents of the calorimeter and the surroundings. Within the calorimeter chemical reactions may occur or heat may pass from one part of the contents to another, but no heat flows into or out of the calorimeter from or to the surroundings.

A. Specific Heat

When heat flows into a substance, the temperature of that substance will increase. The quantity of heat q required to cause a temperature change Δt of any substance is proportional to the mass m of the substance and the temperature change, as shown in Equation 1. The proportionality constant is called the specific heat, $S.H.$, of that substance.

$$q = (\text{specific heat}) \times m \times \Delta t = S.H. \times m \times \Delta t \tag{1}$$

The specific heat can be considered to be the amount of heat required to raise the temperature of one gram of the substance by 1°C (if you make m and Δt in Equation 1 both equal to 1, then q will equal $S.H.$). Amounts of heat are measured in either joules or calories. To raise the temperature of 1 g of water by 1°C, 4.18 joules of heat must be added to the water. The specific heat of water is therefore 4.18 joules/g°C. Since 4.18 joules equals 1 calorie, we can also say that the specific heat of water is 1 calorie/g°C. Ordinarily heat flow into or out of a substance is determined by the effect that that flow has on a known amount of water. Because water plays such an important role in these measurements, the calorie, which was the unit of heat most commonly used until recently, was actually defined to be equal to the specific heat of water.

The specific heat of a metal can readily be measured in a calorimeter. A weighed amount of metal is heated to some known temperature and is then quickly poured into a calorimeter that contains a measured amount of water at a known temperature. Heat flows from the metal to the water, and the two equilibrate at some temperature between the initial temperatures of the metal and the water.

Assuming that no heat is lost from the calorimeter to the surroundings, and that a negligible amount of heat is absorbed by the calorimeter walls, the amount of heat that flows from the metal as it cools is equal to the amount of heat absorbed by the water.

In thermodynamic terms, the heat flow for the metal is equal in magnitude but opposite in direction, and hence in sign, to that for the water. For the heat flow q,

$$q_{H_2O} = -q_{metal} \tag{2}$$

If we now express heat flow in terms of Equation 1 for both the water and the metal M, we get

$$q_{H_2O} = S.H._{H_2O} m_{H_2O} \Delta t_{H_2O} = -S.H._M m_M \Delta t_M \tag{3}$$

In this experiment we measure the masses of water and metal and their initial and final temperatures. (Note that $\Delta t_M < 0$ and $\Delta t_{H_2O} > 0$, since $\Delta t = t_{final} - t_{initial}$.) Given the specific heat of water, we can find the positive specific heat of the metal by Equation 3. We will use this procedure to obtain the specific heat of an unknown metal.

The specific heat of a metal is related in a simple way to its molar mass. Dulong and Petit discovered many years ago that about 25 joules were required to raise the temperature of one mole of many metals by 1°C. This relation, shown in Equation 4, is known as the Law of Dulong and Petit:

$$MM \cong \frac{25}{S.H.(J/g°C)} \tag{4}$$

where *MM* is the molar mass of the metal. Once the specific heat of the metal is known, the approximate molar mass can be calculated by Equation 4. The Law of Dulong and Petit was one of the few rules available to early chemists in their studies of molar masses.

B. Heat of Reaction

When a chemical reaction occurs in aqueous solution, the situation is similar to that which is present when a hot metal sample is put into water. With such a reaction there is an exchange of heat between the reaction mixture and the solvent, water. As in the specific heat experiment, the heat flow for the reaction mixture is equal in magnitude but opposite in sign to that for the water. The heat flow associated with the reaction mixture is also equal to the enthalpy change, ΔH, for the reaction, so we obtain the equation

$$q_{reaction} = \Delta H_{reaction} = -q_{H_2O} \tag{5}$$

By measuring the mass of the water used as solvent, and by observing the temperature change that the water undergoes, we can find q_{H_2O} by Equation 1 and ΔH by Equation 5. If the temperature of the water goes up, heat has been *given off* by the reaction mixture, so the reaction is *exo*thermic; q_{H_2O} is *positive* and ΔH is *negative*. If the temperature of the water goes down, the reaction mixture has *absorbed heat from* the water and the reaction is *endo*thermic. In this case q_{H_2O} is *negative* and ΔH is *positive*. Both exothermic and endothermic reactions are observed.

One of the simplest reactions that can be studied in solution occurs when a solid is dissolved in water. As an example of such a reaction note the solution of NaOH in water:

$$NaOH(s) \rightarrow Na^+(aq) + OH^-(aq); \quad \Delta H = \Delta H_{solution} \tag{6}$$

When this reaction occurs, the temperature of the solution becomes much higher than that of the NaOH and water that were used. If we dissolve a known amount of NaOH in a measured amount of water in a calorimeter, and measure the temperature change that occurs, we can use Equation 1 to find q_{H_2O} for the reaction and use Equation 5 to obtain ΔH. Noting that ΔH is directly proportional to the amount of NaOH used, we can easily calculate $\Delta H_{solution}$ for either a gram or a mole of NaOH. In the second part of this experiment you will measure $\Delta H_{solution}$ for an unknown ionic solid.

Chemical reactions often occur when solutions are mixed. A precipitate may form, in a reaction opposite in direction to that in Equation 6. A very common reaction is that of neutralization, which occurs when an acidic solution is mixed with one that is basic. In the last part of this experiment you will measure the heat effect when a solution of HCl, hydrochloric acid, is mixed with one containing NaOH, sodium hydroxide, which is basic. The heat effect is quite large, and is the result of the reaction between H$^+$ ions in the HCl solution with OH$^-$ ions in the NaOH solution:

$$H^+(aq) + OH^-(aq) \rightarrow H_2O \quad \Delta H = \Delta H_{neutralization} \tag{7}$$

Experimental Procedure

A. Specific Heat

From the stockroom obtain a calorimeter, a sensitive thermometer, a sample of metal in a large stoppered test tube, and a sample of unknown solid. (The thermometer is expensive, so be careful when handling it.)

The calorimeter consists of two nested expanded polystyrene coffee cups fitted with a Styrofoam cover. There are two holes in the cover: one for a thermometer and the other for a glass stirring rod with a perpendicular loop on one end. Assemble the experimental setup as shown in Figure 14.1.

Fill a 400-cm³ beaker two thirds full of water and begin heating it to boiling. While the water is heating, weigh your sample of unknown metal in the large stoppered test tube to the nearest 0.1 g on a top-loading or triple-beam balance. Pour the metal into a dry container and weigh the empty test tube and stopper. Replace the metal in the test tube and put the *loosely* stoppered tube into the hot water in the beaker. The water level in the beaker should be high enough that the top of the metal is below the water surface. Continue heating the metal in the water for at least 10 minutes after the water begins to boil, to ensure that the metal attains the temperature of the boiling water. Add small amounts of water as necessary to maintain the water level.

While the water is boiling, weigh the calorimeter to 0.1 g. Place about 40 cm³ of water in the calorimeter and weigh again. Insert the stirrer and thermometer into the cover and put it on the calorimeter. If a glass thermometer is being used, the thermometer bulb should be completely immersed.

Measure the temperature of the water in the calorimeter to 0.1°C. Take the test tube out of the beaker of boiling water, remove the stopper, and pour the metal into the water in the calorimeter. Be careful that no water adhering to the outside of the test tube runs into the calorimeter when you are pouring the metal. Replace the calorimeter cover and agitate the water as best you can with the glass stirrer. Record to 0.1°C the maximum temperature reached by the water. Repeat the experiment, using about 50 cm³ of water in the calorimeter. Be sure to dry your metal before reusing it; this can be done by heating the metal briefly in the test tube in boiling water and then pouring the metal onto a paper towel to drain. You can dry the hot test tube with a little compressed air.

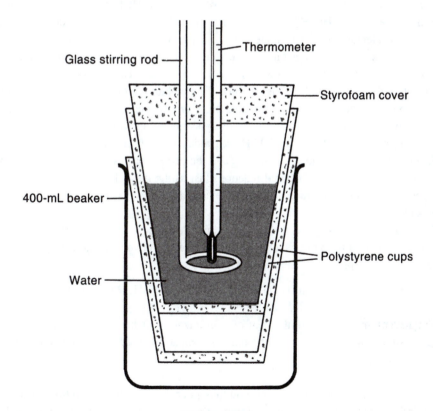

Glass stirring rod

Thermometer

Styrofoam cover

400-mL beaker

Polystyrene cups

Water

Figure 14.1

B. Heat of Solution

Place about 50 cm³ of distilled water in the calorimeter and weigh as in the previous procedure. Measure the temperature of the water to 0.1°C. The temperature should be within a degree or two of room temperature. In a small beaker weigh out about 5 g of the solid compound assigned to you. Make the weighing of the beaker and of the beaker plus solid to 0.1 g. Add the compound to the calorimeter. Stirring continuously and occasionally swirling the calorimeter, determine to 0.1°C the maximum or minimum temperature reached as the solid dissolves. Check to make sure that *all* the solid is dissolved. A temperature change of at least 5 degrees should be obtained in this experiment. If necessary, repeat the experiment, increasing the amount of solid used.

DISPOSAL OF REACTION PRODUCTS.   Dispose of the solution in Part B as directed by your instructor.

C. Heat of Neutralization

Rinse out your calorimeter with distilled water, pouring the rinse into the sink. In a graduated cylinder, measure out 25.0 cm³ of 2.00 M HCl; pour that solution into the calorimeter. Rinse out the cylinder with distilled water, and measure out 25.0 cm³ of 2.00 M NaOH; pour that solution into a dry 50-mL beaker. Measure the temperature of the acid and of the base to ±0.1°C, making sure to rinse and dry your thermometer before immersing it in the solutions. Put the thermometer back in the calorimeter cover. Pour the NaOH solution into the HCl solution and put on the cover of the calorimeter. Stir the reaction mixture, and record the maximum temperature that is reached by the neutralized solution.

Calculate the heat of neutralization. It is necessary to slightly modify Equation 1 when calculating q_{H_2O}. Use a value of 1.02 g/mL as an average density for all solutions in this experiment; thus the mass of the solution is obtained by multiplying 50.0 mL by 1.02 g/mL. We will continue to use 4.18 J/g°C as the specific heat value.

D. Hess' Law   Optional

Using the same procedure as in Part C, measure the heat of neutralization of 2.00 M acetic acid (HAc). You should use 25.0 mL of the acetic acid solution in place of the 2.00 M HCl used in Part C. Calculate the molar heat of solution of acetic acid, $HC_2H_3O_2$, again using 1.02 g/mL as the average density for all solutions. Record your data and calculations on a separate sheet of paper. Write out the net ionic equation for the strong acid-strong base neutralization and the net ionic equation for the acetic acid (weak acid; use HAc to represent undissociated acetic acid)-strong base neutralization reaction. In each case, show the value of the measured molar heat of reaction on the right side of the equation. Remember that the heat flow for an exothermic reaction is negative. Combine (via addition and cancellation of any species that appears on opposite sides of the final equation) the two net ionic equations in such a way as to end up with an equation representing the dissociation reaction of the weak acid HAc to yield H^+(aq) and Ac^-(aq). If you needed to reverse one of the equations, the heat flow value will change signs. As you add the two equations you should add the two heat flow values, to give the final value for the desired reaction. Just as the dissociation of water to form H^+(aq) and OH^-(aq) is expected to be endothermic, so should the corresponding dissociation reaction for a weak acid, such as acetic acid.

If you did the above exercise correctly, you have just illustrated HESS'S LAW. This law states that the value of ΔH for a reaction is the same whether it occurs directly, or in a series of steps. Thus, if a thermochemical equation can be expressed as the sum of two equations,

$$\text{Equation (3)} = \text{Equation (1)} + \text{Equation (2), then } \Delta H_3 = \Delta H_1 + \Delta H_2.$$

Optional Experiment: Measure the heat of neutralization of household vinegar in its reaction with NaOH solution. Assuming that vinegar is about 5 mass percent acetic acid, find the molar heat of neutralization of that acid.

When you have completed the experiment, you may pour the neutralized solution down the sink. Rinse the calorimeter and thermometer with water, and return them, along with the metal sample to the stockroom.

Name _____   Section _____

Experiment 14

Data and Calculations: Heat Effects and Calorimetry

A. Specific Heat

	Trial 1		Trial 2
Mass of stoppered test tube plus metal	_____ g	→	_____ g
Mass of test tube and stopper	_____ g	→	_____ g
Mass of calorimeter	_____ g	→	_____ g
Mass of calorimeter and water	_____ g		_____ g
Mass of water	_____ g		_____ g
Mass of metal	_____ g	→	_____ g
Initial temperature of water in calorimeter	_____ °C		_____ °C
Initial temperature of metal (assume 100°C unless directed to do otherwise)	_____ °C	→	_____ °C
Equilibrium temperature of metal and water in calorimeter	_____ °C		_____ °C
Δt_{water} $(t_{final} - t_{initial})$	_____ °C		_____ °C
Δt_{metal}	_____ °C		_____ °C
q_{H_2O}	_____ J		_____ J
Specific heat of the metal (Eq. 3)	_____ J/g°C		_____ J/g°C
Approximate molar mass of metal	_____		_____
Unknown no.	_____		

B. Heat of Solution

Mass of calorimeter plus water	_____ g
Mass of beaker	_____ g

(continued on following page)

Mass of beaker plus solid   _____ g

Mass of water, m_{H_2O} _____ g

Mass of solid, m_s _____ g

Original temperature   _____ °C

Final temperature   _____ °C

q_{H_2O} for the reaction (Eq. 1) (*S.H.* = 4.18 J/g°C)   _____ joules

ΔH for the reaction (Eq. 5)   _____ joules

The quantity you have just calculated is approximately* equal to the heat of solution of your sample. Calculate the heat of solution per gram of solid sample.

$$\Delta H_{solution} = \text{_____ joules/g}$$

The solution reaction is endothermic exothermic. (Underline correct answer.) Give your reasoning.

Solid unknown no. _____

Optional Formula of compound used (if furnished) _____ Molar mass _____ g

Heat of solution per mole of compound   _____ kJ

C. Heat of Neutralization

Original temperature of HCl solution   _____ °C

Original temperature of NaOH solution   _____ °C

Final temperature of neutralized mixture   _____ °C

Change in temperature. Δt (take average of the
original temperatures of HCl and NaOH)   _____ °C

q_{H_2O} (assume 50.0 mL of solution and use average density of 1.02 g/mL)   _____ J

ΔH for the neutralization reaction   _____ J

ΔH per mole of H^+ and OH^- ions reacting   _____ kJ

*The value of ΔH will be approximate for several reasons. One is that we do not include the amount of heat absorbed by the solute. This effect is smaller than the likely experimental error, and thus we will ignore it.

Experiment 14

Advance Study Assignment: Heat Effects and Calorimetry

1. A metal sample weighing 124.10 g and at a temperature of 99.3°C was placed in 42.87 g of water in a calorimeter at 23.9°C. At equilibrium the temperature of the water and metal was 41.6°C.

   a. What was Δt for the water? ($\Delta t = t_{final} - t_{initial}$)

   _____ °C

   b. What was Δt for the metal?

   _____ °C

   c. How much heat flowed into the water?

   _____ joules

   d. Taking the specific heat of water to be 4.18 J/g°C, calculate the specific heat of the metal, using Equation 3.

   _____ joules/g°C

   e. What is the approximate molar mass of the metal? (Use Eq. 4.)

   _____ g

2. When 5.12 g of NaOH were dissolved in 51.55 g water in a calorimeter at 24.5°C, the temperature of the solution went up to 49.8°C.

   a. Is this solution reaction exothermic? _____ Why?

   b. Calculate q_{H_2O}, using Equation 1.

   _____ joules

   c. Find ΔH for the reaction as it occurred in the calorimeter (Eq. 5).

   $\Delta H = $ _____ joules

(continued on following page)

d. Find ΔH for the solution of 1.00 g NaOH in water.

$\Delta H =$ _____ joules/g

e. Find ΔH for the solution of 1 mole NaOH in water.

$\Delta H =$ _____ joules/mole

f. Given that NaOH exists as Na^+ and OH^- ions in solution, write the equation for the reaction that occurs when NaOH is dissolved in water.

g. Given the following heats of formation, ΔH_f, in kJ per mole, as obtained from a table of ΔH_f data, calculate ΔH for the reaction in Part f. Compare your answer with the result you obtained in Part e. NaOH(s), −425.6; Na^+(aq), −240.1; OH^-(aq), −230.0

$\Delta H =$ _____ kJ

Experiment 15

The Vapor Pressure and Heat of Vaporization of a Liquid*

If we pour a liquid into an open container, which we then close, we find that some of the liquid will evaporate into the air in the container. The vapor will after a short time establish a partial pressure that holds constant as long as the temperature does not change. That pressure is called the vapor pressure of the liquid. It does not change if there are no other gases present.

In the air we breathe, there is some water vapor, usually with a partial pressure well below the vapor pressure. The relative humidity is equal to 100% times the partial pressure of water vapor divided by the vapor pressure at that temperature. On a hot, sticky day the relative humidity is high, perhaps 80 or 90%, and the air is nearly saturated with water vapor; we feel uncomfortable.

If we raise the temperature of a liquid, its vapor pressure increases, ever more rapidly. At the normal boiling point of the liquid, the vapor pressure becomes one atmosphere. That would be the pressure at 100°C in a closed container in which there is just water and its vapor and no air.

At high temperatures, the vapor pressure can become very large, reaching about 5 atm for water at 150°C and 15 atm at 200°C. A liquid that boils at a low temperature, say −50°C, will have a very large vapor pressure at room temperature, and confining it at 25°C in a closed container may well cause an explosion. Failure to recognize this fact has caused several scientists of the authors' acquaintance to have serious accidents.

There is an equation relating the vapor pressure of a liquid to the Kelvin temperature. It is called the Clapeyron Equation and has the form:

$$\ln P = -\Delta H/RT + \text{a constant} \tag{1}$$

If we plot $\ln P$ vs the reciprocal of the Kelvin temperature we obtain a straight line, the slope of which equals $-\Delta H/R$. In the equation $\ln P$ is the natural logarithm of P, and the units of ΔH and R must match. P may be in atm, mm Hg, bars, or any other pressure unit.

There are many methods for measuring vapor pressure, most of which are not intuitively obvious. In this experiment we will use one of the simplest, and yet most accurate, methods. We will inject a measured sample of dry air into a graduated pipet containing the liquid we are studying. We measure the volume of the air-vapor bubble that forms, which will be larger than that of the injected air, since some of the liquid vaporizes. If we raise the temperature the volume will go up as more and more liquid vaporizes. The total pressure in the bubble remains equal to the barometric pressure, so if we can calculate the partial pressure of air in the bubble, we can find the vapor pressure of the liquid at that temperature by simple subtraction. (There is no danger of an explosion in this experiment, in case you are worried.)

WEAR YOUR SAFETY GLASSES WHILE PERFORMING THIS EXPERIMENT

Experimental Procedure

You may work in pairs on this experiment. Both members of the pair must submit a report. From the stockroom obtain a modified 1-mL graduated pipet, a Pasteur pipet, a digital thermometer, and a sample of a liquid unknown, whose density will be given to you.

---

*The general method used in this experiment is one described by DeMuro, Margarian, Mkhikian, No, and Peterson, J Chem Ed 76:1113–1116, 1999.

Setting the Stage

The modified graduated pipet is the instrument you will use; let's call it the device. Weigh the empty dry device accurately on the analytical balance, and record its mass. Using the Pasteur pipet, carefully fill the device completely with your liquid unknown. Dry off the outside and reweigh; record that mass. **Record *all* your data in this experiment on the blank Data and Calculations page, noting in each case what it was you measured. Underline all of the quantities you actually measured.**

Using a syringe, your instructor will inject 0.200 mL of dry air at the closed end of the inverted device, forming a bubble. If, after the syringe is removed, the device is not full at the open end, add unknown with your Pasteur pipet until it is full. Then carefully dry and reweigh the device, and once again record the mass.

Place the device, bubble up, in a vertical position, while holding it near the open end; wait until the liquid level becomes steady. Read and record the level of the liquid meniscus in the bubble as accurately as possible (this is a crucial step, so take your time, and try to get the reading to the nearest 0.002 mL; note that the numbers on the pipet are in tenths of milliliters, and the distance between graduations is 0.01 mL). Record the barometric pressure in mm Hg and the temperature in the lab in °C.

Finding the Vapor Pressure of the Liquid at Room Temperature

From the data you have obtained so far you can calculate the vapor pressure of your unknown at the temperature in the lab. Using the mass measurements you can find the mass of the liquid in the full pipet and the mass of liquid driven out. The volume of the bubble is equal to the mass of liquid driven out divided by its density. This volume is larger than that of the air injected, since some of the liquid vaporizes into the bubble. The partial pressure of the air in the bubble is equal, by Boyle's Law, to the barometric pressure times the initial volume of the air divided by the volume of the bubble. The vapor pressure of the liquid is obtained by subtracting the partial pressure of the air from the barometric pressure. Record all your results. You can make these calculations while performing the rest of the experiment. There are several steps in finding the vapor pressure, so make sure you know what you are doing before making a calculation.

Finding the Vapor Pressure at Other Temperatures

We will use a water bath to study the vapor pressure as a function of temperature. Set up the bath as shown in Figure 15.1. Use a 1000-mL beaker that you can heat with a Bunsen burner or a hot plate. Fill the beaker to near the top with warm water from the tap (between 40° and 45°C). Fill a regular (18 × 150 mm) test tube with your unknown liquid to within 2 cm of the top, and gently drop your device, bubble up, into the tube, first making sure that no air bubble remains at the open end. Clamp the test tube as shown in Figure 15.1, so that the water level in the bath is as high as the liquid level in the test tube. Insert the digital thermometer in the test tube, holding it in position with a loosely-fitting stopper. Lay your lab thermometer in the water in the beaker. If you are heating your bath with a burner, put a stirring rod in the water. With a hot plate use a magnetic stirrer.

The temperature of the water bath should be about 40°C for the first measurement. Heat or cool the bath if necessary, with stirring, to get to within a few degrees of that value. It should stabilize within a few minutes. When the temperature on the digital thermometer holds steady for at least a minute, record the temperature. Note how closely the temperature matches that of the water bath; they should be close to one another when steady. Read the meniscus level in the graduated pipet as carefully as you can. Record the level and the temperature.

Heat the water in the large beaker, with stirring. When the total volume of the bubble has gone up from its last value by about 0.05 mL, as noted by the change in meniscus levels, stop heating. (The temperature at that point will be roughly 10°C above the first value.) The temperature will continue to rise for a few minutes as you stir, but will finally level off. The temperature on the digital thermometer and the bath thermometer should be about equal, and remain constant for at least a minute. Then read the level of the meniscus as accurately as you can, and record its value and the temperature.

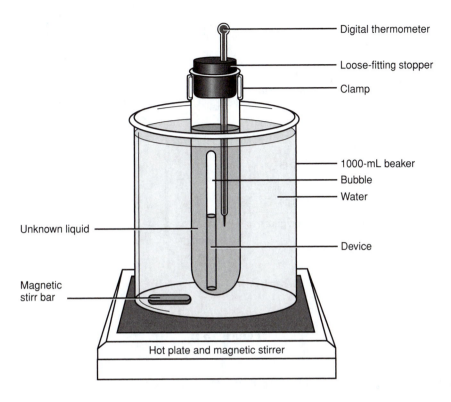

Digital thermometer

Loose-fitting stopper

Clamp

1000-mL beaker
Bubble
Water

Unknown liquid

Device

Magnetic
stirr bar

Hot plate and magnetic stirrer

Figure 15.1

Repeat the measurements you have just made at two higher temperatures, at points where the volume has increased by another 0.05 mL, and then by 0.10 mL more. Each time stop heating at that point, let the temperature level off, and when both temperature and bubble volume are steady for at least a minute, record the meniscus level and the temperature. Do not, do not, do not heat the sample to the point that the bubble volume exceeds the volume of the device. If that appears to be likely to happen, add some cold water to the bath and stir. When things have calmed down, proceed to make a measurement.

The data you have obtained will allow you to calculate the vapor pressure of the liquid at each temperature you used. The method is as noted earlier. First find the volume that the air sample would occupy at each temperature and barometric pressure (use Charles' Law). The partial pressure of air in the bubble will equal barometric pressure times the ratio of the volume you just calculated to the total bubble volume, which you find by noting how much the volume increased over the value at lab temperature (the difference between meniscus levels at the two temperatures). Calculate and record the partial pressures of air and vapor pressures as you perform the next experiment.

Direct Measurement of the Boiling Point

In the last part of this experiment you will measure the boiling point of the liquid at the barometric pressure. Pour the liquid sample in the regular test tube into a large test tube. Determine the boiling point of the liquid using the apparatus shown in Figure 15.2. The thermometer bulb should be just above the liquid surface. Heat the water in the bath until the liquid in the tube boils gently, with the vapor condensing *at least 5 cm below* the top of the test tube. Boiling chips may help in keeping the liquid boiling smoothly. As the boiling proceeds there will be some condensation on the thermometer and droplets will be falling from the thermometer bulb. After a minute or two the temperature should become reasonably steady. Record the temperature. The liquid you are using may be flammable and toxic, so you should not inhale the vapor unnecessarily. *Do not* heat the water bath so strongly that condensation of vapors occurs only at the top of the test tube.

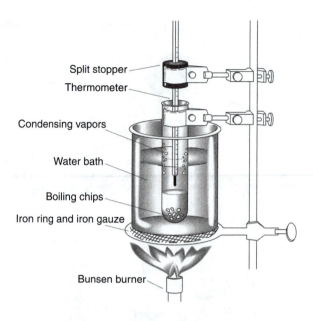

Figure 15.2

Finding the Heat of Vaporization of the Liquid

When you have completed the calculations described in the previous sections, you should have a list of the vapor pressures of the liquid at each of the temperatures you used. To find the molar heat of vaporization of the liquid we use Equation 1. First make a table listing ln P (the natural logarithm of P), the temperature T in Kelvin, and $1/T$ as found from your data. Make a graph of ln P vs $1/T$, using the graph paper provided or software such as Excel. The slope of the line, which should be straight, is equal to $-\Delta H/R$, the molar heat of vaporization divided by the gas constant R. If you want to express ΔH in joules/mole, R will equal 8.314 J/mole K.

Compare the value of the boiling point as measured directly with that you can calculate from your graph. The value of $1/T$ at the boiling point will occur where ln P equals ln $P_{barometric}$.

Finally, make a graph of the vapor pressure as a function of the temperature in °C. Use Excel or the graph paper provided. Connect the points with a smooth curve. Comment on how VP varies with t.

Include your graphs with your report. When you are finished with the experiment, pour the liquid from the test tube back into its container and return it to the stockroom along with the digital thermometer, the device, and the Pasteur pipet.

Experiment 15

Data and Calculations: Vapor Pressure and Heat of Vaporization

For this experiment you will be responsible for recording and processing your data. For each entry note the units and the quantity that was measured, along with temperature and pressure if necessary. Lay the page out so that you can find needed items easily. Put all the data at the top of the page, leaving the bottom part for calculations. Show the equations you used to determine calculated values, first with symbols and then with observed data. Make a table of the observed vapor pressures and temperatures, and the values of ln P and 1/T. Report the value of ΔH as calculated, and the boiling point as found directly and as calculated from the graph you prepared. Use the back of this page if necessary. Include your graph in your report.

Excel can be used for all calculations and graphs. Your report can be the spreadsheet.

Experiment 15

Vapor Pressure and Heat of Vaporization of Liquids

(Data and Calculations)

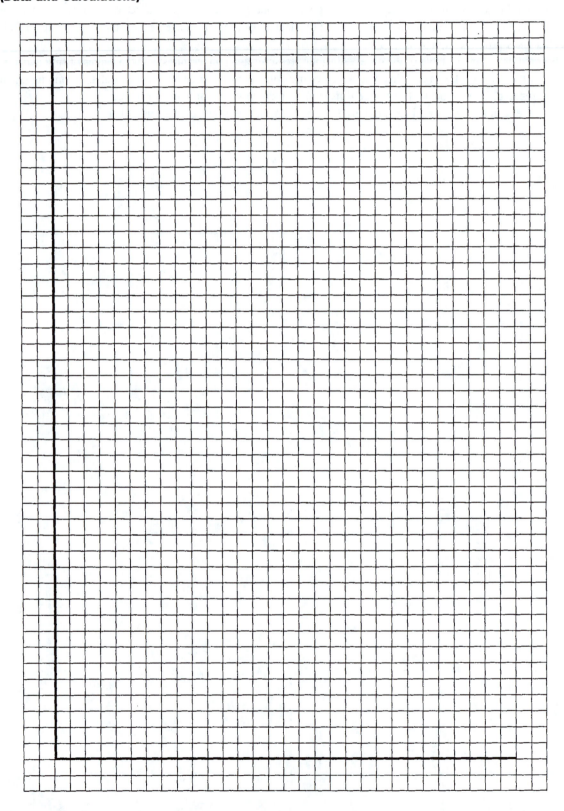

Experiment 15

Advance Study Assignment: Vapor Pressure and Heat of Vaporization
of Liquids

1. In an experiment to measure the vapor pressure of methanol, the following data were obtained:

Mass of empty device	2.6460 g	P_{bar} = 749 mm Hg
Mass full of methanol	3.2035 g	t = 23.0°C
Mass after adding 0.200 mL air	3.0170 g	density of methanol = 0.790 g/mL
Bubble meniscus reading	0.132 mL	

How many grams of methanol were in the full device?   _____ g

How many grams were driven out by the air?   _____ g

What is the volume of the methanol driven out?   _____ mL

What is the volume of the bubble?   _____ mL

Find the partial pressure of air in the bubble.
(Note that the mass of air remains the same, but it occupies a larger volume than 0.200 mL since some vapor entered the bubble. Total P is 749 mm Hg.)

_____ mm Hg

What is the partial pressure of vapor in the bubble?
(This is the vapor pressure of methanol at 23.0°C!)

_____ mm Hg

2. When the device is heated to 42.5°C, the meniscus reading increases to 0.250 mL.

a. What is the new volume of the bubble?   _____ mL

b. Why does the volume increase? (Two reasons)

c. What is the vapor pressure of ethanol at 42.5°C?

_____ mm Hg

Experiment 16

The Structure of Crystals—
An Experiment Using Models

If one examines the crystals of an ordinary substance like table salt, using a magnifying glass or a microscope, one finds many cubic particles, among others, in which planes at right angles are present. This is the situation with many common solids. The regularity we see implies a deeper regularity in the arrangement of atoms or ions in the solid. Indeed, when we study crystals by X-ray diffraction, we find that the atomic nuclei are present in remarkably symmetrical arrays, which continue in three dimensions for thousands or millions of units. Substances having a regular arrangement of atom-size particles in the solid are called crystalline, and the solid material consists of crystals. This experiment deals with some of the simpler arrays in which atoms or ions occur in crystals, and what these arrays can tell us about such properties as atomic sizes, densities of solids, and the efficiency of packing of particles.

Many crystals are complex, almost beyond belief. We will limit ourselves to the simplest crystals, those which have cubic structures. They are by far the easiest to understand, yet they exhibit many of the interesting properties of more complicated structures. We will further limit our discussion to substances containing only one or two kinds of atoms, so we will be working with the crystals of some of the elements and some binary compounds.

The atoms in crystals occur in planes. Sometimes a crystal will cleave readily at such planes. This is the reason the cubic structure of salt is so apparent. Let us begin our study of crystals with a simple example, a two-dimensional crystal in which all of the atoms lie in a square array in a plane, in the manner shown below:

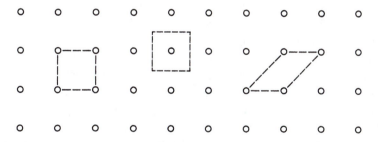

The first thing that you need to realize is that the array goes on essentially forever, in both directions. In this array $\bigcirc$ is always the same kind of atom, so all the atoms are equivalent. The distance, d_o, between atoms in the vertical direction is the same as in the horizontal, so if you knew either distance you could generate the array completely. In crystal studies we select a small section of the array, called the unit cell, that represents the array in the sense that by moving the unit cell repeatedly in either the x or y direction, a distance d_o at a time, you could generate the entire array. In the sketch we have used dotted lines to indicate several possible unit cells. All the cells have the same area, and by moving them up or down or across we could locate all the sites in the array. Ordinarily we select the unit cell on the left, because it includes four of the atoms and has its edges along the natural axes, x and y, for the array, but that is not necessary. In fact, the middle cell has the advantage that it clearly tells us that the number of $\bigcirc$ atoms in the cell is equal to 1. The number of atoms in whatever cell we choose must also equal 1, but that is not so apparent in the cell on the left. However, once you realize that only 1/4 of each atom on the corners of the cell actually belongs to that cell, because it is shared by three other cells, the number of atoms in the whole cell becomes equal to 1/4 × 4, or 1, which is what we got by drawing the cell in a different position.

When we extend the array to three dimensions, the same ideas regarding unit cells apply. The unit cell is the smallest portion of the array that could be used to generate the array. With cubic cells the unit cell is usually chosen to have edges parallel to the x, y, and z axes we could put on the array.

Experimental Procedure

In this experiment we will mix the discussion with the experimental procedure, since one supports and illustrates the other. We will start with the simplest possible cubic crystal, deal with its properties, and then go on to the next more complex example, and the next, and so on. Each kind of crystal will be related in some ways to the earlier ones but will have its own properties as well. So get out your model set and let's begin.

Work in pairs, and complete each section before going on to the next, unless directed otherwise by your instructor.

The Simple Cubic Crystal

You can probably guess the form of the array in the crystal structure we call simple cubic (SC). It is shown on the Calculations page at the end of this section.

The unit cell is a cube with an edge length equal to the distance from the center of one atom to the center of the next. The cell edge is usually given the symbol d_o. The volume of the unit cell is d_o^3, and is very small, since d_o is of the order of 0.5 nm. Using X-rays we can measure d_o to four significant figures quite easily. The number of atoms in the unit cell in a simple cubic crystal is equal to 1. Can you see why? Only 1/8 of each corner atom actually belongs to the cell, since it is shared equally by eight cells.

Using your model set, assemble three attached unit cells having the simple cubic structure shown in the sketch. Use the short bonds (no. 6) and the gray atoms. Each bond goes into a square face on the atom. An actual crystal with this structure would have many such cells, in three dimensions.

If you extended the model you have made further, you would find that each atom would be connected to six others. We say that the coordination number of the atoms in this structure is 6. Only the closest atoms are considered to be bonded to each other. In the other models we will be making, *only* those atoms that are *bonded*, by covalent or ionic bonds, will be connected to one another, so the number of bonds to an atom in a large model will be equal to the coordination number.

Although the model has an open structure to help us see relationships better, in an actual crystal we consider that the atoms that are closest are touching. It is on this assumption that we determine atomic radii. In this SC crystal, if we know d_o we can find the atomic radius r of the atoms, since, if the atoms are touching, d_o must equal $2r$. Another property we can calculate knowing d_o is the density, given the nature of the atoms in the crystal. From d_o we can easily find the volume V of the unit cell. Since there is one atom per cell, a mole will contain Avogadro's number of cells. Given the molar mass of the element, we can find the mass m of a cell. The density is simply m/V. With more complex crystals we can make these same calculations, but must take account of the fact that the number of atoms or ions per unit cell may not be equal to one.

Essentially no elements crystallize in an SC structure. The reason is that SC packing is inefficient, in that the atoms are farther apart than they need be. There is, as you can see, a big hole in the middle of each unit cell that is begging for an atom to go there, and atoms indeed do. Before we go into that, let's calculate the fraction of the volume of the unit cell that is actually occupied by atoms. This is easy to do. Make the calculation of that fraction on the Calculations page.

That fraction is indeed pretty small; only about 52 percent of the cell volume is occupied by atoms. Most of the empty space is in that hole. You can calculate by simple geometry that an atom having a radius equal to 73 percent of that of the atom on a corner would fit into the hole. Or, putting it another way, an atom bigger than that would have to push the corner atoms back to fit in. If the particles on the corners were anions, and the atom in the hole were a cation, the cation in the hole would push the anions apart, decreasing the repulsion between them and maximizing the attraction between anions and cations. We will have more to say about this when we deal with binary salts, but for now you need to note that if $r_+/r_- > 0.732$, the cation will not fit in the cubic hole formed by anions on the corners of an SC cell.

Body-Centered Cubic Crystals

In a body-centered cubic (BCC) crystal, the unit cell still contains the corner atoms present in the SC structure, but in the center of the cell there is another atom of the same kind. The unit cell is shown on the Calculations page.

Using your model set, assemble a BCC crystal. Use the blue balls. Put the short bonds in the eight holes in the triangular faces on one blue ball. Attach a blue ball to each bond, again using the holes in a triangular face. Those eight blue balls define the unit cell in this structure. Add as many atoms to the structure as you have available, so that you can see how the atoms are arranged. In a BCC crystal each atom is bonded to eight others, so the coordination number is 8. Verify this with your model. (There are no bonds along the edges of the unit cell, since the corner atoms are not as close to one another as they are to the central atom.)

The BCC lattice is much more stable than the SC one, in part at least because of the higher coordination number. Many metals crystallize in a BCC lattice—including sodium, chromium, tungsten, and iron—when these metals are at room temperature.

There are several properties of the BCC structure that you should note. The number of atoms per unit cell is two, one from the corner atoms and one from the atom in the middle of the cell, where it is unshared. As with SC cells, there is a relation between the unit cell size and the atom radius. Given that in sodium metal the cell edge is 0.429 nm, calculate the radius of a sodium atom on the Calculations page. When you are finished, calculate the density of sodium metal.

The fraction of the volume that is occupied by atoms in a BCC crystal is quite a bit larger than with SC crystals. Using the same kind of procedure that we did with the SC structure, we can show that in BCC crystals about 68 percent of the cell volume is occupied.

Close-Packed Structures

Although many elements form BCC crystals, still more prefer structures in which the atoms are close packed. In such structures there are layers of atoms in which each atom is in contact with six others, as in the sketch below:

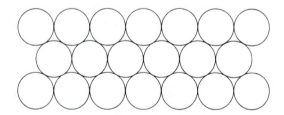

This is the way that billiard balls lie in a rack, or the honeycomb cells are arranged in a bees' nest. It is the most efficient way one can pack spheres, with about 74 percent of the volume in a close-packed structure filled with atoms. It turns out that there is more than one close-packed crystal structure. The layers all have the same structure, but they can be stacked on one another in two different ways. This is certainly not apparent at first sight. The first close-packed crystal we will examine is cubic, amazingly enough, considering all those triangles in the layers of atoms.

Face-Centered Cubic Crystals

In the face-centered cubic (FCC) unit cell there are atoms on the corners and an atom at the center of each face. There is no atom in the center of the cell (see the sketch on the Calculations page, where we show the bonding on only the three exposed faces).

Constructing an FCC unit cell, using bonds between closest atoms, is not as simple as you might think, so let's do it the easy way. This time select the large gray balls, the ones with the most holes. Make the bottom face of the unit cell by attaching four gray balls to a center ball, using short bonds, and holes that are in rectangular faces. You should get the structure shown on the following page:

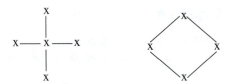

To make the middle layer of atoms, connect four of the balls in a square, again using short bonds and the holes in rectangular faces, as in the right sketch. The top layer is made the same way as the bottom one. Then connect the layers to one another, again using holes in the rectangular faces. You should then have an FCC cubic unit cell. Because the atoms in the middle of adjacent faces are as close to one another as they are to the corner atoms, there should be bonds between them. There are quite a few bonds in the final cell, 32 in all. Put them all in. As with BCC there are no bonds along the edges of the cell.

The FCC, or cubic close-packed, structure is a common one. Among the metals with this type of crystal are copper, silver, nickel, and calcium.

In close-packed structures the coordination number is the largest that is possible. It can be found by looking at a face-centered atom, say on the top face of the cell. That atom has eight bonds to it and would also have bonds to atoms in the cell immediately above, four of them. So its coordination number, like that of every atom in the cell, is 12.

Having seen how to deal with other structures, you should now be able to find the number of atoms in the FCC unit cell and the radius of an atom as related to d_o. From these relationships and the measured density, you can calculate Avogadro's number. On the Calculations page, find these quantities, using copper metal, which has a unit cell edge d_o equal to 0.361 nm and a density equal to 8.92 g/cm^3.

In the center of the unit cell there is a hole. It is smaller than a cubic hole but of significant size nonetheless. It is called an octahedral hole, because the six atoms around it define an octahedron. In an ionic crystal, the anions often occupy the sites of the atoms in your cell, and there is a cation in the center of the octahedral hole. On the Calculations page, find the maximum radius, r_+, of the cation that would just fit in an octahedral hole surrounded by anions of radius r_-. You should find that the r_+/r_- ratio turns out to be 0.414.

There is another kind of hole in the FCC lattice that is important. See if you can find it; there are eight of them in the unit cell. If you look at an atom on the corner of the cell, you can see that it lies in a tetrahedron. In the center of the tetrahedron there is a tetrahedral hole, which is small compared to an octahedral or cubic hole. However, in some crystals small cations are found in some or all of the tetrahedral holes, again in a close-packed anion lattice. The cation-anion radius ratio at which cations would just fit into a tetrahedral hole is not so easy to find; r_+/r_- turns out to be 0.225.

The close-packed layers of atoms in the FCC structure are not parallel to the unit cell faces, but rather are perpendicular to the cell diagonal. If you look down the cell diagonal, you see six atoms in a close-packed triangle in the layer immediately behind the corner atom, and another layer of close-packed atoms below that, followed by another corner atom. The layers are indeed closely packed, and, as one goes down the diagonal of this and succeeding cells, the layers repeat their positions in the order ABCABC…, meaning that atoms in every fourth layer lie below one another.

Clearly there is another way we could stack the layers. The first and second will always be in the same relative positions, but the third layer could be below the first one if it were shifted properly. So we can have a close-packed structure in which the order of the layers is ABABAB…. The crystal obtained from this arrangement of layers is not cubic, but hexagonal. It too is a common structure for metals. Cadmium, zinc, and manganese have this structure. As you might expect, the stability of this structure is very similar to that of FCC crystals. We find that simply changing the temperature often converts a metal from one form to another. Calcium, for example, is FCC at room temperature, but if heated to 450°C it converts to close-packed hexagonal.

Crystal Structures of Some Common Binary Compounds

We have now dealt with all of the possible cubic crystal structures for metals. It turns out that the structures of binary ionic compounds are often related to these metal structures in a very simple way. In many ionic crystals the anions, which are large compared with cations, are essentially in contact with each other, in either an

SC or FCC structure. The cations go into the cubic, or octahedral, or tetrahedral holes, depending on the cation-anion radius ratios that we calculated. The idea is that the cation will tend to go into a hole in which it will not quite fit. This increases the unit cell size from the value it would have if the anions were touching, which reduces the repulsion energy due to anion-anion interaction and increases to the maximum the cation-anion attraction energy, producing the most stable possible crystal structure. According to the so-called radius-ratio rule, large cations go into cubic holes, smaller ones into octahedral holes, and the smallest ones into tetrahedral holes.

The deciding factor for which hole is favored is given by the radius ratio:

$$\text{If } r_+/r_- > 0.732 \qquad \text{cations go into cubic holes}$$
$$\text{If } 0.732 > r_+/r_- > 0.414 \qquad \text{cations go into octahedral holes}$$
$$\text{If } 0.414 > r_+/r_- > 0.225 \qquad \text{cations go into tetrahedral holes}$$

The NaCl Crystal

To apply the radius-ratio rule to NaCl, we simply need to find r_+/r_-, using the data in Table 16.1. Since $r_{Na+} = 0.095$ nm, and $r_{Cl-} = 0.181$ nm, the radius ratio is 0.095/0.181, or 0.525. That value is less than 0.732 and greater than 0.414, so the sodium ions should go into octahedral holes. We saw the octahedral hole in the center of the FCC unit cell, so Na^+ ions go there, and the Cl^- ions have the FCC structure. Actually, there are 12 other octahedral holes associated with the cell, one on each edge, which would be apparent if we had been able to make more cells. An Na^+ ion goes into each of these holes, giving the classic NaCl structure shown on the Calculations page. (Again, only the ions and bonds on the exposed faces are shown.)

Make a model of the unit cell for NaCl, using the large gray balls for the Cl^- ions and the blue ones for Na^+ ions. Clearly the Na^+ ion at the center of the cell is in an octahedral hole, but so are all of the other sodium ions, because if you extend the lattice, every Na^+ ion will be surrounded by six Cl^- ions. The coordination number of Na^+, and of Cl^-, ions is 6. The crystal is FCC in Cl^- and also in Na^+, because you could put Na^+ ions on the corners of the unit cell and maintain the same structure. The unit cell extends from the center of one Cl^- ion to the center of the next Cl^- along the cell edge; or, from the center of one Na^+ ion to the center of the next Na^+ ion.

On the Calculations page find the number of Na^+ and of Cl^- ions in the unit cell.

The CsCl Crystal

Cesium chloride has the same type of formula as NaCl, 1:1. The Cs^+ ion, however, is larger than Na^+, and has a radius equal to 0.169 nm. This makes r_+/r_- equal to 0.933. Because this value is greater than 0.732, we would expect that in the CsCl crystal the Cs^+ ions will fill cubic holes, and this is what is observed. The structure of CsCl will look like that of the BCC unit cell you made earlier, except that the ion in the center will be Cs^+ and those on the corners Cl^-. If you put gray balls in the center of each BCC unit cell you made from the blue balls, you would have the CsCl structure. This structure is *not* BCC, because the corner and center atoms are not the same. Rather it consists of two interpenetrating simple cubic lattices, one made from Cl^- ions and the other from Cs^+ ions.

The Zinc Sulfide Crystal

Zinc sulfide is another 1:1 compound, but its crystal structure is not that of NaCl or of CsCl. In ZnS, the Zn^{2+} ions have a radius of 0.074 nm and the S^{2-} ions a radius of 0.184 nm, making r_+/r_- equal to 0.402. By the radius-ratio rule, ZnS should have close-packed S^{2-} ions, with the Zn^{2+} ions in tetrahedral holes. In the ZnS unit cell, we find that this is indeed the case; the S^{2-} ions are FCC, and alternate tetrahedral holes in the unit cell are occupied by Zn^{2+} ions, which themselves form a tetrahedron.

To make a model for ZnS, first assemble an SC unit cell, using blue balls (b) and the long bonds. Use gray balls (g) for the zinc ions, and assemble the unit shown on the following page, using the short bonds and the holes in triangular faces:

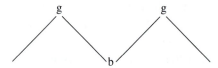

Attach this unit to two corner atoms on the diagonal of the bottom face. Make another unit and attach it to the two corner atoms in the upper face that lie on the face diagonal that is not parallel to the bottom face diagonal. The gray balls should then form a tetrahedron. Then attach the four other FCC blue balls to the gray ones. In this structure the coordination number is four (the long bonds do not count, they just keep the unit cell from falling apart). If the gray balls in the unit cell are replaced with blue ones, so that all atoms are of the same element, we obtain the diamond crystal structure.

These are the three common cubic structures of 1:1 compounds. The radius-ratio rule allows us to predict which structure a given compound will have. It does not always work, but it is correct most of the time. On the Calculations page, use the rule to predict the cubic structures that crystals of the following substances will have: KI, CuBr, and TlBr.

The Calcium Fluoride Crystal

Calcium fluoride is a 1:2 compound, so it cannot have the structure of any crystal we have discussed so far. The radii of Ca^{2+} and F^- are 0.099 and 0.136 nm, respectively, so r_+/r_- is 0.727. This makes CaF_2 on the boundary between compounds with cations in cubic holes or octahedral holes. It turns out that in CaF_2, the F^- ions have a simple cubic structure, with half of the cubic holes filled by Ca^{2+} ions. This produces a crystal in which the Ca^{2+} ions lie in an FCC lattice, with 8 F^- ions in a cube inside the unit cell.

Use your model set to make a unit cell for CaF_2. Use gray balls (g) for the Ca^{2+} ions and red ones (r) for F^-. The cations and anions are linked by short bonds that go into holes in nonadjacent triangular faces. The Ca^{2+} ions on each face diagonal are linked to F^- ions as shown below:

To complete the bottom face, attach two red and two gray balls to the initial line, so that the four red balls form a square and the gray ones an FCC face. The top of the unit cell is made the same way. Attach the two assemblies through four gray balls, which lie at the centers of the other faces. When you are done, the red balls should form a cube inside an FCC unit cell made from gray balls. Extend the model into an adjacent unit cell to show that every other cube of F^- ions has a Ca^{2+} ion at its center.

When you have completed this part of the experiment, your instructor may assign you a model with which to work. Report your results as directed. Then disassemble the models you made and pack the components in the box.

Experiment 16

Data and Calculations: The Structure of Crystals

Table 16.1

Atom	Molar Mass, g/mole	Atomic Radius, nm	Ionic Radius, nm
Na	22.99	—	0.095
Cu	63.55	—	0.096
Cl	35.45	0.099	0.181
Cs	132.9	0.262	0.169
Zn	65.38	0.133	0.074
S	32.06	0.104	0.184
K	39.10	0.231	0.133
I	126.9	0.133	0.216
Br	79.90	0.114	0.195
Tl	204.4	0.171	0.147

Some Cubic Unit Cells

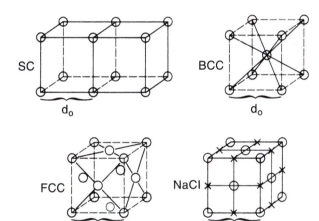

Simple Cubic Crystal

Fraction of volume of unit cell occupied by atoms

Volume of unit cell in terms of d_o _____

Number of atoms per unit cell   _____

Radius r of atom in terms of d_o _____

Volume of atom in terms of d_o $\left(V = \dfrac{4\pi}{3} r^3 \right)$ _____

Volume of atom/volume of cell   _____ = _____

(continued on following page)

Body-Centered Cubic Crystal

a.   Radius of a sodium atom

In BCC atoms touch along the cube diagonal
(see your model)

Length of cube diagonal if $d_o = 0.429$ nm   _____ nm

$4 \times r_{Na} =$ length of cube diagonal $r_{Na} =$ _____ nm

b.   Density of sodium metal

Length of unit cell, d_o, in cm (1 cm $= 10^7$ nm)   _____ cm

Volume of unit cell, V _____ cm^3

Number of atoms per unit cell   _____

Number of unit cells per mole Na   _____

Mass of a unit cell, m _____ g

Density of sodium metal, m/V (Observed value $= 0.97$ g/cm^3)   _____ g/cm^3

Face-Centered Cubic Crystal

a.   Number of atoms per unit cell

Number of atoms on corners _____ Shared by _____ cells _____ × _____ = _____

Number of atoms on faces _____ Shared by _____ cells _____ × _____ = _____

Total atoms per cell _____

b.   Radius of a Cu atom

In FCC atoms touch along face diagonal (see your model and ASA)

Length of face diagonal in Cu, where $d_o = 0.361$ nm   _____ nm

Number of Cu atom radii on face diagonal _____ $r_{Cu} =$ _____ nm

(continued on following page)

c.   Avogadro's Number, N

Length of cell edge, d_o, in cm, in Cu metal   _____ cm

Volume of a unit cell in Cu metal   _____ cm^3

Volume of a mole of Cu metal (V = mass/density)   _____ cm^3

Number of unit cells per mole Cu   _____

Number of atoms per unit cell   _____

Number of atoms per mole, N _____

d.   Size of an octahedral hole

Below is a sketch of the front face of a face-centered cubic cell with the NaCl structure. The ions are drawn so that they are just touching their nearest neighbors. The cations are in octahedral holes.

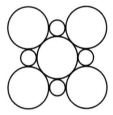

Show the length of d_o on the sketch. (See your model.)

What is the relationship between r_- and d_o?

$r_- =$ _____ $\times d_o$

What is the equation relating r_+, r_-, d_o?

$d_o =$ _____

What is the relationship between r_+ and d_o?

$r_+ =$ _____ $\times d_o$

What is the value of the radius ratio r_+/r_-? Express the ratio to three significant figures.

$r_+/r_- =$ _____

(continued on following page)

NaCl Crystal

Number of Cl^- ions in the unit cell (FCC) _____

Number of Na^+ ions on edges of cell _____ Shared by _____ cells

Number of Na^+ ions in center of cell _____ Shared by _____ cell(s)

Total no. Na^+ ions in unit cell _____

Radius-ratio rule (data in Table 16.1)

KI r_+/r_- _____ Predicted structure _____ Obs. NaCl

CuBr r_+/r_- _____ Predicted structure _____ Obs. ZnS

TlBr r_+/r_- _____ Predicted structure _____ Obs. CsCl

Report on unknown

Experiment 16

Advance Study Assignment: The Structure of Crystals

1. Many substances crystallize in a cubic structure. The unit cell for such crystals is a cube having an edge with a length equal to d_o.

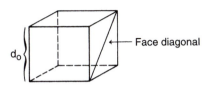

    a. What is the length, in terms of d_o, of the face diagonal, which runs diagonally across one face of the cube?

    _____

    b. What is the length, again in terms of d_o, of the cube diagonal, which runs from one corner, through the center of the cube, to the other corner? Hint: Make a right triangle having a face diagonal and an edge of the cube as its sides, with the hypotenuse equal to the cube diagonal.

    _____

2. In an FCC structure, the atoms are found on the corners of the cubic unit cell and in the centers of each face. The unit cell has an edge whose length is the distance from the center of one corner atom to the center of another corner atom on the same edge. The atoms on the diagonal of any face are touching. One of the faces of the unit cell is shown below.

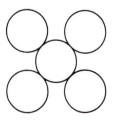

    a. Show the distance d_o on the sketch. Draw the boundaries of the unit cell.

    b. What is the relationship between the length of the face diagonal and the radius of the atoms in the cell?

    Face diagonal = _____

(continued on following page)

c. How is the radius of the atoms related to d_o?

$r =$ _____

d. Silver metal crystals have an FCC structure. The unit cell edge in silver is 0.4086 nm long. What is the radius of a silver atom?

_____nm

Experiment 17

Classification of Chemical Substances

Depending on the kind of bonding present in a chemical substance, the substance may be called ionic, molecular, or metallic.

In a solid ionic compound there are ions; the large electrostatic forces between the positively and negatively charged ions are responsible for the bonding which holds these particles together.

In a molecular substance the bonding is caused by the sharing of electrons by atoms. When the stable aggregates resulting from covalent bonding contain relatively small numbers of atoms, they are called molecules. If the aggregates are very large and include essentially all the atoms in a macroscopic particle, the substance is called macromolecular.

Metals are characterized by a kind of bonding in which the electrons are much freer to move than in other kinds of substances. The metallic bond is stable but is probably less localized than other bonds.

The terms ionic, molecular, macromolecular, and metallic are somewhat arbitrary, and some substances have properties that would place them in a borderline category, somewhere between one group and another. It is useful, however, to consider some of the general characteristics of typical ionic, molecular, macromolecular, and metallic substances, since many very common substances can be readily assigned to one of these categories or another.

Ionic Substances

Ionic substances are all solids at room temperature. They are typically crystalline but may exist as fine powders as well as clearly defined crystals. While many ionic substances are stable up to their melting points, some decompose on heating. It is very common for an ionic crystal to release loosely bound water of hydration at temperatures below 200°C. Anhydrous (dehydrated) ionic compounds have high melting points, usually above 300°C but below 1000°C. They are not readily volatilized and boil only at very high temperatures (Table 17.1).

When molten, ionic compounds conduct an electric current. In the solid state they do not conduct electricity. The conductivity in the molten liquid is attributed to the freedom of motion of the ions, which arises when the crystal lattice is no longer present.

Ionic substances are frequently but not always appreciably soluble in water. The solutions produced conduct the electric current rather well. The conductivity of a solution of a slightly soluble ionic substance is often several times that of the solvent water. Ionic substances are usually not nearly so soluble in other liquids as they are in water. For a liquid to be a good solvent for ionic compounds it must be highly polar, containing molecules with well-defined positive and negative regions with which the ions can interact.

Molecular Substances

All gases and essentially all liquids at room temperature are molecular in nature. If the molar mass of a molecular substance is over about 100 grams, it may be a solid at room temperature. The melting points of molecular substances are usually below 300°C; these substances are relatively volatile, but a good many will decompose before they boil. Most molecular substances do not conduct the electric current either when solid or when molten.

Organic compounds, which contain primarily carbon and hydrogen, often in combination with other nonmetals, are essentially molecular in nature. Since there are a great many organic substances, it is true that most substances are molecular. If an organic compound decomposes on heating, the residue is frequently a

Table 17.1

Physical Properties of Some Representative Chemical Substances						

Substance	M.P., °C	B.P., °C	Solubility		Electrical Conductance	Classification
			Water	Hexane		
NaCl	801	1413	Sol	Insol	High in melt and in soln	Ionic
MgO	2800	—	Sl sol	Insol	Low in sat'd soln	Ionic
CoCl$_2$	Sublimes	1049	Sol	Insol	High in soln	Ionic
CoCl$_2$ · 6H$_2$O	86	Dec	Sol	Insol	High in soln	Ionic hydrate, −H$_2$O at 110°C
C$_{10}$H$_8$	70	255	Insol	Sol	Zero in melt	Molecular
C$_6$H$_5$COOH	122	249	Sl sol	Sl sol	Low in sat'd soln	Molecular-ionic
FeCl$_3$	282	315	Sol	Sl sol	High in soln and in melt	Molecular-ionic
SnI$_4$	144	341	Dec	Sol	~Zero in melt	Molecular
SiO$_2$	1600	2590	Insol	Insol	Zero in solid	Macromolecular
Fe	1535	3000	Insol	Insol	High in solid	Metallic

Key: Sol = at least 0.1 mole/L; Sl sol = appreciable solubility but <0.1 mole/L; Insol = essentially insoluble; Dec = decomposes.

black carbonaceous material. Reasonably large numbers of inorganic substances are also molecular; those that are solids at room temperature include some of the binary compounds of elements in Groups 14, 15, 16, and 17.

Molecular substances are usually soluble in at least a few organic solvents, with the solubility being enhanced if the substance and the solvent are similar in molecular structure.

Some molecular compounds are markedly polar, which tends to increase their solubility in water and other polar solvents. Such substances may ionize appreciably in water, or even in the melt, so that they become conductors of electricity. Usually the conductivity is considerably lower than that of an ionic material. Most polar molecular compounds in this category are organic, but a few, including some of the salts of the transition metals, are inorganic.

Macromolecular Substances

Macromolecular substances are all solids at room temperature. They have very high melting points, usually above 1000°C, and low volatility. They are typically very resistant to thermal decomposition. They do not conduct electric current and are often good insulators. They are not soluble in water or any organic solvents. They are frequently chemically inert and may be used as abrasives or refractories.

Metallic Substances

The properties of metals appear to derive mainly from the freedom of movement of their bonding electrons. Metals are good electrical conductors in the solid form, and have characteristic luster and malleability. Most metals are solid at room temperature and have melting points that range from below 0°C to over 2000°C. They are not soluble in water or organic solvents. Some metals are prepared as gray or black powders, which may not appear to be electrical conductors. However, if one measures the conductance while the powder is under pressure, its metallic character is revealed.

Experimental Procedure

In this experiment you will investigate the properties of several substances with the purpose of determining whether they are molecular, ionic, macromolecular, or metallic. In some cases the classification will be very straightforward. In others the assignment to a class will not be so easy and you may find that the substance has a behavior associated with more than one class.

As the discussion indicates, there are several properties we can use to find out to which class a substance belongs. In this experiment we will use the melting point, solubility in water and organic solvents, and electrical conductivity of the aqueous solution, the solid, and the melt in making the classification.

The substances to be studied in the first part of the experiment are on the laboratory tables along with two organic solvents, one polar and one nonpolar. You need only carry out enough tests on each substance to establish the class to which it belongs, so you will not need to perform every test on every substance. You may, however, carry out any extra tests you wish, if only to satisfy your curiosity. Follow the directions for each test as given below.

Melting Point

Approximate melting points of substances can be determined rather easily. Substances with low melting points, less than 100°C, will melt readily when warmed gently in a small test tube. A sample the size of a pea will suffice. If the sample melts between 100° and 300°C, it will take more than gentle warming, but will melt before the test tube imparts a yellow-orange color to the Bunsen flame. Above 300°C, there will be increasing color; up to about 500°C one can still use a test tube and a strong burner flame; but at about 550°C the pyrex tube will begin to soften. In this experiment we will not attempt to measure any melting points above 500°C.

While heating a sample, keep the tube *loosely* stoppered with a cork. *Do not breathe* any vapors that are given off, and *do not continue to heat* a sample after it has melted. As you heat the sample, look for evidence of decomposition, sublimation, or evolution of water.

Solubility and Conductance of Solutions

In testing for solubility, again use a sample about the size of a pea, this time in a regular (18 mm × 150 mm) test tube. Use about 2 mL of solvent, enough to fill the tube to a depth of about 1 cm. Stir well, using a clean stirring rod. Some samples will dissolve completely almost immediately; some are only slightly soluble and may produce a cloudy suspension; others are completely insoluble. Make solubility tests with distilled water and the two organic solvents and record your results. Use fresh distilled water in your wash bottle for the aqueous solubility tests.

Conductance measurements need only be made on water solutions. We will use portable ohm meters for this purpose. Your instructor may make the measurements for you, but may have you make them. An ohm-meter measures the electrical resistance of a sample in ohms, Ω. A solution with a high resistance has a low electrical conductance, and vice versa. Some of your solutions will have a low resistance, on the order of 1000Ω or less; these are good conductors. Distilled water has a high resistance; with your meter it will probably have a resistance of $50,000\Omega$ or greater. Small amounts of contaminants can lower the resistance of a solution very markedly, particularly if the main solute shows high resistance.

Measure the resistance of any of the aqueous solutions containing soluble or slightly soluble substances. Between tests, rinse the electrodes in a beaker filled with distilled water. For our purposes, a solution with a resistance less than about 2000Ω is a good conductor, denoted "G". Between 2000 and $20,000\Omega$ it is a weak conductor, denoted "W". Above $20,000\Omega$ we will consider it to be essentially nonconducting, and denote it with an "N". Record the resistances you observe in ohms. Then note, with a G, W, or N, whether the solution is a good, weak, or poor conductor.

Electrical Conductance of Solids and Melts

Some substances conduct electricity in the solid state. If the sample contains large crystals, the conductance test is very easy. Select a crystal, put it on the lab bench, and touch it with the two wires on the ohm-meter probe. Metals have a very low resistance. In powder form most substances, metals included, appear to have essentially infinite resistance. However, under pressure, metal powders, unlike those of other substances, show good conductance. To test a powder for conductance, put a penny on the lab bench. On it place a small rubber washer from the box on the bench. Fill the hole in the washer with the powder, and put another penny on top of the washer. Put the whole sandwich between the jaws of a pair of insulated pliers. Touch the electrodes from the ohm-meter probe to the pennies, one electrode to each penny, and squeeze the pliers. If the powder is a metal, the resistance will gradually fall from infinity to a small value. Make sure any drop in resistance is not caused by the pennies touching each other. Record your results.

To check the conductance of a melt, put a pea-size sample in a dry, regular test tube and melt it. Heat the electrodes on the probe for a few seconds in the Bunsen flame and touch them to the melt. Heat gently to ensure that no solid is crystallized on the electrodes. Many melts are good conductors. After testing a melt, clean the electrodes by washing them with water or an organic solvent, or, if necessary, scraping them off with a spatula.

Having made the tests we have described, you should be able to assign each substance to its class, or, possibly, to one or both of two classes. Make this classification for each substance, and give your reasons for doing so.

When you have classified each substance, report to your laboratory supervisor, who will assign you two unknowns for characterization.

DISPOSAL OF REACTION PRODUCTS. All residues from your tests should be discarded in the waste crock unless directed otherwise by your instructor.

Experiment 17

Observations and Conclusions: Classification of Chemical Substances

Table 17.2

Substance No.	Approx. Melting Pt., °C (<100, 100–300, 300–500, >500)	Solubility			Electrical Resistance, Ω			Classification and Reason
		H₂O	Nonpolar Organic	Polar Organic	Solution in H₂O	Solid	Melt	
I								
II								
III								
IV								
V								
VI								
Unknown no.								
_____								
_____								

Key: Sol = soluble; Sl sol = slightly soluble; Insol = insoluble; G = good electrical conductor, R < 2000Ω; W = weak electrical conductor, 2000Ω < R < 20,000Ω; N = nonconductor, R > 20,000Ω.

Experiment 17

Advance Study Assignment: Classification of Chemical Substances

1. List the properties of a substance that would definitely establish that the material is molecular.

2. If we classify substances as ionic, molecular, macromolecular, or metallic, in which if any categories are all the members

    a. soluble in water?

    b. electrical conductors in the melt?

    c. insoluble in all common solvents?

    d. solids at room temperature?

3. A given substance is a white, granular solid at 25°C that does not conduct electricity. It melts at 750°C without decomposing, and the melt conducts an electric current. It dissolves readily in water, to produce an electrically conductive solution. What would be the classification of the substance, based on this information?

4. A lustrous grey-white solid melts at 1650°C. It is electrically conductive as both a solid and a liquid, but not soluble in either water or any organic solvent. Classify the substance as best you can from these properties.

Experiment 18

Some Nonmetals and Their Compounds—Preparations and Properties

Some of the most commonly encountered chemical substances are nonmetallic elements or their simple compounds. O_2 and N_2 in the air, CO_2 produced by combustion of oil, coal, or wood, and H_2O in rivers, lakes, and air are typical of such substances. Substances containing nonmetallic atoms, whether elementary or compound, are all molecular, reflecting the covalent bonding that holds their atoms together. They are often gases, due to weak intermolecular forces. However, with high molecular masses, hydrogen bonding, or macromolecular structures one finds liquids, like H_2O and Br_2, and solids, like I_2 and graphite.

Several of the common nonmetallic elements and some of their gaseous compounds can be prepared by simple reactions. In this experiment you will prepare some typical examples of such substances and examine a few of their characteristic properties.

WEAR YOUR SAFETY GLASSES WHILE PERFORMING THIS EXPERIMENT

Experimental Procedure

In several of the experiments you perform, you will prepare gases. In general we will not describe in detail should be observed, so perform each preparation carefully and report what you actually observe, not what you think we expect you to observe.

There are several tests we will make on the gases you prepare, and the way each of these should be carried out is summarized below.

Test for Odor. To determine the odor of a gas, first pass your hand across the end of the tube, bringing the gas toward your nose. If you don't detect an odor, sniff near the end of the tube, first at some distance and then gradually closer. Don't just put the tube at the end of your nose immediately and take a deep breath. Some of the gases you will make have no odor, and some will have very impressive ones. Some of the gases are very toxic and, even though we will be making only small amounts, caution in testing for odor is important.

Test for Support of Combustion. A few gases will support combustion, but most will not. To make the test, ignite a wood splint with a Bunsen flame, blow out the flame, and put the glowing, but not burning, splint into the gas in the test tube. If the gas supports combustion, the splint will glow more brightly, or may make a small popping noise. If the gas does not support combustion, the splint will go out almost instantly. You can use the same splint for all the combustion tests.

Test for Acid-Base Properties. Many gases are acids. This means that if the gas is dissolved in water it will produce some H^+ ions. A few gases are bases; in water solution such gases produce OH^- ions. Other gases do not interact with water and are neutral. It is easy to establish the acidic or basic nature of a gas by using a chemical indicator. One of the most common acid-base indicators is litmus, which is red in acidic solution and blue in basic solution. To test whether a gas is an acid, moisten a piece of blue litmus paper with water from your wash bottle, and put the paper down in the test tube in which the gas is present. If the gas is an acid, the paper will turn red. Similarly, to test if the gas is a base, moisten a piece of red litmus paper, and hold it down in the tube. A color change to blue will occur if the gas forms a basic solution. Since you may have used an acid or a base in making the gas, do not touch the walls of the test tube with the paper. The color change will occur fairly quickly and smoothly over the surface of the paper if the gas is acidic or basic. It is not necessary to use a new piece of litmus for each test. Start with a piece of blue and a piece of red litmus. If you need to regenerate the blue paper, hold it over an open bottle of 6 M NH_3, whose vapor is basic. If you need to make red litmus, hold the moist paper over the acidic vapor of an open bottle of 6 M acetic acid.

A. Preparation and Properties of Nonmetallic Elements: O_2, N_2, Br_2, I_2

1. Oxygen, O_2. Oxygen can be easily prepared in the laboratory by the decomposition of H_2O_2, hydrogen peroxide, in aqueous solution. Hydrogen peroxide is not very stable and will break down to water and oxygen gas on addition of a suitable catalyst, particularly MnO_2:

$$2\ H_2O_2(aq) \xrightarrow{\ MnO_2\ } O_2(g) + 2\ H_2O \tag{1}$$

Add 1 mL 3% H_2O_2 solution in water to a small test tube. Pick up a small amount (~0.1 g) of MnO_2 on the tip of your spatula and add it to the liquid in the tube. Hold your finger over the end of the tube to help confine the O_2. Test the evolved gas for odor. Test the gas for any acid-base properties, using moist blue and red litmus paper. Test the gas for support of combustion. Record your observations.

2. Nitrogen, N_2. Sodium sulfamate in water solution will react with nitrite ion to produce nitrogen gas:

$$NO_2^-(aq) + NH_2SO_3^-(aq) \rightarrow N_2(g) + SO_4^{2-}(aq) + H_2O \tag{2}$$

Add about 1 mL 1 M KNO_2, potassium nitrite, to a small test tube. Then add 10 to 12 drops 0.5 M $NaNH_2SO_3$, sodium sulfamate, and place the test tube in a hot-water bath made from a 250-mL beaker half full of water. Bubbles of nitrogen should form within a few moments. Confine the gas for a few seconds with a stopper. Cautiously test the evolved gas for odor. Carry out the tests for acid-base properties and for support of combustion. If you need to generate more N_2 to complete the tests, add sodium sulfamate solution as necessary, 5 to 10 drops at a time. Report your observations.

3. Iodine, I_2. The halogen elements are most readily prepared from their sodium or potassium salts. The reaction involves an oxidizing agent, which can remove electrons from the halide ions, freeing the halogen. The reaction that occurs when a solution of potassium iodide, KI, is treated with 6 M HCl and a little MnO_2 is

$$2\ I^-(aq) + 4\ H^+(aq) + MnO_2(s) \rightarrow I_2(aq) + Mn^{2+}(aq) + 2\ H_2O \tag{3}$$

At 25°C I_2 is a solid, with relatively low solubility in water, and appreciable volatility. I_2 can be extracted from aqueous solution into organic solvents, particularly heptane, C_7H_{16} (HEP). The solid, vapor, and solutions in water and HEP all have characteristic colors.

Put two drops 1 M KI into a *regular* (18 × 150 mm) test tube. Add six drops 6 M HCl and a tiny amount of manganese dioxide, MnO_2. Swirl the mixture and note any changes that occur. Put the test tube into a hot-water bath. After a minute or two a noticeable amount of I_2 vapor should be visible above the liquid. Remove the test tube from the water bath and add 10 mL distilled water. Stopper the tube and shake. Note the color of I_2 in the solution. Sniff the vapor above the solution. Decant the liquid into another regular test tube and add 3 mL heptane, C_7H_{16}. Stopper and shake the tube. Observe the color of the HEP layer and the relative solubility of I_2 in water and HEP. Record your observations.

4. Bromine, Br_2. Bromine can be made by the same reaction as is used to make iodine, substituting bromide ion for iodide ion. Bromide ion is less easily oxidized than iodide ion. Bromine at 25°C is a liquid. Br_2 can be extracted from water solution into heptane. Its liquid, vapor, and solutions in water and HEP are colored.

To two or three drops 1 M NaBr in a regular test tube, add six drops 6 M HCl and a small amount of MnO_2, about the size of a small pea. Swirl to mix the reagents, and observe any changes. Heat the tube in the water bath for a minute or two. Try to detect Br_2 vapor above the liquid by observing its color. Add 10 mL water to the tube, stopper, and shake. Note the color of Br_2 in the solution. Sniff the vapor, cautiously. Decant the liquid into a regular test tube. Add 3 mL HEP, stopper, and shake. Note the color of Br_2 in HEP. Record your observations.

B. Preparation and Properties of Some Nonmetallic Oxides: CO_2, SO_2, NO, and NO_2

1. Carbon Dioxide, CO_2. Carbon dioxide, like several nonmetallic oxides, is easily made by treating an oxyanion with an acid. With carbon dioxide the oxyanion is CO_3^{2-}, carbonate ion, which is present in solutions of carbonate salts, such as Na_2CO_3. The reaction is

$$CO_3^{2-}(aq) + 2\,H^+(aq) \rightarrow (aq) \rightarrow H_2CO_3(aq) \rightarrow CO_2(g) + H_2O \tag{4}$$

Carbon dioxide is not very soluble in water, and on acidification, carbonate solutions will tend to effervesce as CO_2 is liberated. Nonmetallic oxides in solution are often acidic but never basic.

To 1 mL M Na_2CO_3 in a small test tube add six drops 3 M H_2SO_4. Test the gas for odor, acidic properties, and ability to support combustion.

2. Sulfur Dioxide, SO_2. Sulfur dioxide is readily prepared by acidification of a solution of sodium sulfite, containing sulfite ion, SO_3^{2-}. As you can see, the reaction that occurs is very similar to that with carbonate ion.

$$SO_3^{2-}(aq) + 2\,H^+(aq) \rightarrow H_2SO_3(aq) \rightarrow SO_2(g) + H_2O \tag{5}$$

Sulfur dioxide is considerably more soluble in water than is carbon dioxide. Some effervescence may be observed on acidification of concentrated sulfite solutions; effervescence will increase if the solution is heated in a water bath.

To 1 mL M Na_2SO_3 in a small test tube add six drops 3 M H_2SO_4. *Cautiously* test the evolved gas for odor. Test its acidic properties and its ability to support combustion. Put the test tube into the water bath for a few seconds to see if effervescence occurs if the solution is hot.

3. Nitrogen Dioxide, NO_2, and Nitric Oxide, NO. If a solution containing nitrite ion, NO_2^-, is treated with acid, two oxides are produced, NO_2 and NO. In solution these gases are combined in the form of N_2O_3, which is colored. When these gases come out of solution, the mixture contains NO and NO_2; the latter is colored. NO is colorless and reacts readily with oxygen in the air to form NO_2. The preparation reaction is

$$2\,NO_2^-(aq) + 2\,H^+(aq) \rightarrow N_2O_3(aq) + H_2O \rightarrow NO(g) + NO_2(g) + H_2O \tag{6}$$

To 1 mL of 1 M KNO_2 in a small test tube, add six drops 3 M H_2SO_4. Swirl the mixture for a few seconds and note the color of the solution. Warm the tube in the water bath for a few seconds to increase the rate of gas evolution. Note the color of the gas that is given off. Cautiously test the odor of the gas. Test its acidic properties and its ability to support combustion. Record your observations.

C. Preparation and Properties of Some Nonmetallic Hydrides: NH_3, H_2S

1. Ammonia, NH_3. A 6 M solution of NH_3 in water is a common laboratory reagent. You may have been using it in this experiment to make your litmus paper turn blue. Ammonia gas can be made by simply heating 6 M NH_3. It can also be prepared by addition of a strongly basic solution to a solution of an ammonium salt, such as NH_4Cl. The latter solution contains NH_4^+ ion. On treatment with OH^- ion, as in a solution of NaOH, the following reaction occurs:

$$NH_4^+(aq) + OH^-(aq) \rightarrow NH_3(aq) + H_2O \rightarrow NH_3(g) + H_2O \tag{7}$$

The odor of NH_3 is characteristic. NH_3 is very soluble in water, so effervescence is not observed, even on heating concentrated solutions.

Add 1 mL 1 M NH_4Cl to a small test tube. Add 1 mL 6 M NaOH. Swirl the mixture and test the odor of the evolved gas. Put the test tube into the water bath for a few moments to increase the amount of NH_3 in the gas phase. Test the gas with moistened blue and red litmus paper. Test the gas for support of combustion.

2. Hydrogen Sulfide, H_2S. Hydrogen sulfide can be made by treating some solid sulfides, particularly FeS, with an acid such as HCl or H_2SO_4. With FeS the reaction is

$$FeS(s) + 2\,H^+(aq) \rightarrow H_2S(g) + Fe^{2+}(aq) \tag{8}$$

This reaction was used for many years to make H_2S in the laboratory in courses in qualitative analysis. In this experiment we will employ the method currently used in such courses for H_2S generation. This involves the decomposition of thioacetamide, CH_3CSNH_2, which occurs in solution on treatment with acid and heat. The reaction is

$$CH_3CSNH_2(aq) + 2\,H_2O \rightarrow H_2S(g) + CH_3COO^-(aq) + NH_4^+(aq) \tag{9}$$

The odor of H_2S is notorious. H_2S is moderately soluble in water and, since it is produced reasonably slowly in Reaction 9, there will be little if any effervescence.

To 1 mL of 1 M thioacetamide in a small test tube add six drops 3 M H_2SO_4. Put the test tube in a boiling-water bath for about 1 minute. There may be some cloudiness due to formation of free sulfur. Carefully smell the gas in the tube; H_2S is toxic. Test the gas for acidic and basic properties and for the ability to support combustion.

D. Identification of Unknown Solution   Optional

In this experiment you prepared nine different species containing nonmetallic elements. In each case the source of the species was in solution. In this part of the experiment we will give you an unknown solution that can be used to make one of the nine species. It will be a solution used in preparing one of the species, but it will not be an acid or a base. Identify by suitable tests the species that can be made from your unknown and the substance that is present in your unknown solution. ▪

CAUTION: *The following gases prepared in this experiment are toxic: Br_2, SO_2, NO_2, NH_3, and H_2S. Do not inhale these gases unnecessarily when testing their odors. One small sniff will be sufficient and will not be harmful. It is good for you to know these odors, in case you encounter them in the future.*

> **DISPOSAL OF REACTION PRODUCTS.** Most of the chemicals used in this experiment can be discarded down the sink drain. Pour the solutions from Sections 3 and 4 of Part A, containing I_2 and Br_2 in heptane, into a waste crock, unless directed otherwise by your instructor.

Experiment 18

Observations and Conclusions: Some Nonmetals and Their Compounds

A.

	Element Prepared	Degree of Effervescence	Odor	Acid-Base Tests	Support of Combustion Test
1.	O_2	_____	_____	_____	_____
2.	N_2	_____	_____	_____	_____

		Odor	Vapor	Color In H_2O	In HEP
3.	I_2	_____	_____	_____	_____
4.	Br_2	_____	_____	_____	_____

B.

	Oxide Prepared	Degree of Effervescence	Odor	Acid Test	Color	Support of Combustion Test
1.	CO_2	_____	_____	_____	_____	_____
2.	SO_2	_____	_____	_____	_____	_____
3.	$NO_2 + NO$	_____	_____		_____ soln	_____
					_____ gas	

C.

	Hydride Prepared	Degree of Effervescence	Odor	Acid-Base Tests	Support of Combustion Test
1.	NH_3	_____	_____	_____	_____
2.	H_2S	_____	_____	_____	_____

D. Properties of Unknown Solution   Optional

Nonmetal species that can be made from unknown _____

Identity of unknown solution _____

Unknown no. _____

Experiment 18

Advance Study Assignment: Some Nonmetals and Their Compounds

In this experiment nine species are prepared and studied. For each of the species, list the reagents that are used in its preparation and the reaction that occurs.

	Reagents Used	**Reaction**
1. O_2		
2. N_2		
3. I_2		
4. Br_2		
5. CO_2		
6. SO_2		
7. $NO_2 + NO$		
8. NH_3		
9. H_2S		

Experiment 19

Molar Mass Determination by Depression of the Freezing Point

The most common liquid we encounter in our daily lives is water. In this experiment we will study the equilibria that can exist between pure water and its aqueous solutions, and ice, the solid form of water. (Water is one of very few substances for which we have a separate name for the solid.)

If we take some ice cubes from the refrigerator and put them into a glass of water from the tap, we find that the water temperature falls and some ice melts. This occurs because heat will always tend to flow from a higher to a lower temperature. Heat from the water flows into the ice. It takes heat, called the heat of fusion, to melt ice. If there is enough ice present, the water temperature will ultimately fall to 0°C, and stay there. At that point, ice and water are in equilibrium, at the freezing point of water. At the freezing point, T_f°, the vapor pressures of ice and water must be equal, and that condition fixes the temperature. (See Fig. 19.1.)

Now let us consider what happens if we add a soluble liquid or solid to the equilibrium mixture of ice and water. Rather surprisingly, we find that the temperature of the ice and the solution falls as equilibrium is reestablished. The reason this happens is that in the solution the vapor pressure of water at 0°C is less than that of the pure liquid. So the vapor pressure of ice at 0°C is higher than that of the solution, and some ice melts. This requires heat, which comes from the solution and from the ice, and the temperature falls. The vapor pressures of the ice and the solution both fall, but that of the ice falls faster, and at some temperature T_f below 0°C equilibrium is established at the new freezing point. The situation is shown in Figure 19.1.

The change in the freezing point that is observed is called the freezing point depression, ΔT_f, equal to $T_f^\circ - T_f$. It is observed with solutions of any solvent. The freezing point depression is one of the colligative properties of solutions. Others are the boiling point elevation for non-volatile solutes, the osmotic pressure, and the vapor pressure lowering. The colligative properties of solutions depend on the number of solute particles present in a given amount of solvent and not on the kinds of particles dissolved, be they molecules, atoms, or ions.

When working with colligative properties it is convenient to express the solute concentration in terms of its molality m as defined by the equation:

$$\text{molality of } A = m_A = \frac{\text{moles } A \text{ dissolved}}{\text{kg solvent in the solution}} \tag{1}$$

For this unit of concentration, the boiling point elevation, $T_b - T_b^\circ$, or ΔT_b, and the freezing point depression, $T_f^\circ - T_f$, or ΔT_f, in °C at very low concentrations are given by the equations:

$$\Delta T_b = k_b m \quad \text{and} \quad \Delta T_f = k_f m \tag{2}$$

where k_b and k_f are characteristic of the solvent used. For water, $k_b = 0.52$ and $k_f = 1.86$. For benzene, $k_b = 2.53$ and $k_f = 5.10$. In this experiment we will assume that Equation 2 is valid, even though our solutions are moderately concentrated.

One of the classic uses of colligative properties was in connection with finding molar masses of unknown substances. With organic molecules, molar masses by FP depression agreed with those found by other methods. With ionic salts, like NaCl, molar masses were lower than the formula masses. On the basis of such experiments, Arrhenius suggested that ionic substances exist as ions in aqueous solution, consistent with the observation that such solutions conduct an electric current. Arrhenius' general idea turned out to be correct, and is now a basic part of modern chemical theory.

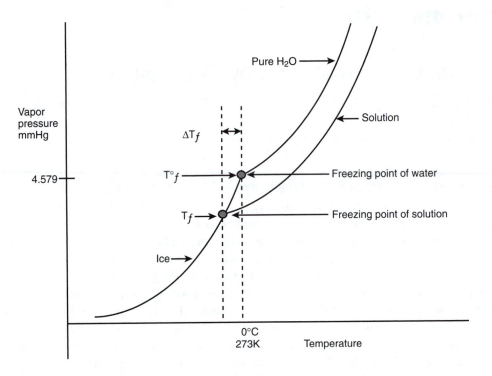

Figure 19.1

Discussion of the Method

In this experiment we will study the freezing point behavior of some aqueous solutions. First you will measure the freezing point of pure water, using a slurry of ice and water. Then you'll add a known mass of an unknown to the slurry, and find the freezing point of the solution, thus determining the freezing point depression, ΔT_f. That allows you to find the molality of the solution. Finally, you will separate the solution from the ice in the mixture, and weigh the solution, which will contain all of the solute. This information furnishes you with the composition of the solution, and lets you calculate the mass of solute in 1 kg of water. The molar mass follows from Equations 1 and 2.

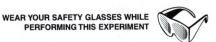

WEAR YOUR SAFETY GLASSES WHILE
PERFORMING THIS EXPERIMENT

Experimental Procedure

You may work in pairs on this experiment. Both members of the pair must submit a report. From the stock room obtain a digital thermometer, an insulated cup, and two unknowns, one liquid and one solid. The actual molar mass of the solid will be furnished to you.

A. Finding the Freezing Point of Water

You will first need to determine the freezing point of pure water. Prepare a water-ice mixture in your insulated cup, using more ice than water. Stir well, and record the lowest temperature you observe for that mixture. (Since your thermometer may not be properly calibrated, that temperature may not be 0.0°C.)

B. Finding the Freezing Point of a Solution of Liquid Unknown

To the water-ice mixture you need to add a known amount of one of your unknowns. Let's work first with the liquid unknown. To estimate the mass of your liquid to add to the mixture, assume that you will have about 100 g of water in the final solution, that the molality of the solute will be about 2 m, and that the molar mass of the solute is 50 g. Tare the insulated cup on the top loading-balance, and add the estimated mass by pouring it down a stirring rod on to the center of the ice-water mix. Record the mass. Then record the lowest temperature you observe after stirring the ice + solution mixture. You should obtain a freezing point depression of at least 4°C. If you need to, add more of the liquid unknown, weighing the amount added. Stir well and measure the lowest temperature of the mixture that you observe. Then promptly separate the solution from the ice by pouring it through a wire screen into a tared beaker on the top-loading balance. (Place the screen over the cup, and pour out the solution, keeping a corner of the screen down.) Record the mass of the solution.

Make a second molar mass measurement for your liquid unknown. This is easily done if you pour the decanted solution back into the cup once you have found its mass. Add some water and ice to the ice + solution mixture, thereby lowering the molality of the solute. Measure the new freezing point for that mixture. Pour off and weigh the solution.

When you have finished this part of the experiment, dispose of the mixture of ice and solution as directed by your instructor.

Determine molar masses as described in the Discussion.

C. Finding the Freezing Point of a Solution of Solid Unknown

With solid unknowns, you can use the same general procedure as with liquids. First, make a solution of the solid in water, using the same total mass of solid as you did with the liquid. You can do this by taring a small dry beaker on the top-loading balance, and adding the solid, noting its mass. Add a minimum amount of water to dissolve the solid (use at least 20 g of water). Stir until dissolution is complete. Add the entire solution of the solid to a new ice + water mix, noting the lowest temperature you obtain. Decant and weigh the solution as you did with the liquid unknown.

Again, make measurements at two different solute concentrations.

If the molar mass you find for your solid is less than the actual molar mass, you have an ionic solid. The ratio of the true molar mass to the value you find is equal to a quantity called the van't Hoff factor, i. This factor can be related to the apparent percentage dissociation of the salt you are studying. For a 1:1 salt,

$$\% \text{ dissociation} = (i - 1) \times 100\% \tag{3}$$

It turns out that Equation 3 works well for weak electrolytes like acetic acid, but is not correct for salts like NaCl, which modern theory assumes are completely ionized in aqueous solution.

In your report, note all the masses observed and the temperatures of the equilibrium mixtures. Report the average molar masses you calculate and show how you obtained them. If you have an ionic solid, find i and the % dissociation as predicted by the Arrhenius theory.

When you are finished with the experiment, pour the solution and the contents of the cup into the sink. Return the digital thermometer and the insulated cup to the stock room.

Optional Experiment: Determine the largest possible freezing point depression for an NaCl solution. Find the composition of that solution. What phases are in equilibrium in the final mixture?

Experiment 19

Data and Calculations: Molar Mass Determination by Depression
of the Freezing Point

For this experiment you will be responsible for recording and processing your data. In the upper part of the page, note the value of each quantity that was measured, along with its units. Below all the experimental data, make the calculations called for in the procedure, and list your results. Include all the equations that you used, first in words, and then with numerical values.

Experiment 19

Advance Study Assignment: Determination of Molar Mass by Depression
of the Freezing Point

1. A student determines the molar mass of methanol, CH_3OH, by the method used in this experiment. She found that the equilibrium temperature of a mixture of ice and pure water was 0.4°C on her thermometer. When she added 10.0 g of her sample to the mixture, the temperature, after thorough stirring, fell to −5.4°C. She then poured off the solution through a screen into a beaker. The mass of the solution was 101.8 g.

   a. What was the freezing point depression? _____ °C

   b. What was the molality of the methanol? _____ m

   c. How much methanol was in the decanted solution? _____ g

   d. How much water was in the decanted solution? _____ g

   e. How much methanol would there be in a solution containing 1 kg of water, with methanol at the same concentration as she had in her experiment?

   _____ g methanol

   f. What did she find to be the molar mass of methanol, assuming she made the calculation properly?

   _____ g

Experiment 20

Rates of Chemical Reactions, I. The Iodination of Acetone

The rate at which a chemical reaction occurs depends on several factors: the nature of the reaction, the concentrations of the reactants, the temperature, and the presence of possible catalysts. All of these factors can markedly influence the observed rate of reaction.

Some reactions at a given temperature are very slow indeed; the oxidation of gaseous hydrogen or wood at room temperature would not proceed appreciably in a century. Other reactions are essentially instantaneous; the precipitation of silver chloride when solutions containing silver ions and chloride ions are mixed and the formation of water when acidic and basic solutions are mixed are examples of extremely rapid reactions. In this experiment we will study a reaction that, in the vicinity of room temperature, proceeds at a moderate, relatively easily measured rate.

For a given reaction, the rate typically increases with an increase in the concentration of any reactant. The relation between rate and concentration is a remarkably simple one in many cases, and for the reaction

$$aA + bB \rightarrow cC$$

the rate can usually be expressed by the equation

$$\text{rate} = k(A)^m(B)^n \tag{1}$$

where m and n are generally, but not always, integers, 0, 1, 2, or possibly 3; (A) and (B) are the concentrations of A and B (ordinarily in moles per liter); and k is a constant, called the *rate constant* of the reaction, which makes the relation quantitatively correct. The numbers m and n are called the *orders of the reaction* with respect to A and B. If m is 1, the reaction is said to be *first order* with respect to the reactant A. If n is 2, the reaction is *second order* with respect to reactant B. The overall order is the sum of m and n. In this example the reaction would be *third order* overall.

The rate of a reaction is also significantly dependent on the temperature at which the reaction occurs. An increase in temperature increases the rate, an often-cited rule being that a 10°C rise in temperature will double the rate. This rule is only approximately correct; nevertheless, it is clear that a rise of temperature of say 100°C could change the rate of a reaction very appreciably.

As with the concentration, there is a quantitative relation between reaction rate and temperature, but here the relation is somewhat more complicated. This relation is based on the idea that to react, the reactant species must have a certain minimum amount of energy present at the time the reactants collide in the reaction step; this amount of energy, which is typically furnished by the kinetic energy of motion of the species present, is called the *activation energy* for the reaction. The equation relating the rate constant k to the absolute temperature T and the activation energy E_a is called the Arrhenius equation:

$$\ln k = \frac{-E_a}{RT} + \text{constant} \tag{2}$$

where $\ln k$ is the natural logarithm of k, and R is the gas constant (8.31 joules/mole K for E_a in joules per mole). This equation is identical in form to Equation 1 in Experiment 15. By measuring k at different temperatures we can determine graphically the activation energy for a reaction.

In this experiment we will study the kinetics of the reaction between iodine and acetone:

$$CH_3\!-\!\overset{\displaystyle O}{\overset{\|}{C}}\!-\!CH_3(aq) + I_2(aq) \rightarrow CH_3\!-\!\overset{\displaystyle O}{\overset{\|}{C}}\!-\!CH_2I(aq) + H^+(aq) + I^-(aq)$$

The rate of this reaction is found to depend on the concentration of hydrogen ion in the solution as well as presumably on the concentrations of the two reactants. By Equation 1, the rate law for this reaction is

$$\text{rate} = k(\text{acetone})^m(\text{I}_2)^n(\text{H}^+)^p \tag{3}$$

where m, n, and p are the orders of the reaction with respect to acetone, iodine, and hydrogen ion, respectively, and k is the rate constant for the reaction.

The rate of this reaction can be expressed as the (small) change in the concentration of I_2, $\Delta(\text{I}_2)$, that occurs, divided by the time interval Δt required for the change:

$$\text{rate} = \frac{-\Delta(\text{I}_2)}{\Delta t} \tag{4}$$

The minus sign is to make the rate positive [as $\Delta(\text{I}_2)$ is negative]. Ordinarily, since rate varies as the concentrations of the reactants according to Equation 3, in a rate study it would be necessary to measure, directly or indirectly, the concentration of each reactant as a function of time; the rate would typically vary markedly with time, decreasing to very low values as the concentration of at least one reactant becomes very low. This makes reaction rate studies relatively difficult to carry out and introduces mathematical complexities that are difficult for beginning students to understand.

The iodination of acetone is a rather atypical reaction, in that it can be easily investigated experimentally. First of all, iodine has color, so that one can readily follow changes in iodine concentration visually. A second and very important characteristic of this reaction is that it turns out to be zero order in I_2 concentration. This means (see Equation 3) that the rate of the reaction does not depend on (I_2) at all; $(\text{I}_2)^0 = 1$, no matter what the value of (I_2) is, as long as it is not itself zero.

Because the rate of the reaction does not depend on (I_2), we can study the rate by simply making I_2 the limiting reagent present in a large excess of acetone and H^+ ion. We then measure the time required for a known initial concentration of I_2 to be used up completely. If both acetone and H^+ are present at much higher concentrations than that of I_2, their concentrations will not change appreciably during the course of the reaction, and the rate will remain, by Equation 3, effectively constant until all the iodine is gone, at which time the reaction will stop. Under such circumstances, if it takes t seconds for the color of a solution having an initial concentration of I_2 equal to $(\text{I}_2)_0$ to disappear, the rate of the reaction, by Equation 4, would be

$$\text{rate} = \frac{-\Delta(\text{I}_2)}{\Delta t} = \frac{(\text{I}_2)_0}{t} \tag{5}$$

Although the rate of the reaction is constant during its course under the conditions we have set up, we can vary it by changing the initial concentrations of acetone and H^+ ion. If, for example, we should *double* the initial concentration of *acetone* over that in Mixture 1, keeping (H^+) and (I_2) at the *same* values they had previously, then the rate of Mixture 2 would, according to Equation 3, be different from that in Mixture 1:

$$\text{rate } 2 = k(2A)^m(\text{I}_2)^0(\text{H}^+)^p \tag{6a}$$

$$\text{rate } 1 = k(A)^m(\text{I}_2)^0(\text{H}^+)^p \tag{6b}$$

Dividing the first equation by the second, we see that the k's cancel, as do the terms in the iodine and hydrogen ion concentrations, since they have the same values in both reactions, and we obtain simply

$$\frac{\text{rate } 2}{\text{rate } 1} = \frac{(2A)^m}{(A)^m} = \left(\frac{2A}{A}\right)^m = 2^m \tag{6}$$

Having measured both rate 2 and rate 1 by Equation 5, we can find their ratio, which must be equal to 2^m. We can then solve for m either by inspection or using logarithms and so find the *order* of the reaction with respect to acetone.

By a similar procedure we can measure the order of the reaction with respect to H^+ ion concentration and also confirm the fact that the reaction is zero order with respect to I_2. Having found the order with respect to each reactant, we can then evaluate k, the rate constant for the reaction.

The determination of the orders m and p, the confirmation of the fact that n, the order with respect to I_2, equals zero, and the evaluation of the rate constant k for the reaction at room temperature comprise your assignment in this experiment. You will be furnished with standard solutions of acetone, iodine, and hydrogen ion, and with the composition of one solution that will give a reasonable rate. The rest of the planning and the execution of the experiment will be your responsibility.

An optional part of the experiment is to study the rate of this reaction at different temperatures to find its activation energy. The general procedure here would be to study the rate of reaction in one of the mixtures at room temperature and at two other temperatures, one above and one below room temperature. Knowing the rates, and hence the k's, at the three temperatures, you can then find E_a, the energy of activation for the reaction, by plotting $\ln k$ vs. $1/T$. The slope of the resultant straight line, by Equation 2, must be $-E_a/R$.

Experimental Procedure

WEAR YOUR SAFETY GLASSES WHILE
PERFORMING THIS EXPERIMENT

Select two regular (18×150 mm) test tubes; when filled with distilled water, they should appear to have identical color when you view them down the tubes against a white background.

Draw 50 mL of each of the following solutions into clean, dry, 100-mL beakers, one solution to a beaker: 4 M acetone, 1 M HCl, and 0.005 M I_2. Cover each beaker with a watch glass.

With your graduated cylinder, measure out 10.0 mL of the 4 M acetone solution and pour it into a clean 125-mL Erlenmeyer flask. Then measure out 10.0 mL 1 M HCl and add that to the acetone in the flask. Add 20.0 mL distilled H_2O to the flask. Drain the graduated cylinder, shaking out any excess water, and then use the cylinder to measure out 10.0 mL 0.005 M I_2 solution. Be careful not to spill the iodine solution on your hands or clothes.

Noting the time on your wristwatch or the wall clock to 1 second, pour the iodine solution into the Erlenmeyer flask and quickly swirl the flask to mix the reagents thoroughly. The reaction mixture will appear yellow because of the presence of the iodine, and the color will fade slowly as the iodine reacts with the acetone. Fill one of the test tubes 3/4 full with the reaction mixture, and fill the other test tube to the same depth with distilled water. Look down the test tubes toward a well-lit piece of white paper, and note the time the color of the iodine just disappears. Measure the temperature of the mixture in the test tube.

Repeat the experiment, using as a reference the reacted solution instead of distilled water. The amount of time required in the two runs should agree within about 20 seconds.

The rate of the reaction equals the initial concentration of I_2 *in the reaction mixture* divided by the elapsed time. Since the reaction is zero order in I_2, and since both acetone and H^+ ion are present in great excess, the rate is constant throughout the reaction and the concentrations of both acetone and H^+ remain essentially at their initial values in the reaction mixture.

Having found the reaction rate for one composition of the system, think for a moment about what changes in composition you might make to decrease the time and hence increase the rate of reaction. In particular, how could you change the composition in such a way as to allow you to determine how the rate depends upon acetone concentration? If it is not clear how to proceed, reread the discussion preceding Equation 6. In your new mixture you should keep the total volume at 50 mL, and be sure that the concentrations of H^+ and I_2 are the *same* as in the first experiment. Carry out the reaction twice with your new mixture; the times should not differ by more than about 15 seconds. The temperature should be kept within about a degree of that in the initial run. Calculate the rate of the reaction. Compare it with that for the first mixture, and then calculate the order of the reaction with respect to acetone, using a relation similar to Equation 6. First, write an equation like 6a for the second reaction mixture, substituting in the values for the rate as obtained by Equation 5 and the initial concentration of acetone, I_2, and H^+ in the reaction mixture. Then write an equation like 6b for the first reaction mixture, using the observed rate and the initial concentrations in that mixture. Obtain an equation like 6 by dividing Equation 6a by Equation 6b. Solve Equation 6 for the order m of the reaction with respect to acetone.

Again change the composition of the initial reaction mixture so that this time a measurement of the reaction rate will give you information about the order of the reaction with respect to H^+. Repeat the experiment with this mixture to establish the time of reaction to within 15 seconds, again making sure that the temperature is within about a degree of that observed previously. From the rate you determine for this mixture find p, the order of the reaction with respect to H^+.

Finally, change the reaction mixture composition in such a way as to allow you to show that the order of the reaction with respect to I_2 is zero. Measure the rate of the reaction twice, and calculate n, the order with respect to I_2.

Having found the order of the reaction for each species on which the rate depends, evaluate k, the rate constant for the reaction, from the rate and concentration data in each of the mixtures you studied. If the temperatures at which the reactions were run are all equal to within a degree or two, k should be about the same for each mixture. Calculate the mean value of k and the standard deviation for the set of values. (See Appendix VIII.)

Optional   As a final reaction, make up a mixture using reactant volumes that you did not use in any previous experiments. Using Equation 3, the values of concentrations in the mixtures, the orders, and the rate constant you calculated from your experimental data, predict how long it will take for the I_2 color to disappear from your mixture. Measure the time for the reaction and compare it with your prediction. If time permits, select one of the reaction mixtures you have already used that gave a convenient time, and use that mixture to measure the rate of reaction at about 10°C and at about 40°C. From the two rates you find, plus the rate at room temperature, calculate the energy of activation for the reaction, using Equation 2.

DISPOSAL OF REACTION PRODUCTS.   Dispose of your solutions from this experiment as directed by your laboratory supervisor.

Experiment 20

Data and Calculations: The Iodination of Acetone

A. Reaction Rate Data

Mixture	Volume in mL 4.0 M Acetone	Volume in mL 1.0 M HCl	Volume in mL 0.0050 M I_2	Volume in mL H_2O	Time for Reaction in sec 1st Run	2nd Run	Temp. in °C
I	10	10	10	20	_____	_____	_____
II	_____	_____	_____	_____	_____	_____	_____
III	_____	_____	_____	_____	_____	_____	_____
IV	_____	_____	_____	_____	_____	_____	_____

B. Determination of Reaction Orders with Respect to Acetone, H⁺ Ion, and I_2

$$\text{rate} = k(\text{acetone})^m(I_2)^n(H^+)^p \tag{3}$$

Calculate the *initial* concentrations of acetone, H⁺ ion, and I_2 in each of the mixtures you studied. Use Equation 5 to find the rate of each reaction.

Mixture	(Acetone)	(H⁺)	$(I_2)_0$	Rate $= \dfrac{(I_2)_0}{\text{avg. time}}$
I	0.80 M	0.20 M	0.0010 M	_____
II	_____	_____	_____	_____
III	_____	_____	_____	_____
IV	_____	_____	_____	_____

Substituting the initial concentrations and the rate from the table above, write Equation 3 as it would apply to Reaction Mixture II:

Rate II =

Now write Equation 3 for Reaction Mixture I, substituting concentrations and the calculated rate from the table:

Rate I =

(continued on following page)

Divide the equation for Mixture II by the equation for Mixture I; the resulting equation should have the ratio of Rate II to Rate I on the left side, and a ratio of acetone concentrations raised to the power m on the right. It should be similar in appearance to Equation 6. Put the resulting equation below:

$$\frac{\text{Rate II}}{\text{Rate I}} =$$

The only unknown in the equation is m. Solve for m. $m =$ _____

Now write Equation 3 as it would apply to Reaction Mixture III and as it would apply to Reaction Mixture IV:

Rate III =

Rate IV =

Using the ratios of the rates of Mixtures III and IV to those of Mixtures II or I, find the orders of the reaction with respect to H^+ ion and I_2:

$$\frac{\text{Rate III}}{\text{Rate _____}} =$$

         $p =$ _____

$$\frac{\text{Rate IV}}{\text{Rate _____}} =$$

         $n =$ _____

C. Determination of the Rate Constant k

Given the values of m, p, and n as determined in Part B, calculate the rate constant k for each mixture by simply substituting those orders, the initial concentrations, and the observed rate from the table into Equation 3.

Mixture	I	II	III	IV	Average
k	_____	_____	_____	_____	_____

Standard deviation in k _____

D. Prediction of Reaction Rate   Optional

Reaction mixture

Volume in mL 4.0 M acetone	Volume in mL 1.0 M HCl	Volume in mL 0.0050 M I_2	Volume in mL H_2O
_____	_____	_____	_____

Initial concentrations
(acetone) _____ M (H^+) _____ M $(I_2)_0$ _____ M

Predicted rate _____ (Eq. 3)

(continued on following page)

Predicted time for reaction _____ sec (Eq. 5)

Observed time for reaction _____ sec

E. Determination of Energy of Activation   Optional

Reaction mixture used _____ (same for all temperatures)

Time for reaction at about 10°C _____ sec   Temperature _____ °C; _____ K

Time for reaction at about 40°C _____ sec   Temperature _____ °C; _____ K

Time for reaction at room temp _____ sec   Temperature _____ °C; _____ K

Calculate the rate constant at each temperature from your data, following the procedure in Part C.

	Rate	k	**ln** k	$\dfrac{1}{T(K)}$
~10°C	_____	_____	_____	_____
~40°C	_____	_____	_____	_____
Room temp	_____	_____	_____	_____

Plot ln k vs. 1/T using Excel or the graph paper on the following page. Find the slope of the best straight line through the points. (If you need to, see Appendix V.)

Slope = _____

By Equation 2: $E_a = -8.31 \times$ slope

$E_a =$ _____ joules

Experiment 20

The Iodination of Acetone

(Data and Calculations)

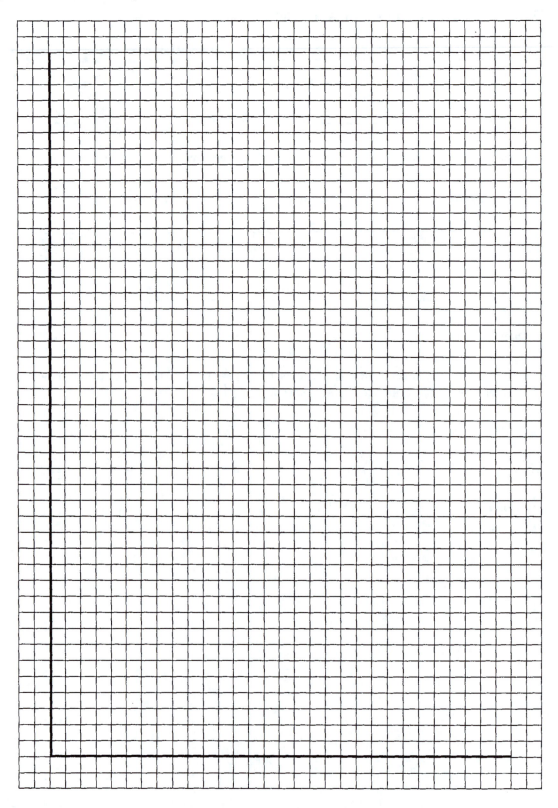

Experiment 20

Advance Study Assignment: The Iodination of Acetone

1. In a reaction involving the iodination of acetone, the following volumes were used to make up the reaction mixture:

    10 mL 4.0 M acetone + 10 mL 1.0 M HCl + 10 mL 0.0050 M I_2 + 20 mL H_2O

    a. How many moles of acetone were in the reaction mixture? Recall that, for a component A, no. moles $A = M_A \times V$, where M_A is the molarity of A and V is the volume in liters of the solution of A that was used.

    _____ moles acetone

    b. What was the molarity of acetone in the *reaction mixture?* The volume of the *mixture* was 50 mL, 0.050 L, and the number of moles of acetone was found in Part (a). Again,

    $$M_A = \frac{\text{no. moles } A}{V \text{ of soln. in liters}}$$

    _____ M acetone

    c. How could you double the molarity of the acetone in the reaction mixture, keeping the total volume at 50 mL and keeping the same concentrations of H^+ ion and I_2 as in the original mixture?

2. Using the reaction mixture in Problem 1, a student found that it took 230 seconds for the color of the I_2 to disappear.

    a. What was the rate of the reaction? Hint: First find the initial concentration of I_2 in the reaction mixture, $(I_2)_0$. Then use Equation 5.

    rate = _____

    b. Given the rate from Part (a), and the initial concentrations of acetone, H^+ ion, and I_2 in the reaction mixture, write Equation 3 as it would apply to the mixture.

    rate =

    c. What are the unknowns that remain in the equation in Part (b)?

    _____ _____ _____ _____

(continued on following page)

3. A second reaction mixture was made up in the following way:

$$10 \text{ mL } 4.0 \text{ M acetone} + 20 \text{ mL } 1.0 \text{ M HCl} + 10 \text{ mL } 0.0050 \text{ M I}_2 + 10 \text{ mL } H_2O$$

a. What were the initial concentrations of acetone, H^+ ion, and I_2 in the reaction mixture?

(acetone) _____ M; (H^+) _____ M; $(I_2)_0$ _____ M

b. It took 120 seconds for the I_2 color to disappear from the reaction mixture when it occurred at the same temperature as the reaction in Problem 2. What was the rate of the reaction?

_____

Write Equation 3 as it would apply to the second reaction mixture:

rate =

c. Divide the equation in Part (b) by the equation in Problem 2b. The resulting equation should have the ratio of the two rates on the left side and a ratio of acetone concentrations raised to the m power on the right. Write the resulting equation and solve for the value of m, the order of the reaction with respect to acetone. (Round off the value of m to the nearest integer.)

$m = $ _____

4. A third reaction mixture was made up in the following way:

$$10 \text{ mL } 4.0 \text{ M acetone} + 20 \text{ mL } 1.0 \text{ M HCl} + 5 \text{ mL } 0.0050 \text{ M I}_2 + 15 \text{ mL } H_2O$$

If the reaction is zero order in I_2, how long would it take for the I_2 color to disappear at the temperature of the reaction mixture in Problem 3?

_____ seconds

Experiment 21

Rates of Chemical Reactions, II. A Clock Reaction

In the previous experiment we discussed the factors that influence the rate of a chemical reaction and presented the terminology used in quantitative relations in studies of the kinetics of chemical reactions. That material is also pertinent to this experiment and should be studied before you proceed further.

This experiment involves the study of the rate properties, or chemical kinetics, of the following reaction between iodide ion and bromate ion under acidic conditions:

$$6\ I^-(aq) + BrO_3^-(aq) + 6\ H^+(aq) \rightarrow 3\ I_2(aq) + Br^-(aq) + 3\ H_2O \tag{1}$$

This reaction proceeds reasonably slowly at room temperature, its rate depending on the concentrations of the I^-, BrO_3^-, and H^+ ions according to the rate law discussed in the previous experiment. For this reaction the rate law takes the form

$$\text{rate} = k(I^-)^m(BrO_3^-)^n(H^+)^p \tag{2}$$

One of the main purposes of this experiment will be to evaluate the rate constant k and the reaction orders m, n, and p for this reaction. We will also investigate the manner in which the reaction rate depends on temperature, and will evaluate the activation energy of the reaction, E_a.

Our method for measuring the rate of the reaction involves what is frequently called a clock reaction. In addition to Reaction 1, whose kinetics we will study, the following reaction will also be made to occur simultaneously in the reaction flask:

$$I_2(aq) + 2\ S_2O_3^{2-}(aq) \rightarrow 2\ I^-(aq) + S_4O_6^{2-}(aq) \tag{3}$$

As compared with Equation 1 this reaction is essentially instantaneous. The I_2 produced in (1) reacts completely with the thiosulfate, $S_2O_3^{2-}$, ion present in the solution, so that until all the thiosulfate ion has reacted, the concentration of I_2 is effectively zero. As soon as the $S_2O_3^{2-}$ is gone from the system, the I_2 produced by (1) remains in the solution and its concentration begins to increase. The presence of I_2 is made strikingly apparent by a starch indicator that is added to the reaction mixture, since I_2 even in small concentrations reacts with starch solution to produce a deep blue color.

By carrying out Reaction 1 in the presence of $S_2O_3^{2-}$ and a starch indicator, we introduce a "clock" into the system. Our clock tells us when a given amount of BrO_3^- ion has reacted (1/6 mole BrO_3^- per mole $S_2O_3^{2-}$), which is just what we need to know, since the rate of reaction can be expressed in terms of the time it takes for a particular amount of BrO_3^- to be used up. In all our reactions, the amount of BrO_3^- that reacts in the time we measure will be constant and small as compared to the amounts of any of the other reactants. This means that the concentrations of all reactants will be essentially constant in Equation 2, and hence so will the rate during each reaction.

In our experiment we will carry out the reaction between BrO_3^-, I^-, and H^+ ions under different concentration conditions. Measured amounts of each of these ions in aqueous solution will be mixed in the presence of a constant, small amount of $S_2O_3^{2-}$. The time it takes for each mixture to turn blue will be measured. The time obtained for each reaction will be inversely proportional to its rate. By changing the concentration of one reactant and keeping the other concentrations constant, we can investigate how the rate of the reaction varies with the concentration of a particular reactant. Once we know the order for each reactant we can determine the rate constant for the reaction.

In the last part of the experiment we will investigate how the rate of the reaction depends on temperature. You will recall that in general the rate increases sharply with temperature. By measuring how the rate varies with temperature we can determine the activation energy, E_a, for the reaction by making use of the Arrhenius equation:

$$\ln k = -\frac{E_a}{RT} + \text{constant} \tag{4}$$

In this equation, k is the rate constant at the Kelvin temperature T, E_a is the activation energy, and R is the gas constant. By plotting $\ln k$ against $1/T$ we should obtain, by Equation 4, a straight line whose slope equals $-E_a/R$. From the slope of that line we can easily calculate the activation energy.

WEAR YOUR SAFETY GLASSES WHILE
PERFORMING THIS EXPERIMENT

Experimental Procedure

A. Dependence of Reaction Rate on Concentration

In Table 21.1 we have summarized the reagent volumes to be used in carrying out the several reactions whose rates we need to know to find the general rate law for Reaction 1. First, measure out 100 mL of each of the listed reagents (except H_2O) into clean, labeled flasks or beakers. Use these reagents in your reaction mixtures.

Table 21.1

	Reaction Mixtures at Room Temperature (Reagent volumes in mL)				
	Reaction Flask I (250 mL)			**Reaction Flask II (125 mL)**	
Reaction Mixture	**0.010 M KI**	**0.0010 M Na$_2$S$_2$O$_3$**	**H$_2$O**	**0.040 M KBrO$_3$**	**0.10 M HCl**
1	10	10	10	10	10
2	20	10	0	10	10
3	10	10	0	20	10
4	10	10	0	10	20
5	8	10	12	5	15

The actual procedure for each reaction mixture will be much the same, and we will describe it now for Reaction Mixture 1.

Since there are several reagents to mix, and since we don't want the reaction to start until we are ready, we will put some of the reagents into one flask and the rest into another, selecting them so that no reaction occurs until the contents of the two flasks are mixed. Using a 10-mL graduated cylinder to measure volumes, measure out 10 mL 0.010 M KI, 10 mL 0.0010 M Na$_2$S$_2$O$_3$, and 10 mL distilled water into a 250-mL Erlenmeyer flask (Reaction Flask I). Then measure out 10 mL 0.040 M KBrO$_3$ and 10 mL 0.10 M HCl into a 125-mL Erlenmeyer flask (Reaction Flask II). To Flask II add three or four drops of starch indicator solution.

Pour the contents of Reaction Flask II into Reaction Flask I and swirl the solutions to mix them thoroughly. Note the time at which the solutions were mixed. Continue swirling the solution. It should turn blue in less than 2 minutes. Note the time at the instant that the blue color appears. Record the temperature of the blue solution to 0.2°C.

Repeat the procedure with the other mixtures in Table 21.1. **Don't forget to add the indicator** before mixing the solutions in the two flasks. The reaction flasks should be rinsed with distilled water between runs. When measuring out reagents, rinse the graduated cylinder with distilled water after you have added the reagents to Reaction Flask I, and **before** you measure out the reagents for Reaction Flask II. Try to keep the temperature just about the same in all the runs. Repeat any experiments that did not appear to proceed properly.

B. Dependence of Reaction Rate on Temperature

In this part of the experiment, the reaction will be carried out at several different temperatures, using Reaction Mixture 1 in all cases. The temperatures we will use will be about 20°C, 40°C, 10°C, and 0°C.

We will take the time at about 20°C to be that for Reaction Mixture 1 as determined at room temperature. To determine the time at 40°C proceed as follows. Make up Reaction Mixture 1 as you did in Part A, including the indicator. However, instead of mixing the solutions in the two flasks at room temperature, put the flasks into water at 40°C, drawn from the hot-water tap into one or more large beakers. Check to see that the water is indeed at about 40°C, and leave the flasks in the water for several minutes to bring them to the proper temperature. Then mix the two solutions, noting the time of mixing. Continue swirling the reaction flask in the warm water. When the color change occurs, note the time and the temperature of the solution in the flask.

Repeat the experiment at about 10°C, cooling all the reactants in water at that temperature before starting the reaction. Record the time required for the color to change and the final temperature of the reaction mixture. Repeat once again at about 0°C, this time using an ice-water bath to cool the reactants.

C. Dependence of the Reaction Rate on the Presence of Catalyst   Optional

Some ions have a pronounced catalytic effect on the rates of many reactions in water solution. Observe the effect on this reaction by once again making up Reaction Mixture 1. Before mixing, add one drop 0.5 M $(NH_4)_2MoO_4$, ammonium molybdate, and a few drops of starch indicator to Reaction Flask II. Swirl the flask to mix the catalyst thoroughly. Then mix the solutions, noting the time required for the color to change. ▢

DISPOSAL OF REACTION PRODUCTS.   The reaction products in this experiment are very dilute and may be poured into the sink as you complete each part of the experiment if so directed by your instructor.

Experiment 21

Data and Calculations: Rates of Chemical Reactions, II. A Clock Reaction

A. Dependence of Reaction Rate on Concentration

$$\text{Reaction: } 6\, I^-(aq) + BrO_3^-(aq) + 6\, H^+(aq) \rightarrow 3\, I_2(aq) + Br^-(aq) + 3\, H_2O \tag{1}$$

$$\text{rate} = k(I^-)^m(BrO_3^-)^n(H^+)^p = -\frac{\Delta(BrO_3^-)}{t} \tag{2}$$

In all the reaction mixtures used in this experiment, the color change occurred when a constant predetermined number of moles of BrO_3^- had been used up by the reaction. The color "clock" allows you to measure the *time required* for this *fixed number of moles of BrO_3^- to react*. The rate of each reaction is determined by the time t required for the color to change; since in Equation 2 the change in concentration of BrO_3^- ion, $\Delta(BrO_3^-)$, is the same in each mixture, the relative rate of each reaction is inversely proportional to the time t. Since we are mainly concerned with relative rather than absolute rate, we will for convenience take all relative rates as being equal to $1000/t$. Fill in the following table, first calculating the relative reaction rate for each mixture.

Reaction Mixture	Time t (sec) for Color to Change	Relative Rate of Reaction $1000/t$	Reactant Concentrations in Reacting Mixture (M)			Temp. in (°C)
			(I^-)	(BrO_3^-)	(H^+)	
1	_____	_____	0.0020	_____	_____	_____
2	_____	_____	_____	_____	_____	_____
3	_____	_____	_____	_____	_____	_____
4	_____	_____	_____	_____	_____	_____
5	_____	_____	_____	_____	_____	_____

The reactant concentrations in the reaction mixture are *not* those of the stock solutions, since the reagents were diluted by the other solutions. The final volume of the reaction mixture is 50 mL in all cases. Since the number of moles of reactant does not change on dilution we can say, for example, for I^- ion, that

$$\text{no. moles } I^- = (I^-)_{stock} \times V_{stock} = (I^-)_{mixture} \times V_{mixture}$$

For Reaction Mixture 1,

$$(I^-)_{stock} = 0.010 \text{ M}, \quad V_{stock} = 10 \text{ mL}, \quad V_{mixture} = 50 \text{ mL}$$

Therefore,

$$(I^-)_{mixture} = \frac{0.010 \text{ M} \times 10 \text{ mL}}{50 \text{ mL}} = 0.0020 \text{ M}$$

Calculate the rest of the concentrations in the table using the same approach.

Determination of the Orders of the Reaction

Given the data in the table, the problem is to find the order for each reactant and the rate constant for the reaction. Since we are dealing with relative rates, we can modify Equation 2 to read as follows:

$$\text{relative rate} = k'(I^-)^m(BrO_3^-)^n(H^+)^p \tag{5}$$

(continued on the following page)

We need to determine the relative rate constant k' and the orders m, n, and p in such a way as to be consistent with the data in the table.

The solution to this problem is quite simple, once you make a few observations on the reaction mixtures. Each mixture (2 to 4) differs from Reaction Mixture 1 in the concentration of only one species (see table). This means that for any pair of mixtures that includes Reaction Mixture 1, there is only one concentration that changes. From the ratio of the relative rates for such a pair of mixtures we can find the order for the reactant whose concentration was changed. Proceed as follows.

Write Equation 5 below for Reaction Mixtures 1 and 2, substituting the relative rates and the concentrations of I^-, BrO_3^-, and H^+ ions from the table you have just completed.

Relative Rate 1 = _____ = $k'($ ___ $)^m($ ___ $)^n($ ___ $)^p$

Relative Rate 2 = _____ = $k'($ ___ $)^m($ ___ $)^n($ ___ $)^p$

Divide the first equation by the second, noting that nearly all the terms cancel out. The result is simply

$$\frac{\text{Relative Rate 1}}{\text{Relative Rate 2}} =$$

If you have done this properly, you will have an equation involving only m as an unknown. Solve this equation for m, the order of the reaction with respect to I^- ion.

$$m = \text{_____} \text{(nearest integer)}$$

Applying the same approach to Reaction Mixtures 1 and 3, find the value of n, the order of the reaction with respect to BrO_3^- ion.

Relative Rate 1 = _____ = $k'($ ___ $)^m($ ___ $)^n($ ___ $)^p$

Relative Rate 3 = _____ = $k'($ ___ $)^m($ ___ $)^n($ ___ $)^p$

Dividing one equation by the other:

$$=$$

$$n = \text{_____}$$

Now that you have the idea, apply the method once again, this time to Reaction Mixtures 1 and 4, and find p, the order with respect to H^+ ion.

Relative Rate 4 = $k'($ ___ $)^m($ ___ $)^n($ ___ $)^p$

Dividing the equation for Relative Rate 1 by that for Relative Rate 4, we get

$$=$$

$$p = \text{_____}$$

Having found m, n, and p (nearest integers), the relative rate constant, k', can be calculated by substitution of m, n, p, and the known rates and reactant concentrations into Equation 5. Evaluate k' for Reaction Mixtures 1 to 4.

Reaction	1	2	3	4	
k'	_____	_____	_____	_____	k'_{ave} _____

Standard deviation in k'_____ (See Appendix VIII)

Why should k' have nearly the same value for each of the above reactions?

Using k'_{ave} in Equation 5, predict the relative rate and time, t_{pred}, for Reaction Mixture 5. Use the concentrations in the table.

Relative rate$_{pred}$ _____ t_{pred} _____ t_{obs} _____

(continued on the following page)

B. Effect of Temperature on Reaction Rate: The Activation Energy

To find the activation energy for the reaction it will be helpful to complete the following table.
    The dependence of the rate constant, k', for a reaction is given by Equation 4:

$$\ln k' = -\frac{E_a}{RT} + \text{constant} \tag{4}$$

Since the reactions at the different temperatures all involve the same reactant concentrations, the rate constants, k', for two different mixtures will have the same ratio as the reaction rates themselves for the two mixtures. This means that in the calculation of E_a, we can use the observed relative rates instead of rate constants. Proceeding as before, calculate the relative rates of reaction in each of the mixtures and enter these values in (c). Take the ln rate for each mixture and enter these values in (d). To set up the terms in $1/T$, fill in (b), (e), and (f) in the table.

	Approximate Temperature in °C			
	20	40	10	0
(a) Time t in seconds for color to appear	____	____	____	____
(b) Temperature of the reaction mixture in °C	____	____	____	____
(c) Relative rate = 1000/t	____	____	____	____
(d) ln of relative rate	____	____	____	____
(e) Temperature T in K	____	____	____	____
(f) $1/T$, K^{-1}	____	____	____	____

To evaluate E_a, make a graph of ln relative rate vs. $1/T$, using Excel or the graph paper provided. (See Appendix V.)
    Find the slope of the line obtained by drawing the best straight line through the experimental points.

Slope = _____

The slope of the line equals $-E_a/R$, where $R = (8.31$ joules/mole $K)$ if E_a is to be in joules per mole. Find the activation energy, E_a, for the reaction.

E_a = _____ joules/mole

C. Effect of a Catalyst on Reaction Rate   Optional

	Reaction 1	Catalyzed Reaction 1
Time for color to appear (seconds)	_____	_____

Would you expect the activation energy, E_a, for the catalyzed reaction to be greater than, less than, or equal to the activation energy for the uncatalyzed reaction? Why?

Experiment 21

Rates of Chemical Reactions, II. A Clock Reaction

(Data and Calculations)

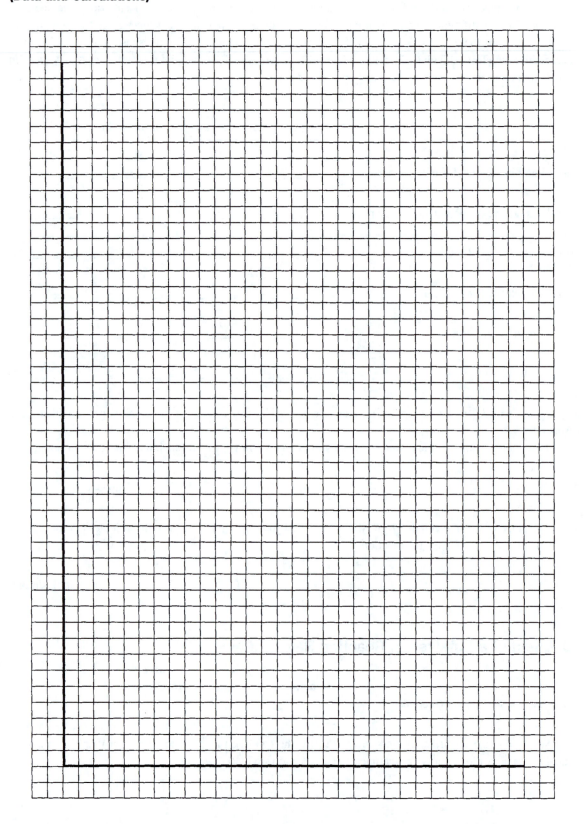

Experiment 21

Advance Study Assignment: Rates of Chemical Reactions,
II. A Clock Reaction

1. A student studied the clock reaction described in this experiment. She set up Reaction Mixture 4 by mixing 10 mL 0.010 M KI, 10 mL 0.001 M $Na_2S_2O_3$, 10 mL 0.040 M $KBrO_3$, and 20 mL 0.10 M HCl using the procedure given. It took about 21 seconds for the color to turn blue.

    a. She found the concentrations of each reactant in the reacting mixture by realizing that the number of moles of each reactant did not change when that reactant was mixed with the others, but that its concentration did. For any reactant A,

    $$\text{no. moles } A = M_{A \text{ stock}} \times V_{\text{stock}} = M_{A \text{ mixture}} \times V_{\text{mixture}}$$

    The volume of the mixture was 50 mL. Revising the above equation, she obtained

    $$M_{A \text{ mixture}} = M_{A \text{ stock}} \times \frac{V_{\text{stock}}(\text{mL})}{50 \text{ mL}}$$

    Find the concentrations of each reactant using the equation above.

    $(I^-) = $ _____ M; $(BrO_3^-) = $ _____ M; $(H^+) = $ _____ M

    b. What was the relative rate of the reaction $(1000/t)$? _____

    c. Knowing the relative rate of reaction for Mixture 4 and the concentrations of I^-, BrO_3^-, and H^+ in that mixture, she was able to set up Equation 5 for the relative rate of the reaction. The only quantities that remained unknown were k', m, n, and p. Set up Equation 5 as she did, presuming she did it properly. Equation 5 is on page 163.

2. For Reaction Mixture 1 the student found that 85 seconds were required. On dividing Equation 5 for Reaction Mixture 1 by Equation 5 for Reaction Mixture 4, and after canceling out the common terms [k' and terms in (I^-) and (BrO_3^-)], she got the following equation:

$$\frac{11.8}{48} = \left(\frac{0.020}{0.040}\right)^p = \left(\frac{1}{2}\right)^p$$

Recognizing that 11.8/48 is about equal to ¼, she obtained an approximate value for m. What was that value?

$m = $ _____

By taking logarithms of both sides of the equation, she got an exact value for m. What was that value?

$m = $ _____

Since orders of reactions are often integers, she rounded her value of m to the nearest integer, and reported that value as the order of the reaction with respect to H^+.

Experiment 22

Properties of Systems in Chemical Equilibrium—Le Châtelier's Principle

When working in the laboratory, one often makes observations that at first sight are surprising and hard to explain. One might add a reagent to a solution and obtain a precipitate. Addition of more of that reagent to the precipitate causes it to dissolve. A violet solution turns yellow on addition of a reagent. Subsequent addition of another reagent brings back first a green solution and then the original violet one. Clearly, chemical reactions are occurring, but how and why they behave as they do is not at once obvious.

In this experiment we will examine and attempt to explain several observations of the sort we have just mentioned. Central to our explanation will be recognition of the fact that chemical systems tend to exist in a state of equilibrium. If one disturbs the equilibrium in one way or another, the reaction may shift to the left or right, producing the kinds of effects we have mentioned. If one can understand the principles governing the equilibrium system, it is often possible to see how one might disturb the system, such as by adding a particular reagent or heat, and so cause it to change in a desirable way.

Before proceeding to specific examples, let us examine the situation in a general way, noting the key principle that allows us to make a system in equilibrium behave as we wish. Consider the reaction

$$A(aq) \rightleftharpoons B(aq) + C(aq) \tag{1}$$

where A, B, and C are molecules or ions in solution. If we have a mixture of these species in equilibrium, it turns out that their concentrations are not completely unrelated. Rather, there is a condition that those concentrations must meet, namely that

$$\frac{[B] \times [C]}{[A]} = K_c \tag{2}$$

where K_c is a constant, called the equilibrium constant for the reaction. For a given reaction at any given temperature, K_c has a particular value.

When we say that K_c has a particular value, we mean just that. For example, we might find that, for a given solution in which Reaction 1 can occur, when we substitute the equilibrium values for the molarities of A, B, and C into Equation 2, we get a value of 10 for K_c. Now, suppose we add more of species A to that solution. What will happen? Remember, K_c can't change. If we substitute the new higher molarity of A into Equation 2 we get a value that is smaller than K_c. This means that the system is not in equilibrium, and *must* change in some way to get back to equilibrium. How can it do this? It can do this by shifting to the right, producing more B and C and using up some A. It *must* do this, and *will*, until the molarities of C, B, and A reach values that, on substitution into Equation 2, equal 10. At that point the system is once again in equilibrium. In the new equilibrium state, [B] and [C] are greater than they were initially, and [A] is larger than its initial value but smaller than if there had been no forced shift to the right.

The conclusion you should reach on reading the last paragraph is that *one can always cause a reaction to shift to the right by increasing the concentration of a reactant*. An *increase* in concentration of a *product* will force a *shift* to the *left*. By a similar argument we find that a *decrease* in *reactant* concentration causes a *shift* to the *left*; a *decrease* in *product* concentration produces a *shift to* the *right*. This is all true because K_c does not change (unless you change the temperature). The changes in concentration that one can produce by adding particular reagents may be simply enormous, so the shifts in the equilibrium system may also be enormous. Much of the mystery of chemical behavior disappears once you understand this idea.

Another way one might disturb an equilibrium system is by changing its temperature. When this happens, the value of K_c changes. It turns out that the change in K_c depends upon the enthalpy change, ΔH, for the reaction. If ΔH is positive, greater than zero (an "endothermic reaction"), K_c increases with increasing T. If ΔH is negative (an "exothermic" reaction), K_c decreases with an increase in T. Let us return to our original equilibrium between A, B, and C, where K_c equals 10. Let us assume that ΔH for Reaction 1 is -40 kJ. If we raise the temperature, K_c will go down ($\Delta H < 0$), say to a value of 1. This means that the system will no longer be in equilibrium. Substitution of the initial values of [A], [B], and [C] into Equation 2 produces a value that is too big, 10 instead of 1. How can the system change itself to regain equilibrium? It must of necessity shift to the left, lowering [B] and [C] and raising [A]. This will make the expression in Equation 2 smaller. The shift will continue until the concentrations of A, B, and C, on substitution into Equation 2, give the expression a value of 1.

From the discussion in the previous paragraph, you should be able to conclude that an equilibrium system will shift to the left on being heated if the reaction is exothermic ($\Delta H < 0$, K_c decreases with temperature). It will shift to the right if the reaction is endothermic ($\Delta H > 0$, K_c increases with temperature). Again, since we can change temperatures very markedly, we can shift equilibria a long, long way. An endothermic reaction that at 25°C has an equilibrium state that consists mainly of reactants might, at 1000°C, exist almost completely as products.

The effects of concentration and temperature on systems in chemical equilibrium are often summarized by Le Châtelier's principle. The principle states that:

If you attempt to change a system in chemical equilibrium, it will react in such a way as to counteract the change you attempted.

If you think about the principle for a while, you will see that it predicts the same kind of behavior as we did by using the properties of K_c. Increasing the concentration of a reactant will, by the principle, cause a change that decreases that concentration; that change must be a shift to the right. Increasing the temperature of a reaction mixture will cause a change that tends to absorb heat; that change must be a shift in the endothermic direction. The principle is an interesting one, but does require more careful reasoning in some cases than the more direct approach we employed. For the most part we will find it more useful to base our arguments on the properties of K_c.

In working with aqueous systems, the most important equilibrium is often that which involves the dissociation of water into H^+ and OH^- ions:

$$H_2O \rightleftharpoons H^+(aq) + OH^-(aq) \quad K_c = [H^+][OH^-] = 1 \times 10^{-14} \tag{3}$$

In this reaction the concentration of water is very high and is essentially constant at about 55 M; it is incorporated into K_c. The value of K_c is very small, which means that in **any** water system the product of [H^+] and [OH^-] must be very small. In pure water, [H^+] equals [OH^-] equals 1×10^{-7} M.

Although the product, [H^+] $\times$ [OH^-] is small, that does not mean that both concentrations are necessarily small. If, for example, we dissolve HCl in water, the HCl in the solution will dissociate completely to H^+ and Cl^- ions; in 1 M HCl, [H^+] will become 1 M, and there is nothing that Reaction 3 can do about changing that concentration appreciably. Rather, Reaction 3 must occur in such a direction as to maintain equilibrium. It does this by lowering [OH^-] by reaction to the left; this uses up a little bit of H^+ ion and drives [OH^-] to the value it must have when [H^+] is 1 M, namely, 1×10^{-14} M. In 1 M HCl, [OH^-] is a factor of ten million *smaller* than it is in water. This makes the properties of 1 M HCl quite different from those of water, particularly where H^+ and OH^- ions are involved.

If we take 1 M HCl and add a solution of NaOH to it, an interesting situation develops. Like HCl, NaOH is completely dissociated in solution, so in 1 M NaOH, [OH^-] is equal to 1 M. If we add 1 M NaOH to 1 M HCl, we will initially raise [OH^-] ions way above 1×10^{-14} M. However, Reaction 3 cannot be in equilibrium when both [H^+] and [OH^-] are high; reaction must occur to re-establish equilibrium. The added OH^- ions react with H^+ ions to form H_2O, decreasing both concentrations until equilibrium is established. If only a small amount of OH^- ion is added, it will essentially all be used up; [H^+] will remain high, and [OH^-] will still be very small, but somewhat larger than 10^{-14} M. If we add OH^- ion until the amount added equals in moles the amount of H^+ originally present, then Reaction 3 will go to the left until [H^+] equals [OH^-] equals 1×10^{-7} M, and both concentrations will be very small. Further addition of OH^- ion will raise [OH^-] to much higher values, easily as high as 1 M. In such a solution, [H⁺] would be very low, 1×10^{-14} M. So, in aqueous solution, depending on the solutes present, we can have [H⁺] and [OH⁻] range from about 1 M to 10^{-14} M, or 14 orders of magnitude. This will have a tremendous effect on **any** other equilibrium system in which [H⁺] or [OH⁻] ions are reactants. Similar situations arise in other equilibrium systems in which the concentration of a reactant or product can be changed significantly by adding a particular reagent.

In many equilibrium systems, several equilibria are present simultaneously. For example, in aqueous solution, Reaction 3 must **always** be in equilibrium. There may, in addition, be equilibria between the solutes in the aqueous solution. Some examples are those in Reactions 4, 5, 7, 8, 9, and 10 in the Experimental Procedure section. In some of those reactions, H⁺ and OH⁻ ions appear; in others, they do not. In Reaction 4, for example, H⁺ ion is a product. The molarity of H⁺ in Reaction 4 is *not* determined by the indicator HMV, since it is only present in a tiny amount. Reaction 4 will have an equilibrium state that is fixed by the state of Reaction 3, which as we have seen depends markedly on the presence of solutes such as HCl or NaOH. Reactions 8 and 9 can similarly be controlled by Reaction 3. Reactions 5 and 7, which do not involve H⁺ or OH⁻ ions, are not dependent on Reaction 3 for their equilibrium state. Reaction 10 is sensitive to NH_3 concentration and can be driven far to the right by addition of a reagent such as 6 M NH_3.

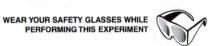

WEAR YOUR SAFETY GLASSES WHILE
PERFORMING THIS EXPERIMENT

Experimental Procedure

In this experiment we will work with several equilibrium systems, each of which is similar to the A-B-C system we discussed. We will alter these systems in various ways, forcing shifts to the right and left by changing concentrations and temperature. You will be asked to interpret your observations in terms of the principles we have presented.

A. Acid-Base Indicators

There is a large group of chemical substances, called acid-base indicators, which change color in solution when [H⁺] changes. A typical substance of this sort is called methyl violet, which we will give the formula HMV. In solution HMV dissociates as follows:

$$HMV(aq) \rightleftharpoons H^+(aq) + MV^-(aq) \tag{4}$$
$$\text{yellow} \qquad\qquad\qquad \text{violet}$$

HMV has an intense yellow color, while the anion MV⁻ is violet. The color of the indicator in solution depends very strongly on [H⁺].

Step 1 Add about 5 mL of distilled water to a regular (18×150 mm) test tube. Add a few drops of methyl violet indicator. Report the color of the solution on the Data page.

Step 2 How could you force the equilibrium system to go to the other form (color)? Select a reagent that should do this and add it to the solution, drop by drop, until the color change is complete. If your reagent works, write its formula on the Data page. If it doesn't, try another until you find one that does. Work with 6 M reagents if they are available.

Step 3 Equilibrium systems are reversible. That is, the reaction can be driven to the left and right many times by changing the conditions in the system. How can you force the system in Step 2 to revert to its original color? Select a reagent that should do this and add it drop by drop until the color has become the original one. Again, if your first choice was incorrect, try another reagent. On the Data page write the formula of the reagent that was effective. Answer all the questions for Part A before going on to Part B.

B. Solubility Equilibrium; Finding a Value for K_{sp}

Many ionic substances have limited water solubility. A typical example is $PbCl_2$, which dissolves to some extent in water according to the reaction

$$PbCl_2(s) \rightleftharpoons Pb^{2+}(aq) + 2\ Cl^-(aq) \tag{5}$$

The equilibrium constant for this reaction takes the form

$$K_c = [Pb^{2+}] \times [Cl^-]^2 = K_{sp} \tag{6}$$

The $PbCl_2$ does not enter into the expression because it is a solid, and so has a constant effect on the system, independent of its amount. The equilibrium constant for a solubility equilibrium is called the solubility product, and is given the symbol K_{sp}.

For the equilibrium in Reaction 5 to exist, there *must* be some solid $PbCl_2$ present in the system. If there is no solid, there is no equilibrium; Equation 6 is not obeyed, and $[Pb^{2+}] \times [Cl^-]^2$ must be *less* than the value of K_{sp}. If the solid is present, even in a tiny amount, then the values of $[Pb^{2+}]$ and $[Cl^-]$ are subject to Equation 6.

Step 1   Set up a hot-water bath, using a 400-mL beaker half full of water. Start heating the water with a burner while proceeding with Step 2.

Step 2   To a regular test tube add 5.0 mL 0.30 M $Pb(NO_3)_2$. In this solution $[Pb^{2+}]$ equals 0.30 M. Add 5.0 mL 0.30 M HCl to a 10-mL graduated cylinder. In this solution $[Cl^-]$ equals 0.30 M. Add 1 mL of the HCl solution to the $Pb(NO_3)_2$ solution. Stir, and wait about 15 seconds. What happens? Record your result.

Step 3   To the $Pb(NO_3)_2$ solution add the HCl in 1-mL increments until a noticeable amount of white solid $PbCl_2$ is present after stirring. Record the total volume of HCl added at that point.

Step 4   Put the test tube with the precipitate of $PbCl_2$ into the hot-water bath. Stir for a few moments. What happens? Record your observations. Cool the test tube under the cold-water tap. What happens?

Step 5   Rinse out your graduated cylinder and then add about 5.0 mL of distilled water to the cylinder. Add water in 1-mL increments to the mixture in the test tube, stirring well after each addition. Record the volume of water added when the precipitate just dissolves. Answer the questions and do the calculations in Part B before proceeding.

C. Complex Ion Equilibria

Many metallic ions in solution exist not as simple ions but, rather, as complex ions in combination with other ions or molecules, called ligands. For example, the Co^{2+} ion in solution exists as the pink $Co(H_2O)_6^{2+}$ complex ion, and Cu^{2+} as the blue $Cu(H_2O)_4^{2+}$ complex ion. In both of these ions the ligands are H_2O molecules. Complex ions are reasonably stable but may be converted to other complex ions on addition of ligands that form more stable complexes than the original ones. Among the common ligands that may form complex species, OH^-, NH_3, and Cl^- are important.

An interesting Co(II) complex is the $CoCl_4^{2-}$ ion, which is blue. This ion is stable in concentrated Cl^- solutions. Depending upon conditions, Co(II) in solution may exist as either $Co(H_2O)_6^{2+}$ or as $CoCl_4^{2-}$. The principles of chemical equilibrium can be used to predict which ion will be present:

$$Co(H_2O)_6^{2+}(aq) + 4\ Cl^-(aq) \rightleftharpoons CoCl_4^{2-}(aq) + 6\ H_2O \tag{7}$$

Step 1   Put a few small crystals (~0.1 g) of $CoCl_2 \cdot 6\ H_2O$ in a regular test tube. Add 2 mL of 12 M HCl. **CAUTION:** *This is a concentrated solution of an acidic gas; avoid contact with it and its fumes. This is a concentrated strong acid with a choking odor.* Stir to dissolve the crystals. Record the color of the solution.

Step 2   Add 2-mL portions of distilled water, stirring after each dilution, until no further color change occurs. Record the new color.

Step 3   Place the test tube into the hot-water bath and note any change in color. Cool the tube under the water tap and report your observations. Complete the questions in Part C before continuing.

D. Dissolving Insoluble Solids

We saw in Part B that we can dissolve more $PbCl_2$ by either heating its saturated solution or by simply adding water. In most cases these procedures won't work very well on other solids because they are typically much less soluble than $PbCl_2$.

There are, however, some very powerful methods for dissolving solids that depend on their effectiveness upon the principles of equilibrium. As an example of an insoluble substance, we might consider $Zn(OH)_2$:

$$Zn(OH)_2(s) \rightleftharpoons Zn^{2+}(aq) + 2\ OH^-(aq) \qquad K_{sp} = 5 \times 10^{-17} = [Zn^{2+}] \times [OH^-]^{-2} \qquad (8)$$

The equilibrium constant for Reaction 8 is very small, which tells us that the reaction does not go very far to the right or, equivalently, that $Zn(OH)_2$ is almost completely insoluble in water. Adding a few drops of a solution containing OH^- ion to one containing Zn^{2+} ion will cause precipitation of $Zn(OH)_2$.

At first sight you might well wonder how one could possibly dissolve, say, 1 mole of $Zn(OH)_2$ in an aqueous solution. If, however, you examine Equation 8, you can see, from the equation for K_{sp}, that in the saturated solution $[Zn^{2+}] \times [OH^-]^2$ must equal 5×10^{-17}. If, by some means, we can lower that product to a value *below* 5×10^{-17}, then $Zn(OH)_2$ will dissolve, until the product becomes equal to K_{sp}, where equilibrium will again exist. To lower the product, we need to lower the concentration of either Zn^{2+} or OH^- very drastically. This turns out to be very easy to do. To lower $[OH^-]$ we can add H^+ ions from an acid. If we do that, we drive Reaction 3 to the left, making $[OH^-]$ very, very small—small enough to dissolve substantial amounts of $Zn(OH)_2$.

To lower $[Zn^{2+}]$ we can take advantage of the fact that zinc(II) forms stable complex ions with both OH^- and NH_3:

$$Zn^{2+}(aq) + 4\ OH^-(aq) \rightleftharpoons Zn(OH)_4^{2-}\ (aq) \quad K_1 = 3 \times 10^{15} \qquad (9)$$

$$Zn^{2+}(aq) + 4\ NH_3(aq) \rightleftharpoons Zn(NH_3)_4^{2+}\ (aq) \quad K_2 = 1 \times 10^9 \qquad (10)$$

In high concentrations of OH^- ion, Reaction 9 is driven strongly to the right, making $[Zn^{2+}]$ very low. The same thing would happen in solutions containing high concentrations of NH_3. In both media we would therefore expect that $Zn(OH)_2$ might dissolve, since if $[Zn^{2+}]$ is very low, Reaction 8 must go to the right.

Step 1   To each of three small test tubes add about 2 mL 0.1 M $Zn(NO_3)_2$. In this solution $[Zn^{2+}]$ equals 0.1 M. To each test tube add one drop 6 M NaOH and stir. Report your observations.

Step 2   To the first tube add 6 M HCl drop by drop, with stirring. To the second add 6 M NaOH, again drop by drop. To the third add 6 M NH_3. Note what happens in each case.

Step 3   Repeat Steps 1 and 2, this time using a solution of 0.1 M $Mg(NO_3)_2$. Record your observations. Answer the questions in Part D.

DISPOSAL OF REACTION PRODUCTS.   The residues from Parts B and C should be poured in the waste crock. Those from Parts A and D may be poured down the sink.

Experiment 22

Data and Observations: Properties of Systems in Chemical Equilibrium

A. Acid-Base Indicators

1. Color of methyl violet in water _____

2. Reagent causing color change _____

3. Reagent causing shift back _____. Explain, by considering how changes in $[H^+]$ will cause Reaction 4 to shift to right and left, why the reagents in Steps 2 and 3 caused the solution to change color. Note that Reactions 3 and 4 must both go to equilibrium after a reagent is added.

B. Solubility Equilibrium; Finding a Value for K_{sp}

2. Vol. 0.30 M $Pb(NO_3)_2$ = 5.0 mL No. moles Pb^{2+} = $[M \times V(\text{stock})]$ = 1.5×10^{-3} moles

    Observations:

3. Vol. 0.30 M HCl used _____ mL No. moles Cl^- added _____ moles

4. Observations: in hot water _____ in cold water _____

5. Volume H_2O added to dissolve $PbCl_2$ _____ mL

    Total volume of solution _____ mL

    a. Explain why $PbCl_2$ did not precipitate immediately on addition of HCl. (What condition must be met by $[Pb^{2+}]$ and $[Cl^-]$ if $PbCl_2$ is to form?)

    b. Explain your observations in Step 4. (In which direction did Reaction 5 shift when heated? What must have happened to the value of K_{sp} in the hot solution? What does this tell you about the sign of ΔH in Reaction 5?)

    c. Explain why the $PbCl_2$ dissolved when water was added in Step 5. (What was the effect of the added water on $[Pb^{2+}]$ and $[Cl^-]$? In what direction would such a change drive Reaction 5?)

(continued on following page)

d. Given the numbers of moles of Pb^{2+} and Cl^- in the final solution in Step 5, and the volume of that solution, calculate $[Pb^{2+}]$ and $[Cl^-]$ in that solution.

$[Pb^{2+}]$ _____ M; $[Cl^-]$ _____ M

Noting that the molarities just calculated are essentially those in equilibrium with solid $PbCl_2$, calculate $[Pb^{2+}] \times [Cl^-]^2$. This is equal to K_{sp} for $PbCl_2$.

K_{sp} = _____

C. Complex Ion Equilibria

1. Color of $CoCl_2 \cdot 6\,H_2O$ _____

    Color in solution in 12 M HCl   _____

2. Color in diluted solution   _____

3. Color of hot solution   _____

    Color of cooled solution   _____

    Formula of Co(II) complex ion present in solution in

    a.   12 M HCl   _____

    b.   diluted solution   _____

    c.   hot solution   _____

    d.   cooled solution   _____

    Explain the color change that occurred when

    a.   water was added in Step 2. (Consider how a change in $[Cl^-]$ and $[H_2O]$ will shift Reaction 7.)

    b.   the diluted solution was heated. (How would Reaction 7 shift if K_c went up? How did increasing the temperature affect the value of K_c? What is the sign of ΔH in Reaction 7?)

(continued on following page)

D. Dissolving Insoluble Solids

1. Observations on addition of a drop of 6 M NaOH to $Zn(NO_3)_2$ solution:

2. Effect on solubility of $Zn(OH)_2$:

   a. of added HCl solution

   b. of added NaOH solution

   c. of added NH_3 solution

3. Observations on addition of one drop of 6 M NaOH to $Mg(NO_3)_2$ solution:

   Effect on solubility of $Mg(OH)_2$:

   a. of added HCl solution

   b. of added NaOH solution

   c. of added NH_3 solution

Explain your observations in Step 1. (Consider Reaction 8; how is it affected by addition of OH^- ion?)

In Step 2(a), how does an increase in $[H^+]$ affect Reaction 3? (What does that do to Reaction 8?) Explain your observations in Step 2(a).

In Step 2(b), how does an increase in $[OH^-]$ affect Reaction 9? What does that do to Reaction 8? Explain your observations in Step 2(b).

In Step 2(c), how does an increase in $[NH_3]$ affect Reaction 10? What does that do to Reaction 8? Explain your observations in Step 2(c).

In Step 3, you probably found that $Mg(OH)_2$ was similar in some ways in its behavior to that of $Zn(OH)_2$, but different in others.

   a. How was it similar? Explain that similarity. (In particular, why would any insoluble hydroxide tend to dissolve in acidic solution?)

   b. How was it different? Explain that difference. (In particular, does Mg^{2+} appear to form complex ions with OH^- and NH_3? What would we observe if it did? If it did not?)

Experiment 22

Advance Study Assignment: Properties of Systems in Chemical Equilibrium

1. Methyl orange, HMO, is a common acid-base indicator. In solution it ionizes according to the equation:

$$HMO(aq) \rightleftharpoons H^+(aq) + MO^-(aq)$$
$$\text{red} \qquad\qquad\qquad \text{yellow}$$

If methyl orange is added to distilled water, the solution turns yellow. If a drop or two of 6 M HCl is added to the yellow solution, it turns red. If to that solution one adds a few drops of 6 M NaOH the color reverts to yellow.

a. Why does adding 6 M HCl to the yellow solution of methyl orange tend to cause the color to change to red? (Note that in solution HCl exists as H⁺ and Cl⁻ ions.)

b. Why does adding 6 M NaOH to the red solution tend to make it turn back to yellow? (Note that in solution NaOH exists as Na⁺ and OH⁻ ions. How does increasing [OH⁻] shift Reaction 3 in the discussion section? How would the resulting change in [H⁺] affect the dissociation reaction of HMO?)

2. Magnesium hydroxide is only very slightly soluble in water. The reaction by which it goes into solution is:

$$Mg(OH)_2(s) \rightleftharpoons Mg^{2+}(aq) + 2\ OH^-(aq)$$

a. Formulate the expression for the equilibrium constant, K_{sp}, for the above reaction.

b. It is possible to dissolve significant amounts of $Mg(OH)_2$ in solutions in which the concentration of either Mg^{2+} or OH^- is very, very small. Explain, using K_{sp}, why this is the case.

c. Explain why $Mg(OH)_2$ might have very appreciable solubility in 1 M HCl. (Consider the effect of Reaction 3 on the $Mg(OH)_2$ solution reaction.)

Experiment 23

Determination of the Equilibrium Constant for a Chemical Reaction

When chemical substances react, the reaction typically does not go to completion. Rather, the system goes to some intermediate state in which both the reactants and products have concentrations that do not change with time. Such a system is said to be in chemical equilibrium. When in equilibrium at a particular temperature, a reaction mixture obeys the Law of Chemical Equilibrium, which imposes a condition on the concentrations of reactants and products. This condition is expressed in the equilibrium constant K_c for the reaction.

In this experiment we will study the equilibrium properties of the reaction between iron(III) ion and thiocyanate ion:

$$Fe^{3+}(aq) + SCN^-(aq) \rightleftharpoons FeSCN^{2+}(aq) \tag{1}$$

When solutions containing Fe^{3+} ion and thiocyanate ion are mixed, Reaction 1 occurs to some extent, forming the $FeSCN^{2+}$ complex ion, which has a deep red color. As a result of the reaction, the equilibrium amounts of Fe^{3+} and SCN^- will be less than they would have been if no reaction occurred; for every mole of $FeSCN^{2+}$ that is formed, one mole of Fe^{3+} and one mole of SCN^- will react. According to the Law of Chemical Equilibrium, the equilibrium constant expression K_c for Reaction 1 is formulated as follows:

$$\frac{[FeSCN^{2+}]}{[Fe^{3+}][SCN^-]} = K_c \tag{2}$$

The value of K_c in Equation 2 is constant at a given temperature. This means that mixtures containing Fe^{3+} and SCN^- will react until Equation 2 is satisfied, so that the same value of the K_c will be obtained no matter what initial amounts of Fe^{3+} and SCN^- were used. Our purpose in this experiment will be to find K_c for this reaction for several mixtures made up in different ways, and to show that K_c indeed has the same value in each of the mixtures. The reaction is a particularly good one to study because K_c is of a convenient magnitude and the color of the $FeSCN^{2+}$ ion makes for an easy analysis of the equilibrium mixture.

The mixtures will be prepared by mixing solutions containing known concentrations of iron(III) nitrate, $Fe(NO_3)_3$, and potassium thiocyanate, KSCN. The color of the $FeSCN^{2+}$ ion formed will allow us to determine its equilibrium concentration. Knowing the initial composition of a mixture and the equilibrium concentration of $FeSCN^{2+}$, we can calculate the equilibrium concentrations of the rest of the pertinent species and then determine K_c.

Since the calculations required in this experiment may not be apparent, we will go through a step-by-step procedure by which they can be made. As a specific example, let us assume that we prepare a mixture by mixing 10.0 mL of 2.00×10^{-3} M $Fe(NO_3)_3$ with 10.0 mL of 2.00×10^{-3} M KSCN. As a result of Reaction 1, some red $FeSCN^{2+}$ ion is formed. By the method of analysis described later, its concentration at equilibrium is found to be 1.50×10^{-4} M. Our problem is to find K_c for the reaction from this information. To do this we first need to find the initial number of moles of each reactant in the mixture. Second, we determine the number of moles of product that were formed at equilibrium. Since the product was formed at the expense of reactants, we can calculate the amount of each reactant that was used up. In the third step we find the number of moles of each reactant remaining in the equilibrium mixture. Fourth, we determine the concentration of each reactant. Finally, in the fifth step we evaluate K_c for the reaction.

Step 1 **Finding the Initial Number of Moles of Each Reactant.** This requires relating the volumes and concentrations of the reagent solutions that were mixed to the numbers of moles of each reactant species in those solutions. By the definition of the molarity, M_A, of a species A,

$$M_A = \frac{\text{no. moles } A}{\text{no. liters of solution, } V} \quad \text{or} \quad \text{no. moles } A = M_A \times V \tag{3}$$

Using Equation 3, it is easy to find the initial number of moles of Fe^{3+} and SCN^-. For each solution the volume used was 10.0 mL, or 0.0100 L. The molarity of each of the solutions was 2.00×10^{-3} M, so $M_{Fe^{3+}} = 2.00 \times 10^{-3}$ M and $M_{SCN^-} = 2.00 \times 10^{-3}$ M. Therefore, in the reagent solutions, we find that

initial no. moles $Fe^{3+} = M_{Fe^{3+}} \times V = 2.00 \times 10^{-3}$ M $\times$ 0.0100 L $= 20.0 \times 10^{-6}$ moles

initial no. moles $SCN^- = M_{SCN^-} \times V = 2.00 \times 10^{-3}$ M $\times$ 0.0100 L $= 20.0 \times 10^{-6}$ moles

Step 2 **Finding the Number of Moles of Product Formed.** Here again we can use Equation 3 to advantage. The concentration of $FeSCN^{2+}$ was found to be 1.50×10^{-4} M at equilibrium. The volume of the mixture at equilibrium is the *sum* of the two volumes that were mixed, and is 20.0 mL, or 0.0200 L. So,

no. moles $FeSCN^{2+} = M_{FeSCN^{2+}} \times V = 1.50 \times 10^{-4}$ M $\times$ 0.0200 L $= 3.00 \times 10^{-6}$ moles

The number of moles of Fe^{3+} and SCN^- that were *used up* in producing the $FeSCN^{2+}$ must also both be equal to 3.00×10^{-6} moles since, by Equation 1, it takes *one mole* Fe^{3+} *and one mole* SCN^- to make each mole of $FeSCN^{2+}$.

Step 3 **Finding the Number of Moles of Each Reactant Present at Equilibrium.** In Step 1 we determined that initially we had 20.0×10^{-6} moles Fe^{3+} and 20.0×10^{-6} moles SCN^- present. In Step 2 we found that in the reaction 3.00×10^{-6} moles Fe^{3+} and 3.00×10^{-6} moles SCN^- were used up. The number of moles present at equilibrium must equal the number we started with minus the number that reacted. Therefore, *at equilibrium*,

no. moles at equilibrium = initial no. moles − no. moles used up

equil. no. moles $Fe^{3+} = 20.0 \times 10^{-6} - 3.00 \times 10^{-6} = 17.0 \times 10^{-6}$ moles $\tag{4}$

equil. no. moles $SCN^- = 20.0 \times 10^{-6} - 3.00 \times 10^{-6} = 17.0 \times 10^{-6}$ moles

Step 4 **Find the Concentrations of All Species at Equilibrium.** Experimentally, we obtained the concentration of $FeSCN^{2+}$ directly. $[FeSCN^{2+}] = 1.50 \times 10^{-4}$ M. The concentrations of Fe^{3+} and SCN^- follow from Equation 3. The number of moles of each of these species at equilibrium was obtained in Step 3. The volume of the mixture being studied was 20.0 mL, or 0.0200 L. So, *at equilibrium*,

$$[Fe^{3+}] = M_{Fe^{3+}} = \frac{\text{no. moles } Fe^{3+}}{\text{volume of solution}} = \frac{17.0 \times 10^{-6} \text{ moles}}{0.0200 \text{ L}} = 8.50 \times 10^{-4} \text{ M}$$

$$[SCN^-] = M_{SCN^-} = \frac{\text{no. moles } SCN^-}{\text{volume of solution}} = \frac{17.0 \times 10^{-6} \text{ moles}}{0.0200 \text{ L}} = 8.50 \times 10^{-4} \text{ M}$$

Step 5 **Finding the Value of K_c for the Reaction.** Once the equilibrium concentrations of all the reactants and products are known, one needs merely to substitute into Equation 2 to determine K_c:

$$K_c = \frac{[FeSCN^{2+}]}{[Fe^{3+}][SCN^-]} = \frac{1.50 \times 10^{-4}}{(8.50 \times 10^{-4}) \times (8.50 \times 10^{-4})} = 208$$

Step 5   Continued. In this experiment you will obtain data similar to that shown in this example. The calculations involved in processing that data are completely analogous to those we have made. (Actually, your results will differ from the ones we obtained, since the data in our example were obtained at a different temperature and so relate to a different value of K_c.)

In carrying out this analysis we made the assumption that the reaction which occurred was given by Equation 1. There is no inherent reason why the reaction might not have been

$$Fe^{3+}(aq) + 2\ SCN^-(aq) \rightleftharpoons Fe(SCN)_2{}^+(aq) \tag{5}$$

If you are interested in matters of this sort, you might ask how we know whether we are actually observing Reaction 1 or Reaction 5. The line of reasoning is the following. If Reaction 1 is occurring, K_c for that reaction as we calculate it should remain constant with different reagent mixtures. If, however, Reaction 5 is going on, K_c as calculated for that reaction should remain constant. In the optional part of the Data and Calculations section, we will assume that Reaction 5 occurs and make the analysis of K_c on that basis. The results of the two sets of calculations should make it clear that Reaction 1 is the one that we are studying.

Two analytical methods can be used to determine $[FeSCN^{2+}]$ in the equilibrium mixtures. The more precise method uses a spectrophotometer, which measures the amount of light absorbed by the red complex at 447 nm, the wavelength at which the complex most strongly absorbs. The absorbance, A, of the complex is proportional to its concentration, M, and can be measured directly on the spectrophotometer:

$$A = kM \tag{6}$$

Your instructor will show you how to operate the spectrophotometer, if one is available in your laboratory, and will provide you with a calibration curve or equation from which you can find $[FeSCN^{2+}]$ once you have determined the absorbance of your solutions. See Appendix IV for information about spectrophotometers.

In the other analytical method a solution of known concentration of $FeSCN^{2+}$ is prepared. The $[FeSCN^{2+}]$ concentrations in the solutions being studied are found by comparing the color intensities of these solutions with that of the known. The method involves matching the color intensity of a given depth of unknown solution with that for an adjusted depth of known solution. The actual procedure and method of calculation are discussed in the Experimental Procedure section.

In preparing the mixtures in this experiment we will maintain the concentration of H^+ ion at 0.5 M. The hydrogen ion does not participate directly in the reaction, but its presence is necessary to avoid the formation of brown-colored species such as $FeOH^{2+}$, which would interfere with the analysis of $[FeSCN^{2+}]$.

WEAR YOUR SAFETY GLASSES WHILE
PERFORMING THIS EXPERIMENT

Experimental Procedure

Label five regular test tubes (18 mm × 150 mm) 1 to 5, with labels or by noting their positions on your test tube rack. Pour about 30 mL 2.00×10^{-3} M $Fe(NO_3)_3$ in 1 M HNO_3 into a dry 100-mL beaker. Pipet 5.00 mL of that solution into each test tube. Then add about 20 mL 2.00×10^{-3} M KSCN to another dry 100-mL beaker. Pipet 1, 2, 3, 4, and 5 mL from the KSCN beaker into each of the corresponding test tubes labeled 1 to 5. Then pipet the proper number of milliliters of water into each test tube to bring the total volume in each tube to 10.00 mL. The volumes of reagents to be added to each tube are summarized in Table 23.1, which you should complete by filling in the required volumes of water. See Appendix IV for a discussion of the use of pipets.

Mix each solution thoroughly with a glass stirring rod. Be sure to dry the stirring rod after mixing each solution.

Method I. Analysis by Spectrophotometric Measurement

Place a portion of the mixture in tube 1 in a spectrophotometer cell, as demonstrated by your instructor, and measure the absorbance of the solution at 447 nm. Determine the concentration of $FeSCN^{2+}$ from the calibration curve provided for each instrument or from the equation furnished to you. Record the value on the Data

Table 23.1

	Test Tube No.				
	1	**2**	**3**	**4**	**5**
Volume Fe(NO$_3$)$_3$ solution (mL)	5.00	5.00	5.00	5.00	5.00
Volume KSCN solution (mL)	1.00	2.00	3.00	4.00	5.00
Volume H$_2$O (mL)	—	—	—	—	—

page. Repeat the measurement using the mixtures in each of the other test tubes. For a discussion of how absorbance and concentration are related, see Appendix IV.

Method II. Analysis by Comparison with a Standard

Prepare a solution of known [FeSCN^{2+}] by pipetting 10.00 mL 0.200 M Fe(NO$_3$)$_3$ in 1 M HNO$_3$ into a test tube and adding 2.00 mL 0.00200 M KSCN and 8.00 mL water. Mix the solution thoroughly with a stirring rod.

Since in this solution [Fe^{3+}] >> [SCN$^-$], Reaction 1 is driven strongly to the right. You can assume without serious error that essentially all the SCN$^-$ added is converted to FeSCN^{2+}. Assuming that this is the case, calculate [FeSCN^{2+}] in the standard solution and record the value on the Data page.

The [FeSCN^{2+}] in the unknown mixture in test tubes 1 to 5 can be found by comparing the intensity of the red color in these mixtures with that of the standard solution. This can be done by placing the test tube containing Mixture 1 next to a test tube containing the standard. Look down both test tubes toward a well-illuminated piece of white paper on the laboratory bench.

Pour out the standard solution into a dry, clean beaker until the color intensity you see down the tube containing the standard matches that which you see when looking down the tube containing the unknown. Use a well-lit piece of white paper as your background. When the colors match, the following relation is valid:

$$[FeSCN^{2+}]_{unknown} \times \text{depth of unknown solution} = [FeSCN^{2+}] \times \text{depth of standard solution} \qquad \textbf{(7)}$$

Measure the depths of the matching solutions with a rule and record them. Repeat the measurement for Mixtures 2 through 5, recording the depth of each unknown and that of the standard solution which matches it in intensity.

DISPOSAL OF REACTION PRODUCTS.   In this experiment, reactant concentrations are very low. In most localities you can pour the contents of the test tubes down the sink when you have completed your measurements. However, consult your instructor for alternate disposal procedures.

Experiment 23

Data and Calculations: Determination of the Equilibrium Constant for a Chemical Reaction

Mixture	Volume in mL, 2.00×10^{-3} M $Fe(NO_3)_3$	Volume in mL, 2.00×10^{-3} M KSCN	Volume in mL, Water	Method I Absorbance	Method II Depth in mm		$[FeSCN]^{2+}$
					Standard	**Unknown**	
1	5.00	1.00	_____	_____	_____	_____	_____ $\times 10^{-4}$ M
2	5.00	2.00	_____	_____	_____	_____	_____ $\times 10^{-4}$ M
3	5.00	3.00	_____	_____	_____	_____	_____ $\times 10^{-4}$ M
4	5.00	4.00	_____	_____	_____	_____	_____ $\times 10^{-4}$ M
5	5.00	5.00	_____	_____	_____	_____	_____ $\times 10^{-4}$ M

If Method II was used, $[FeSCN^{2+}]_{standard}$ = _____ $\times 10^{-4}$ M; $[FeSCN^{2+}]$ in Mixtures 1 to 5 is found by Equation 7.

Processing the Data

A. Calculation of K_c assuming the reaction:

(1) $$Fe^{3+}(aq) + SCN^-(aq) \rightleftharpoons FeSCN^{2+}(aq)$$

This calculation is most easily done by following Steps 1 through 5 in the discussion. Results are to be entered in the table on the following page. If you are using Excel, set up the table as we have, and follow the directions that follow.

Step 1 Find the initial number of moles of Fe^{3+} and SCN^- in the mixtures in test tubes 1 through 5. Use Equation 3 and enter the values in the first two columns of the table.

Step 2 Enter the experimentally determined value of $[FeSCN^{2+}]$ at equilibrium for each of the mixtures in the next to last column in the table. Use Equation 3 to find the number of moles of $FeSCN^{2+}$ in each of the mixtures, and enter the values in the fifth column of the table. Note that this is also the number of moles of Fe^{3+} and SCN^- that were used up in the reaction.

(continued on following page)

Step 3 From the number of moles of Fe^{3+} and SCN^- initially present in each mixture, and the number of moles of Fe^{3+} and SCN^- used up in forming $FeSCN^{2+}$, calculate the number of moles of Fe^{3+} and SCN^- that remain in each mixture at equilibrium. Use Equation 4. Enter the results in columns 3 and 4 of the table.

Step 4 Use Equation 3 and the results of Step 3 to find the concentrations of all of the species at equilibrium. The volume of the mixture is 10.00 mL, or 0.0100 liter in all cases. Enter the values in columns 6 and 7 of the table.

Step 5 Calculate K_c for the reaction for each of the mixtures by substituting values for the equilibrium concentrations of Fe^{3+}, SCN^-, and $FeSCN^{2+}$ in Equation 2.

Step 6 Calculate the mean value for K_c and the standard deviation. (See Appendix VIII.)

Mixture	Initial No. Moles		Equilibrium No. Moles			Equilibrium Concentrations			K_c
	Fe^{3+}	SCN^-	Fe^{3+}	SCN^-	$FeSCN^{2+}$	$[Fe^{3+}]$	$[SCN^-]$	$[FeSCN^{2+}]$	
1	$\times 10^{-6}$	$\times 10^{-6}$	$\times 10^{-6}$	$\times 10^{-6}$	$\times 10^{-6}$	$\times 10^{-4}$ M	$\times 10^{-4}$ M	$\times 10^{-4}$ M	
2									
3									
4									
5									

Mean value of K_c _____   Standard deviation _____

(continued on following page)

B. In calculating K_c in Part A, we assume, correctly, that the formula of the complex ion is $FeSCN^{2+}$. It is by no means obvious that this is the case and one might have assumed, for instance, that $Fe(SCN)_2^+$ was the species formed. The reaction would then be

$$Fe^{3+}(aq) + 2\ SCN^-(aq) \rightleftharpoons Fe(SCN)_2^+(aq) \tag{5}$$

If we analyze the equilibrium system we have studied, assuming that Reaction 5 occurs rather than Reaction 1, we would presumably obtain nonconstant values of K_c. Using the same kind of procedure as in Part A, calculate K_c for Mixtures 1, 3, and 5 on the basis that $Fe(SCN)_2^+$ is the formula of the complex ion formed by the reaction between Fe^{3+} and SCN^-. As a result of the procedure used for calibrating the system by Method I or Method II, $[Fe(SCN)_2^+]$ will equal *one-half* the $[FeSCN^{2+}]$ obtained for each solution in Part A. Note that *two* moles SCN_2 are needed to form *one* mole $Fe(SCN)_2^+$. This changes the expression for K_c. Also, in calculating the equilibrium number of moles SCN^- you will need to subtract ($2 \times$ number of moles $Fe(SCN)_2^+$) from the initial number of moles SCN^-.

Mixture	Initial No. Moles		Equilibrium No. Moles			Equilibrium Concentrations			K_c
	Fe^{3+}	SCN^-	Fe^{3+}	SCN^-	$Fe(SCN)_2^+$	$[Fe^{3+}]$	$[SCN^-]$	$[Fe(SCN)_2^+]$	
1									
3									
5									

On the basis of the results of Part A, what can you conclude about the validity of the equilibrium concept, as exemplified by Equation 2? What do you conclude about the formula of iron(III) thiocyanate complex ion?

Experiment 23

Advance Study Assignment: Determination of the Equilibrium Constant
for a Chemical Reaction

1. A student mixes 5.00 mL 2.00×10^{-3} M $Fe(NO_3)_3$ with 3.00 mL 2.00×10^{-3} M KSCN. She finds that
    in the equilibrium mixture the concentration of $FeSCN^{2+}$ is 1.28×10^{-4} M. Find K_c for the reaction
    $Fe^{3+}(aq) + SCN^-(aq) \rightleftharpoons FeSCN^{2+}(aq)$.

    Step 1 Find the number of moles Fe^{3+} and SCN^- initially present. (Use Eq. 3.)

    _____ moles Fe^{3+}; _____ moles SCN^-

    Step 2 How many moles of $FeSCN^{2+}$ are in the mixture at equilibrium? What is the volume of the
    equilibrium mixture? (Use Eq. 3.)

    _____ mL; _____ moles $FeSCN^{2+}$

    How many moles of Fe^{3+} and SCN^- are used up in making the $FeSCN^{2+}$?

    _____ moles Fe^{3+}; _____ moles SCN^-

    Step 3 How many moles of Fe^{3+} and SCN^- remain in the solution at equilibrium? (Use Eq. 4 and
    the results of Steps 1 and 2.)

    _____ moles Fe^{3+}; _____ moles SCN^-

    Step 4 What are the concentrations of Fe^{3+}, SCN^-, and $FeSCN^{2+}$ at equilibrium? What is the volume
    of the equilibrium mixture? (Use Eq. 3 and the results of Step 3.)

    $[Fe^{3+}]$ = _____ M; $[SCN^-]$ = _____ M; $[FeSCN^{2+}]$ = _____ M

    _____ mL

    Step 5 What is the value of K_c for the reaction? (Use Eq. 2 and the results of Step 4.)

    K_c = _____

(continued on following page)

2.   Optional   Assume that the reaction studied in Problem 1 is $Fe^{3+}(aq) + 2\ SCN^-(aq) \rightleftharpoons Fe(SCN)_2^+(aq)$. Find K_c for this reaction, given the data in Problem 1, except that the equilibrium concentration of $Fe(SCN)_2^+$ is equal to 0.49×10^{-4} M.

a.   Formulate the expression for K_c for the alternate reaction just cited.

b.   Find K_c as you did in Problem 1; take due account of the fact that two moles SCN^- are used up per mole $Fe(SCN)_2^+$ formed.

Step 1   Results are as in Problem 1.

Step 2   How many moles of $Fe(SCN)_2^+$ are in the mixture at equilibrium? (You should use Eq. 3.)

_____ moles $Fe(SCN)_2^+$

How many moles of Fe^{3+} and SCN^- are used up in making the $Fe(SCN)_2^+$?

_____ moles Fe^{3+};   _____ moles SCN^-

Step 3   How many moles of Fe^{3+} and SCN^- remain in solution at equilibrium? Use the results of Steps 1 and 2, noting that no. moles SCN^- at equilibrium = original no. moles $SCN^- - (2 \times$ no. moles $Fe(SCN)_2^+)$.

_____ moles Fe^{3+};   _____moles SCN^-

Step 4   What are the concentrations of Fe^{3+}, SCN^-, and $Fe(SCN)_2^+$ at equilibrium? (Use Eq. 3 and the results of Step 3.)

$[Fe^{3+}] =$ _____ M; $[SCN^-] =$ _____ M; $[Fe(SCN)_2^+] =$ _____ M

Step 5   Calculate K_c on the basis that the alternate reaction occurs. (Use the answer to Part 2a.)

$K_c =$ _____

Experiment 24

The Standardization of a Basic Solution and the Determination of the Molar Mass of an Acid

When a solution of a strong acid is mixed with a solution of a strong base, a chemical reaction occurs that can be represented by the following net ionic equation:

$$H^+(aq) + OH^-(aq) \rightarrow H_2O$$

This is called a neutralization reaction, and chemists use it extensively to change the acidic or basic properties of solutions. The equilibrium constant for the reaction is about 10^{14} at room temperature, so that the reaction can be considered to proceed completely to the right, using up whichever of the ions is present in the lesser amount and leaving the solution either acidic or basic, depending on whether H^+ or OH^- ion was in excess.

Since the reaction is essentially quantitative, it can be used to determine the concentrations of acidic or basic solutions. A frequently used procedure involves the titration of an acid with a base. In the titration, a basic solution is added from a buret to a measured volume of acid solution until the number of moles of OH^- ion added is just equal to the number of moles of H^+ ion present in the acid. At that point the volume of basic solution that has been added is measured.

Recalling the definition of the molarity, M_A, of species A, we have

$$M_A = \frac{\text{no. moles of A}}{\text{no. liters of solution, } V} \quad \text{or} \quad \text{no. moles of A} = M_A \times V \tag{1}$$

At the end point of the titration,

$$\text{no. moles } H^+ \text{ originally present} = \text{no. moles } OH^- \text{ added} \tag{2}$$

So, by Equation 1,

$$M_{H^+} \times V_{acid} = M_{OH^-} \times V_{base} \tag{3}$$

Therefore, if the molarity of either the H^+ or the OH^- ion in its solution is known, the molarity of the other ion can be found from the titration.

The equivalence point or end point in the titration is determined by using a chemical, called an indicator, that changes color at the proper point. The indicators used in acid-base titrations are weak organic acids or bases that change color when they are neutralized. One of the most common indicators is phenolphthalein, which is colorless in acid solutions but becomes red when the pH of the solution becomes 9 or higher.

When a solution of a strong acid is titrated with a solution of a strong base, the pH at the end point will be about 7. At the end point a drop of acid or base added to the solution will change its pH by several pH units, so that phenolphthalein can be used as an indicator in such titrations. If a weak acid is titrated with a strong base, the pH at the equivalence point is somewhat higher than 7, perhaps 8 or 9, and phenolphthalein is still a very satisfactory indicator. If, however, a solution of a weak base such as ammonia is titrated with a strong acid, the pH will be a unit or two less than 7 at the end point, and phenolphthalein will not be as good an indicator for that titration as, for example, methyl red, whose color changes from red to yellow as the pH changes from about 4 to 6. Ordinarily, indicators will be chosen so that their color change occurs at about the pH at the equivalence point of a given acid-base titration.

In this experiment you will determine the molarity of OH^- ion in an NaOH solution by titrating that solution against a standardized solution of HCl. Since in these solutions one mole of acid in solution furnishes one mole of H^+ ion and one mole of base produces one mole of OH^- ion, $M_{HCl} = M_{H^+}$ in the acid solution, and $M_{NaOH} = M_{OH^-}$ in the basic solution. Therefore, the titration will allow you to find M_{NaOH} as well as M_{OH^-}.

In the second part of this experiment you will use your standardized NaOH solution to titrate a sample of a pure solid organic or inorganic acid. By titrating a weighed sample of the unknown acid with your standardized NaOH solution you can, by Equation 2, find the number of moles H^+ ion that your sample can furnish.

If your acid has one acidic hydrogen atom in the molecule, with formula HB, then the number of moles of acid will equal the number of moles of H^+ that react during the titration. The molar mass of the acid, *MM*, will equal the number of grams of acid that contain one mole of H^+ ion.

$$MM = \frac{\text{no. grams of acid}}{\text{no. moles of } H^+ \text{ ion furnished}} \tag{4}$$

Many acids release one mole of H^+ ion per mole of acid on titration with NaOH solution. Such acids are called monoprotic. Acetic acid, $HC_2H_3O_2$, is a classic example of a monoprotic acid (only the first H atom in the formula is acidic). Like all organic acids, acetic acid is weak, in that it only ionizes to a small extent in water solution.

Some acids contain more than one acidic hydrogen atom in the molecule, and would have the general formula H_2B or H_3B. Sulfurous acid, H_2SO_3, is an example of an inorganic diprotic acid. Maleic acid, $H_2C_4H_2O_4$, is a diprotic organic acid. If we should titrate samples of these acids with a solution of a strong base like NaOH, it would take two moles of OH^- ion, or two moles of NaOH, to neutralize one mole of acid, since each mole of acid would release two moles of H^+ ion. If you don't know the formula of an acid, you can't be sure it is monoprotic, so you can only calculate the mass of acid that will react with one mole of OH^- ion. That mass will contain one mole of H^+ ion and is called the *equivalent* mass of the acid. The molar mass and the equivalent mass are related by simple equations. Since sulfurous acid gives up two moles of H^+ ion in titration with NaOH, the molar mass must equal *twice* the equivalent mass, which gives up one mole of H^+ ion.

To simplify matters, in this experiment we will only use monoprotic acids, with formula HB, so one mole of acid will react with one mole of NaOH, and you can find the molar mass of your acid by Equation 4.

WEAR YOUR SAFETY GLASSES WHILE PERFORMING THIS EXPERIMENT

Experimental Procedure

Note: This experiment is relatively long unless you know precisely what to do. Study the experiment carefully before coming to class, so that you don't have to spend a lot of time finding out what the experiment is all about.

Obtain two burets and a sample of solid unknown acid from the stockroom.

A. Standardization of NaOH Solution

Into a small graduated cylinder draw about 7 mL of the stock 6 M NaOH solution provided in the laboratory and dilute to about 400 mL with distilled water in a 500-mL Florence flask. Stopper the flask tightly and mix the solution thoroughly at intervals over a period of at least 15 minutes before using the solution.

Draw into a clean, *dry* 125-mL Erlenmeyer flask about 75 mL of standardized HCl solution (about 0.1 M) from the stock solution on the reagent shelf. This amount should provide all the standard acid you will need; do not waste it. Record the molarity of the HCl.

Prepare for the titration by using the procedure described in Experiment 7 and Appendix IV. The purpose of this procedure is to make sure that the solution in each buret has the same molarity as it has in the container from which it was poured. Clean the two burets and rinse them with distilled water. Then rinse one buret three times with a few milliliters of the HCl solution, in each case thoroughly wetting the walls of the buret with the solution and then letting it out through the stopcock. Fill the buret with HCl; open the stopcock momentarily to fill the tip. Proceed to clean and fill the other buret with your NaOH solution in a similar manner. Put the acid buret, A, on the left side of your buret clamp, and the base buret, B, on the right side. Check to see that your burets do not leak and that there are no air bubbles in either buret tip. Read and record the levels in the two burets to 0.02 mL.

Draw about 25 mL of the HCl solution from the buret into a clean 250-mL Erlenmeyer flask; add to the flask about 25 mL of distilled H_2O and two or three drops of phenolphthalein indicator solution. Place a white sheet of paper under the flask to aid in the detection of any color change. Add the NaOH solution intermittently from its buret to the solution in the flask, noting the pink phenolphthalein color that appears and disappears as the drops hit the liquid and are mixed with it. Swirl the liquid in the flask gently and continuously as you add the NaOH solution. When the pink color begins to persist, slow down the rate of addition of NaOH. In the final stages of the titration add the NaOH drop by drop until the entire solution just turns a pale pink color that will persist for about 30 seconds. If you go past the end point and obtain a red solution, add a few drops of the HCl solution to remove the color, and then add NaOH a drop at a time until the pink color persists. Carefully record the *final* readings on the HCl and NaOH burets.

To the 250-mL Erlenmeyer flask containing the titrated solution, add about 10 mL more of the standard HCl solution. Titrate this as before with the NaOH to an end point, and carefully record both buret readings once again. To this solution add about 10 mL more HCl and titrate a third time with NaOH.

You have now completed three titrations, with *total* HCl volumes of about 25, 35, and 45 mL. Using Equation 3, calculate the molarity of your base, M_{OH^-} for each of the three titrations. In each case, use the *total volumes* of acid and base that were added up to that point. At least two of these molarities should agree to within 1%. If they do, proceed to the next part of the experiment. If they do not, repeat these titrations until two calculated molarities do agree. Calculate the mean value of M_{OH^-} and the standard deviation. (See Appendix VIII.)

B. Determination of the Molar Mass of an Acid

Weigh the vial containing your solid acid on the analytical balance to ±0.0001 g. Carefully pour out about half the sample into a clean but not necessarily dry 250-mL Erlenmeyer flask. Weigh the vial accurately. Add about 50 mL of distilled water and two or three drops of phenolphthalein to the flask. The acid may be relatively insoluble, so don't worry if it doesn't all dissolve.

Fill your NaOH buret with the (now standardized) NaOH solution. Add the standard HCl to your HCl buret until it is about half full. Read both levels carefully and record them.

Titrate the solution of the solid acid with NaOH. As the acid is neutralized it will tend to dissolve in the solution. If the acid appears to be relatively insoluble, add NaOH until the pink color persists, and then swirl to dissolve the solid. If the solid still will not dissolve, and the solution remains pink, add 25 mL of ethanol to increase the solubility. If you go past the end point, add HCl as necessary. The final pink end point should appear on addition of one drop of NaOH. Record the final levels in the NaOH and HCl burets.

Pour the rest of your acid sample into a clean 250-mL Erlenmeyer flask, and weigh the vial accurately. Titrate this sample of acid as before with the NaOH and HCl solutions.

If you use HCl in these titrations, and you probably will, the calculations needed are a bit more complicated than in the standardization of the NaOH solution. To find the number of moles of H^+ ion in the solid acid, you must subtract the number of moles of HCl used from the number of moles of NaOH. For a back-titration, which is what we have in this case:

$$\text{no. moles } H^+ \text{ in solid acid} = \text{no. moles } OH^- \text{ in NaOH soln.} - \text{no. moles } H^+ \text{ in HCl soln.} \qquad (5)$$

For volumes in milliliters, this equation takes the form:

$$\text{no. moles } H^+ \text{ in solid acid} = \frac{M_{NaOH} \times V_{NaOH}}{1000} - \frac{M_{HCl} \times V_{HCl}}{1000} \qquad (6)$$

C. Determination of K_a for the Unknown Acid   Optional

Your instructor will tell you in advance if you are to do this part of the experiment. If you do this part, you will need to **save** one of the titrated solutions from Part B. Given that titrated solution, it is easy to prepare one in which [H⁺] ion is equal to K_a for your acid.

A weak acid, HB, dissociates in water solution according to the equation:

$$HB(aq) \rightleftharpoons H^+(aq) + B^-(aq) \tag{7}$$

The equilibrium constant for this reaction is called the acid dissociation constant for HB and is given the symbol K_a. A solution of HB will obey the equilibrium condition given by the equation:

$$K_a = \frac{[H^+][B^-]}{[HB]} \tag{8}$$

Your acid is a weak acid, so will follow both of the above equations.

In the titration in Part B you converted a solution of HB into one containing NaB by adding NaOH to the end point. If, to your titrated solution, you now add some HCl, it will convert some of the B⁻ ions in the NaB solution back to HB. If the number of moles of HCl you add equals *one-half* the number of moles of H⁺ in your original sample, then *half* of the B⁻ ions will be converted to HB. In the resulting solution, [HB] will equal [B⁻], and so, by Equation 8,

$$K_a = [H^+] \quad \text{in the half-neutralized solution} \tag{9}$$

Using the approach we have outlined, use your titrated solution to prepare one in which [HB] equals [B⁻]. Measure the pH of that solution with a pH meter, using the procedure described in Appendix IV. From that value, calculate K_a for your unknown acid. ■

Optional Experiment: Using the procedure in this experiment, find the mass percent acetic acid in household vinegar. What is the molarity of the acid?

DISPOSAL OF REACTION PRODUCTS. The reaction products in this experiment may be diluted and poured down the drain.

Experiment 24

Data and Calculations: The Standardization of a Basic Solution and the Determination of the Molar Mass of an Acid

A. Standardization of NaOH Solution

	Trial 1	Trial 2	Trial 3
Initial reading, HCl buret	_____ mL		
Final reading, HCl buret	_____ mL	_____ mL	_____ mL
Initial reading, NaOH buret	_____ mL		
Final reading, NaOH buret	_____ mL	_____ mL	_____ mL

B. Determination of the Molar Mass of an Unknown Acid

Mass of vial plus contents	_____ g
Mass of vial plus contents less Sample 1	_____ g
Mass less Sample 2	_____ g

	Trial 1	Trial 2
Initial reading, NaOH buret	_____ mL	_____ mL
Final reading, NaOH buret	_____ mL	_____ mL
Initial reading, HCl buret	_____ mL	_____ mL
Final reading, HCl buret	_____ mL	_____ mL

Processing the Data

A. Standardization of NaOH Solution

	Trial 1	Trial 2	Trial 3
Total volume HCl	_____ mL	_____ mL	_____ mL
Total volume NaOH	_____ mL	_____ mL	_____ mL

(continued on following page)

Molarity, M_{HCl}, of standardized HCl ⎯⎯⎯⎯⎯⎯ M

Molarity, M_{H^+}, in standardized HCl ⎯⎯⎯⎯⎯⎯ M

By Equation 3,

$$M_{H^+} \times V_{acid} = M_{OH^-} \times V_{base} \quad \text{or} \quad M_{OH^-} = M_{H^+} \times \frac{V_{HCl}}{V_{NaOH}} \tag{3}$$

Use Equation 3 to find the molarity, M_{OH^-}, of the NaOH solution. Note that the volumes do not need to be converted to liters, since we use the volume ratio.

Trial 1	**Trial 2**	**Trial 3**

M_{OH^-} ⎯⎯⎯⎯ M ⎯⎯⎯⎯ M ⎯⎯⎯⎯ M (should agree within 1%)

The molarity of the NaOH will equal M_{OH^-}, since one mole NaOH → one mole OH^-.

Average molarity of NaOH solution, M_{NaOH} ⎯⎯⎯⎯⎯⎯ M   Standard deviation ⎯⎯⎯⎯⎯⎯ M

B. Determination of the Molar Mass of the Unknown Acid

	Trial 1	**Trial 2**
Mass of sample	⎯⎯⎯⎯ g	⎯⎯⎯⎯ g
Volume NaOH used	⎯⎯⎯⎯ mL	⎯⎯⎯⎯ mL
No. moles NaOH $=\dfrac{V_{NaOH} \times M_{NaOH}}{1000}$	⎯⎯⎯⎯	⎯⎯⎯⎯
Volume HCl used	⎯⎯⎯⎯ mL	⎯⎯⎯⎯ mL
No. moles HCl $=\dfrac{V_{HCl} \times M_{HCl}}{1000}$	⎯⎯⎯⎯	⎯⎯⎯⎯
No. moles H^+ in sample (use Eq. 5)	⎯⎯⎯⎯	⎯⎯⎯⎯
$MM = \dfrac{\text{no. grams acid}}{\text{no. moles } H^+}$	⎯⎯⎯⎯ g	⎯⎯⎯⎯ g
Unknown no.		⎯⎯⎯⎯

C. Determination of K_a for the Unknown Acid   Optional

No. moles H^+ in sample (from Part B) ⎯⎯⎯⎯⎯⎯

No. moles HCl to be added ⎯⎯⎯⎯⎯⎯   Volume of HCl added ⎯⎯⎯⎯⎯⎯ mL

pH of half-neutralized solution ⎯⎯⎯⎯⎯⎯ K_a of acid ⎯⎯⎯⎯⎯⎯

Experiment 24

Advance Study Assignment: Molar Mass of an Acid

1. 7.0 mL of 6.0 M NaOH are diluted with water to a volume of 400 mL. You are asked to find the molarity of the resulting solution.

    a. First find out how many moles of NaOH there are in 7.0 mL of 6.0 M NaOH. Use Equation 1. Note that the volume must be in liters.

    _____ moles

    b. Since the total number of moles of NaOH is not changed on dilution, the molarity after dilution can also be found by Equation 1, using the final volume of the solution. Calculate that molarity.

    _____ M

2. In an acid-base titration, 23.91 mL of an NaOH solution are needed to neutralize 24.58 mL of a 0.1002 M HCl solution. To find the molarity of the NaOH solution, we can use the following procedure:

    a. First note the value of M_{H^+} in the HCl solution.

    _____ M

    b. Find M_{OH^-} in the NaOH solution. (Use Eq. 3.)

    _____ M

    c. Obtain M_{NaOH} from M_{OH^-}.

    _____ M

3. A 0.3174 g sample of an unknown acid requires 26.23 mL of 0.1056 M NaOH for neutralization to a phenolphthalein end point. There are 0.27 mL of 0.1096 M HCl used for back-titration.

    a. How many moles of OH^- are used? How many moles of H^+ from HCl?

    _____ moles OH^- _____ moles H^+

    b. How many moles of H^+ are there in the solid acid? (Use Eq. 5.)

    _____ moles H^+ in solid

    c. What is the molar mass of the unknown acid? (Use Eq. 4.)

    _____ g/mol

Experiment 25

pH Measurements—Buffers and Their Properties

O ne of the more important properties of an aqueous solution is its concentration of hydrogen ion. The H^+ or H_3O^+ ion has great effect on the solubility of many inorganic and organic species, on the nature of complex metallic cations found in solutions, and on the rates of many chemical reactions. It is important that we know how to measure the concentration of hydrogen ion and understand its effect on solution properties.

For convenience, the concentration of H^+ ion is frequently expressed as the pH of the solution rather than as molarity. The pH of a solution is defined by the following equation:

$$pH = -\log[H^+] \qquad (1)$$

where the logarithm is taken to the base 10. If $[H^+]$ is 1×10^{-4} moles per liter, the pH of the solution is, by the equation, 4. If the $[H^+]$ is 5×10^{-2} M, the pH is 1.3.

Basic solutions can also be described in terms of pH. In water solutions the following equilibrium relation will always be obeyed:

$$[H^+] \times [OH^-] = K_w = 1 \times 10^{-14} \text{ at } 25°C \qquad (2)$$

In distilled water $[H^+]$ equals $[OH^-]$, so, by Equation 2, $[H^+]$ must be 1×10^{-7} M. Therefore, the pH of distilled water is 7. Solutions in which $[H^+] > [OH^-]$ are said to be acidic and will have a pH < 7; if $[H^+] < [OH^-]$, the solution is basic and its pH > 7. A solution with a pH of 10 will have a $[H^+]$ of 1×10^{-10} M and a $[OH^-]$ of 1×10^{-4} M.

We measure the pH of a solution experimentally in two ways. In the first of these we use a chemical called an indicator, which is sensitive to pH. These substances have colors that change over a relatively short pH range (about two pH units) and can, when properly chosen, be used to determine roughly the pH of a solution. Two very common indicators are litmus, usually used on paper, and phenolphthalein, the most common indicator in acid-base titrations. Litmus changes from red to blue as the pH of a solution goes from about 6 to about 8. Phenolphthalein changes from colorless to red as the pH goes from 8 to 10. A given indicator is useful for determining pH only in the region in which it changes color. Indicators are available for measurement of pH in all the important ranges of acidity and basicity. By matching the color of a suitable indicator in a solution of known pH with that in an unknown solution, one can determine the pH of the unknown to within about 0.3 pH units.

The other method for finding pH is with a device called a pH meter. In this device two electrodes, one of which is sensitive to $[H^+]$, are immersed in a solution. The potential between the two electrodes is related to the pH. The pH meter is designed so that the scale will directly furnish the pH of the solution. A pH meter gives much more precise measurement of pH than does a typical indicator and is ordinarily used when an accurate determination of pH is needed.

Some acids and bases undergo substantial ionization in water, and are called strong because of their essentially complete ionization in reasonably dilute solutions. Other acids and bases, because of incomplete ionization (often only about 1% in 0.1 M solution), are called weak. Hydrochloric acid, HCl, and sodium hydroxide, NaOH, are typical examples of a strong acid and a strong base. Acetic acid, $HC_2H_3O_2$, and ammonia, NH_3, are classic examples of a weak acid and a weak base.

A weak acid will ionize according to the Law of Chemical Equilibrium:

$$HB(aq) \rightleftharpoons H^+(aq) + B^-(aq) \qquad (3)$$

At equilibrium,

$$\frac{[H^+][B^-]}{[HB]} = K_a \qquad (4)$$

K_a is a constant characteristic of the acid HB; in solutions containing HB, the product of concentrations in the equation will remain constant at equilibrium independent of the manner in which the solution was made. A similar relation can be written for solutions of a weak base.

The value of the ionization constant K_a for a weak acid can be found experimentally in several ways. In Experiment 24 we used a very simple method for finding K_a. In general, however, we need to find the concentrations of each of the species in Equation 4 by one means or another. In this experiment we will determine K_a for a weak acid in connection with our study of the properties of those solutions we call buffers.

Salts that can be formed by the reaction of strong acids and bases—such as NaCl, KBr, or $NaNO_3$—ionize completely but do not react with water when in solution. They form neutral solutions with a pH of about 7. When dissolved in water, salts of *weak* acids or *weak* bases furnish ions that tend to react to some extent with water, producing molecules of the weak acid or base and liberating some OH^- or H^+ ion to the solution.

If HB is a weak acid, the B^- ion produced when NaB is dissolved in water will react with water to some extent, according to the equation

$$B^-(aq) + H_2O \rightleftharpoons HB(aq) + OH^-(aq) \tag{5}$$

Solutions of sodium acetate, $NaC_2H_3O_2$, the salt formed by reaction of sodium hydroxide with acetic acid, will be slightly *basic* because of the reaction of $C_2H_3O_2^-$ ion with water to produce $HC_2H_3O_2$ and OH^-. Because of the analogous reaction of the NH_4^+ ion with water to form H_3O^+ ion, solutions of ammonium chloride, NH_4Cl, will be slightly *acidic*.

Salts of most transition metal ions are acidic. A solution of $CuSO_4$ or $FeCl_3$ will typically have a pH equal to 5 or lower. The salts are completely ionized in solution. The acidity comes from the fact that the cation is hydrated (e.g., $Cu(H_2O)_4^{2+}$ or $Fe(H_2O)_6^{3+}$). The large + charge on the metal cation attracts electrons from the O—H bonds in water, weakening them and producing some H^+ ions in solution; with $CuSO_4$ solutions the reaction would be

$$Cu(H_2O)_4^{2-}(aq) \rightleftharpoons Cu(H_2O)_3OH^+(aq) + H^+(aq) \tag{6}$$

Buffers

Some solutions, called buffers, are remarkably resistant to pH changes. Water is not a buffer, since its pH is very sensitive to addition of any acidic or basic species. Even bubbling your breath through a straw into distilled water can lower its pH by at least one pH unit, just due to the small amount of CO_2, which is acidic, in exhaled air. With a good buffer solution, you could blow exhaled air into it for half an hour and not change the pH appreciably. All living systems contain buffer solutions, since stability of pH is essential for the occurrence of many of the biochemical reactions that go on to maintain the living organism.

There is nothing mysterious about what one needs to make a buffer. All that is required is a solution containing a weak acid and its conjugate base. An example of such a solution is one containing the weak acid HB, and its conjugate base, B^- ion, obtained by dissolving the salt NaB in water.

The pH of such a buffer is established by the relative concentrations of HB and B^- in the solution. If we manipulate Equation 4, we can solve for the concentration of H^+ ion:

$$[H^+] = K_a \times [HB]/[B^-] \tag{4a}$$

This equation is often further modified when working with buffers, by taking the negative logarithm to base 10 of both sides:

$$pH = pK_a + \log [B^-]/[HB] \tag{4b}$$

where pK_a equals $-\log K_a$. This equation is called the Henderson-Hasselbach equation, familiar to biochemists. Equation 4b tells us what quantities fix the pH of a buffer.

If we are working with a weak acid HB whose K_a equals 1×10^{-5}, its pK_a equals 5.0. If we mix equal volumes of 0.10 M HB and 0.10 M NaB, the pH equals pK_a equals 5.0, since the concentrations of HB and B⁻ are equal in the final solution, and log 1 equals 0. [H⁺] equals 1×10^{-5} M.

The H⁺ ion in the solution has to come from ionization of HB, but the amount is tiny, so we can assume that **in any buffer the weak acid and its conjugate base do not react appreciably with one another when their solutions are mixed.** Their relative concentrations can be calculated from the way the buffer was put together.

Using Equation 4b, you should be able to answer several questions regarding buffers. For example, how does the pH of buffer solution change if you dilute it with water? How does it change if you add an HB solution? What happens if you add a solution of NaB, containing the conjugate base? What is the pH range over which a buffer would be useful, if you assume that the ratio of [B⁻] to [HB] must lie between 10:1 and 1:10? How sensitive is the pH of water to the addition of a strong acid, like HCl, or addition of a strong base, like NaOH? What happens when you add a strong acid or a strong base to a buffer? Why does the pH of a buffer resist changing when small amounts of a strong acid or strong base are added?

Some of these questions you can answer by just looking at Equation 4b. For others you need to realize that, although a weak acid and its conjugate base don't react with each other, a weak acid like HB will react quantitatively in a buffer with a strong base, like NaOH:

$$\text{HB (aq)} + \text{OH}^- \text{(aq)} \rightarrow \text{B}^- \text{(aq)} + \text{H}_2\text{O} \qquad (7)$$

and a weak base like NaB will react with a strong acid like HCl:

$$\text{B}^- \text{(aq)} + \text{H}^+ \text{(aq)} \rightarrow \text{HB (aq)} \qquad (8)$$

In Reaction 7, a small amount of NaOH will not raise the pH very much, since the OH⁻ ion is essentially soaked up by the acid HB, producing some B⁻ ion and increasing the value of [B⁻]/[HB], but not destroying the buffer.

Reactions 7 and 8 can also be used to **prepare** a buffer from a solution of HB by addition of NaOH, or from a solution of NaB by addition of HCl.

In this experiment you will determine the approximate pH of several solutions by using acid-base indicators. Then you will find the pH of some other solutions with a pH meter. In the rest of the experiment you will carry out some reactions that will allow you to answer all the questions we have raised about buffers. Finally, you will prepare one or two buffers having specific pH values.

Experimental Procedure

WEAR YOUR SAFETY GLASSES WHILE
PERFORMING THIS EXPERIMENT

You may work in pairs on the first three parts of this experiment.

A. Determination of pH by the Use of Acid-Base Indicators

To each of five small test tubes add about 1 mL 0.10 M HCl (about 1/2-inch depth in tube). To each tube add a drop or two of one of the indicators mentioned in Table 25.1, one indicator to a tube. Note the color of the solution you obtain in each case. By comparing the colors you observe with the information in Table 25.1, estimate the pH of the solution to within a range of one pH unit, say 1 to 2, or 4 to 5. In making your estimate, note that the color of an indicator is most indicative of pH in the region where the indicator is changing color.

Repeat the procedure with each of the following solutions:

$$\text{0.10 M NaH}_2\text{PO}_4 \quad \text{0.10 M HC}_2\text{H}_3\text{O}_2 \quad \text{0.10 M ZnSO}_4$$

Record the colors you observe and the pH range for each solution.

Table 25.1

Indicator	Useful pH Range (Approximate)							
	0	1	2	3	4	5	6	7
Methyl violet	yellow [ ] violet							
Thymol blue	red [ ] yellow							
Methyl yellow	red [ ] yellow							
Congo red	violet [ ] orange-red							
Bromcresol green	yellow [ ] blue							

B. Measurement of the pH of Some Typical Solutions

In the rest of this experiment we will use pH meters to find pH. Your instructor will show you how to operate your meter. The electrodes may be fragile, so use due caution when handling the electrode probe. See Appendix IV for a discussion of pH meters.

Using a 25-mL sample in a 150-mL beaker, measure and record the pH of a 0.10 M solution of each of the following substances:

$$NaCl \quad Na_2CO_3 \quad NaC_2H_3O_2 \quad NaHSO_4$$

Rinse the electrode probe in distilled water between measurements. After you have completed a measurement, add a drop or two of bromcresol green to the solution and record the color you obtain.

Some of the solutions are nearly neutral; others are acidic or basic. For each solution having a pH less than 6 or greater than 8, write a net ionic equation that explains qualitatively why the observed pH value is reasonable.

Then write a rationale for the colors obtained with bromcresol green with these solutions.

C. Some Properties of Buffers

On the lab bench we have 0.10 M stock solutions that can be used to make three different common buffer systems. These are

$HC_2H_3O_2 - C_2H_3O_2^-$	$NH_4^+ - NH_3$	$HCO_3^- - CO_3^{2-}$
acetic acid–acetate ion	ammonium ion–ammonia	hydrogen carbonate–carbonate

The sources of the ions will be sodium and ammonium salts containing those ions. Select one of these buffer systems for your experiment.

1. Using a graduated cylinder, measure out 15 mL of the acid component of your buffer into a 100-mL beaker. The acid will be one of the following solutions: 0.10 M $HC_2H_3O_2$, 0.10 M NH_4Cl, or 0.10 M $NaHCO_3$. Rinse out the graduated cylinder with distilled water and use it to add 15 mL of the conjugate base of your buffer. Measure the pH of your mixture and record it on the Data page. Calculate pK_a for the acid.

2. Add 30 mL water to your buffer mixture, mix, and pour half of the resulting solution into another 100-mL beaker. Measure the pH of the diluted buffer. Calculate pK_a once again. Add five drops of 0.10 M NaOH to the diluted buffer and measure the pH again. To the other half of the diluted buffer add 5 drops 0.10 M HCl, and again measure the pH. Record your results.

3. Make a buffer mixture containing 2 mL of the acid component and 20 mL of the solution containing the conjugate base. Mix, and measure the pH. Calculate a third value for pK_a. To that solution add 3 mL 0.10 M NaOH. Measure the pH. Explain your results.

4. Put 25 mL distilled water into a 100-mL beaker. Measure the pH. Add 5 drops 0.10 M HCl and measure the pH again. To that solution add 10 drops 0.10 M NaOH, mix, and measure the pH.

5. Select a pH different from any of those you observed in your experiments. Design a buffer which should have that pH by selecting appropriate volumes of your acidic and basic components. Make up the buffer and measure its pH.

D. Preparation of a Buffer from a Solution of a Weak Acid

So far in this experiment the buffers that we used were made up from a weak acid and its conjugate base. Chemists faced with making a buffer would take a simpler approach. They would start with a solution of a weak acid with a pK_a roughly equal to the pH of the buffer that was needed. To the acid they would slowly add an NaOH solution from a buret, stirring well, while at the same time measuring the pH of the solution. When they got to the desired pH, they would stop adding the NaOH. The buffer would be ready to use.

In this part of the experiment you are to work alone. We will furnish you with an 0.50 M solution of a weak acid with a known pK_a, and the pH of the buffer you will be asked to prepare. First, dilute the acid solution to 0.10 M by adding 10 mL of the acid to 40 mL water in a 100 mL beaker and stirring well.

Using Equation 4b, calculate the ratio of $[B^-]$ to $[HB]$ in the buffer to be prepared. Calculate how much 0.10 M NaOH you will have to add to 20 mL of your 0.10 M acid solution to produce the required ratio. This is easily done. If we add y mL of the NaOH, the value of $[B^-]/[HB]$ will become equal to $y/(20 - y)$. Can you see why? Think about it, and you will soon see that is the case. Then complete the calculation and record the volume of NaOH that should be needed.

Now do the experiment to check your prediction. Use the buret containing 0.10 M NaOH solution that has been set up by the pH meter. Record the volume before starting to add the base, noting the pH of the acid. Slowly add the NaOH to 20 mL of the acid, stirring well, and watching the pH as it slowly goes up. When you obtain a solution of the pH to be prepared, stop adding NaOH. Record the volume reading on the buret. Report the volume required to produce your buffer.

Optional Experiment: Vitamin C (MM = 176 g) is a weak acid, called ascorbic acid. Study the buffering properties of Vitamin C using a solution made from a 500 mg tablet. Find K_a for Vitamin C and the pH range over which it might be useful as a buffer.

   DISPOSAL OF REACTION PRODUCTS.   When you are finished with the experiment, you may dilute the solutions with water and pour them down the drain.

Experiment 25

Observations and Calculations: pH: Buffers and Their Properties

A. Determination of pH by the Use of Acid-Base Indicators

	Color with 0.1 M Solution of			
Indicator	HCl	NaH_2PO_4	$HC_2H_3O_2$	$ZnSO_4$
Methyl violet	_____	_____	_____	_____
Thymol blue	_____	_____	_____	_____
Methyl yellow	_____	_____	_____	_____
Congo red	_____	_____	_____	_____
Bromcresol green	_____	_____	_____	_____
pH range	_____	_____	_____	_____

Circle the observation(s) for each solution that was most useful in estimating the pH range.

B. Measurement of the pH of Some Typical Solutions

Record the pH and the color observed with bromcresol green for each of the 0.1 M solutions that were tested.

	NaCl	Na_2CO_3	$NaC_2H_3O_2$	$NaHSO_4$
pH	_____	_____	_____	_____
Color	_____	_____	_____	_____

For any two solutions having a pH less than 6 or greater than 8, write a net ionic equation to explain qualitatively why the solution has that pH.

Solution _____ Equation _____

Solution _____ Equation _____

Explain why the color observed with bromcresol green for each of the four solutions is reasonable, given the pH.

C. Some Properties of Buffers

Buffer system selected _____ HB is _____ (name the acid)

1. pH of buffer _____ $[H^+]$ _____ M pK_a (by Eq. 4b) _____

2. pH of diluted buffer _____ $[H^+]$ _____ M pK_a _____

    pH after addition of 5 drops NaOH   _____

    pH after addition of 5 drops HCl   _____

    Comment on your observations in Parts 1 and 2.

(continued on following page)

3. pH of buffer in which $[HB]/[B^-] = 0.10$ _____ pK_a _____

 pH after addition of excess NaOH _____

 Explain your observations

4. pH of distilled water _____

 pH after addition of 5 drops HCl _____

 pH after addition of 5 drops NaOH _____

 Explain your observations

5. pH of buffer solution to be prepared _____

 Average value of pK_a (as found in Parts 1, 2, and 3) _____

 $[B^-]/[HB]$ in buffer _____

 Volume 0.10 M NaB/Volume 0.10 M HB needed in buffer _____

 Volume 0.10 M NaB used _____ mL   Volume 0.10 M HB used _____ mL

 pH of buffer you prepared _____

D. Preparing a Buffer from a Solution of a Weak Acid

 pH of buffer to be prepared _____ pK_a of acid _____

 Ratio of $[B^-]/[HB]$ required in buffer _____

 Volume of 0.10 M NaOH calculated _____ mL

 pH of acid solution before titration _____

 Initial volume reading of NaOH _____ mL

 Final volume reading _____ mL

 Volume of NaOH actually required _____ mL

 How does that volume compare to your calculated value?

Experiment 25

Advance Study Assignment: pH, Its Measurement and Applications

1. A solution of a weak acid was tested with the indicators used in this experiment. The colors observed were as follows:

Methyl violet	violet	Congo red	violet
Thymol blue	orange	Bromcresol green	yellow
Methyl yellow	red		

What is the approximate pH of the solution?

_____

2. An aqueous solution of NH_3 has a pH of 11.6. The ammonia (NH_3) molecule is the conjugate base of the NH_4^+ (called "ammonium") ion. Write the net ionic equation for the reaction that makes an aqueous solution of NH_3 basic. (Eq. 5.)

3. The pH of a 0.10 M HOBr solution is 4.8.

   a. What is [H$^+$] in that solution?

   _____ M

   b. What is [OBr$^-$]? What is [HOBr]? (Where do the H$^+$ and OBr$^-$ ions come from?)

   _____ M; _____ M

   c. What is the value of K_a for HOBr? What is the value of pK_a?

   _____   _____

4. Formic acid, HFor, has a K_a value equal to about 1.8×10^{-4}. A student is asked to prepare a buffer having a pH of 3.90 from a solution of formic acid and a solution of sodium formate having the same molarity. How many milliliters of the NaFor solution should she add to 20. mL of the HFor solution to make the buffer? (See discussion of buffers.)

   _____ mL

5. How many mL of 0.10 M NaOH should the student add to 20 mL 0.10 M HFor if she wished to prepare a buffer with a pH of 3.90, the same as in Problem 4?

   _____ mL

Experiment 26

Determination of the Solubility Product of $Ba(IO_3)_2$

In several of the earlier experiments in this manual we used the concept of the solubility product in various applications. Every insoluble inorganic salt has an associated number, called its solubility product, that influences its behavior in mixtures where it is present as a solid. This number puts a condition on the concentrations of the ions in a mixture in which the salt is present. The equation expressing the condition follows from the chemical formula of the solid. For barium iodate the equation for the condition on the solubility product, K_{sp}, takes its form from the equation by which the solid dissolves:

$$Ba(IO_3)_2(s) = Ba^{2+}(aq) + 2\ IO_3^-(aq) \tag{1}$$

where

$$K_{sp} = [Ba^{2+}]\ [IO_3^-]^2 \tag{2}$$

and the concentrations of the barium and iodate ions are in moles per liter in the solution in which solid barium iodate is present. Equation 2 essentially states that in a mixture of barium and iodate ions in equilibrium with solid barium iodate the product of the molarity of the Ba^{2+} ion times the molarity of the IO_3^- ion squared cannot vary, no matter how you change the ion concentrations in the system. If one goes up, the other most go down, keeping the ion product in Equation 2 exactly at its constant value. That's rather amazing. For barium iodate, K_{sp} has a small value, less than 10^{-7}. So, if one ion has a typical concentration, say 0.1 M, the other must have a very low concentration. If you add 0.1 M $BaCl_2$ to a small volume of a solution of 0.05 M KIO_3, you can be sure that a reaction will occur, in which $Ba(IO_3)_2$ precipitates, where the final solution will contain almost no iodate ion, and the product of the ions in Equation 2 has a fixed value.

This behavior is seen in many analyses of practical importance. In Experiment 7, in which a solution of chloride ion is titrated with a solution of $AgNO_3$, containing silver and nitrate ions, when the Ag^+ ions first meet the Cl^- ions, insoluble AgCl precipitates, since the product, K_{sp}, of their concentrations cannot be very large. As the silver nitrate is slowly added, AgCl precipitates until the Cl^- ion is essentially gone, and the Ag^+ ion concentration can finally go up. At the end point, silver dichlorofluorosceinate will form on the surface of the silver chloride particles.

In this experiment we will be examining the solubility behavior of $Ba(IO_3)_2$ in various mixtures of $Ba(NO_3)_2$ and KIO_3 solutions, for the purpose of ultimately measuring the K_{sp} of barium iodate and observing whether it indeed remains constant.

WEAR YOUR SAFETY GLASSES WHILE PERFORMING THIS EXPERIMENT

Experimental Procedure

This is an experiment that can give excellent results if it is done carefully. It is probably the most difficult experiment in the manual to do properly. You will be working with small amounts of the chemical reactants, so small errors will be important.

Preparing the Reaction Mixtures

From the stock solutions that are available, measure out about 35 mL of 0.0200 M $Ba(NO_3)_2$ into a small dry beaker. To a second small beaker add about 20 mL 0.0350 M KIO_3. Use the labeled graduated cylinders next to the reagent bottles for measuring out these solutions.

In the experiment we will precipitate solid $Ba(IO_3)_2$ under several different conditions, with varying proportions of the reagents you will be using. We will analyze the solutions remaining after the precipitation for their iodate ion concentrations at equilibrium, and from those values and the way you made the mixtures, we can find the concentrations of barium ion, and the value of K$_{sp}$ for barium iodate.

Label five regular (18 × 150 mm) dry test tubes, either with labels or by noting their location in your test tube rack.

To Test Tube 1, carefully pipet 1.00 mL 0.0350 M KIO$_3$. Add 2, 3, 4, and 5 mL of that reagent to Test Tubes 2, 3, 4, and 5, respectively. You may use your 10-mL graduated pipet for these additions. (The easiest way to do this is to fill the pipet to the 0.00 level. With the pipet in Test Tube 1, let the level fall to 1.00. In Tube 2, let the level fall to 3.00, thereby adding 2 mL to the Tube. In Tube 3, let the level fall to 6.00, and in Tube 4 let it go to 10.00. Refill the pipet to the 5.00 level and in Tube 5 let it fall to 10.00.)

Then, to each of the test tubes add 5.00 mL of 0.0200 M Ba(NO$_3$)$_2$, using your 5-mL pipet. Finally, pipet 6, 5, 4, 3, and 2 mL of distilled water into Test Tubes 1 through 5, respectively. The final volume in each tube should be 12.00 mL. The compositions of the mixtures are summarized in Table 26.1.

Table 26.1

Volumes of Reagents Used in Precipitating Ba(IO$_3$)$_2$ (mL)			
Test Tube	**0.0200 M Ba(NO$_3$)$_2$**	**0.0350 M KIO$_3$**	**H$_2$O**
1	5.00	1.00	6.00
2	5.00	2.00	5.00
3	5.00	3.00	4.00
4	5.00	4.00	3.00
5	5.00	5.00	2.00

With a stirring rod, stir each solution for 30 seconds, up and down as well as swirling, wiping the rod on a paper towel before proceeding to the next solution. Touch the wall of the tube occasionally as you stir.

Look at each tube. In most of them you should see a crystalline white precipitate, which slowly settles out. It will be most noticeable in the higher number tubes, but it should be present as a light cloudiness in Test Tubes 1 and 2. There will only be a tiny amount of precipitate, only a few mg, so the amount is not impressive. It is most easily seen if you look through the tube toward the light, where you may observe tiny particles, very slowly settling out. The crystallization occurs slowly, so give it time.

Put a cork stopper in each of the tubes in which there is a precipitate, and slowly rotate the tubes so that the liquid goes from end to end twenty times or so to ensure thorough mixing and help bring the system to equilibrium.

(For any tubes in which you observe only a clear liquid, and no solid, repeat the stirring operation, scratching the walls of the tube as you stir, which may get things started. If that still doesn't do the job, touch your stirring rod to the sample of pure solid barium iodate that we have in the lab and put the rod in the reluctant liquid. Insert a cork stopper in the tube and shake with vigor. Sooner or later you will get the cloudiness to form.)

Repeat the rotation step, turning each tube end to end to do what you can to make the liquid homogeneous and in equilibrium with the precipitate. You can't overstir, so be patient. Give yourself five minutes with this. It will be well spent.

Finally, let the test tubes stand for 15 minutes to let the solid precipitate particles settle out completely, leaving a clear solution above the solids.

While you are waiting for the precipitates to settle out, you can begin the calculations you will need later. Fill in as many blanks as you can on the Data and Calculation page, or on your Excel sheet. From the way you made your mixtures, you can determine the numbers of moles of Ba^{2+} and IO$_3^-$ that were initially present in each tube. Once you have completed the analysis of the solutions at equilibrium you can finish the calculations of K$_{sp}$.

Preparing the Equilibrium Mixtures for Analysis

The mixtures in your test tubes at this point should contain essentially clear solutions above the precipitates that formed. You now need to remove an aliquot from each of those solutions, which should now be at equilibrium, without disturbing the solid at the bottom of the test tubes. This is the most difficult part of the experiment, and must be done very carefully. It is done with a 5-mL pipet. First, rinse and drain that pipet with distilled water. Then blow out the water remaining in the tip.

Clamp Test Tube 1 on a ring stand, so that you can work on it without it being able to move. Immerse the 5-mL pipet about half-way down the clear solution, and hold it there with your hand. Compress a suction bulb slightly, and attach it gently to the top of the pipet. Slowly release the pressure on the bulb, and draw the solution into the pipet until it is near the top of the pipet. Remove the bulb and quickly put your finger on the top of the pipet to hold the liquid in place. Let liquid flow back into the test tube, slowly, slowly, until the level is at the marking line on the pipet. Remove the pipet from the test tube and let the contents flow into a clean dry test tube. This operation is easier said than done. If something goes wrong while you are drawing the liquid into the pipet, remove the bulb and let the liquid go back into the test tube. After things settle down, you can try again, until you get it right.

In order to increase your chances of accomplishing this transfer properly, practice doing it with water in a clamped test tube. After about five successful practice runs, you should be able to make the transfer where it counts. So, go ahead and do it, for each of the five test tubes, using five clean dry test tubes to receive the 5-ml aliquots.

Analyzing the Equilibrium Solutions

Having separated 5 mL of each of the five equilibrium solutions, it is now possible to analyze for the concentration of IO$_3^-$ ion in those solutions. We do it by reducing iodate ion to I$_2$, which is colored and can be analyzed spectroscopically. The reaction we will use is:

$$IO_3^-(aq) + 5\ I^-(aq) + 6\ H^+(aq) \rightarrow 3\ I_2(aq) + 3\ H_2O^* \tag{3}$$

To each of the test tubes add, by pipet, 1.00 mL 1 M KI and 1.00 mL 1 M HCl, and 3.00 mL distilled water, bringing the volume of the five solutions to 10.00 mL. During this step the color of the solution will become orange as the reaction proceeds. Stir each mixture well with a stirring rod, drying the rod on a paper towel before you go on to the next tube.

Go to the Spectronic 20 in the lab. The wavelength setting should be 500 nm. If you have not used the Spec 20 before, read up in Appendix IV on how to use it to measure light absorption. Pick up a Spectronic 20 tube, and if it is not dry, shake it to remove as much liquid as you can, and then add about a milliliter of the orange liquid from Tube 1, shake to mix it with the old liquid, and then pour that liquid into a waste beaker. Rinse, and drain again with another 1-mL portion of your solution. Then fill the Spec 20 tube up to the level of the label with the liquid in Tube 1. Set the zero and 100% readings if they have not yet been adjusted; then put your tube in the tube holder and read the absorbance of the solution. The concentration of the IO$_3^-$ ion in the **equilibrium 12-mL mixture** will equal the value on the graph in the lab. (This is drawn to take account of the dilution you made.) Record the absorbance and concentration for Tube 1. Repeat these measurements for the other four tubes you worked with, and record your values for those solutions.

Pour all of the solutions and solids in the ten test tubes you used in the waste crock. Rinse out the tubes with distilled water.

Processing the Data

From the volume and concentration of the KIO$_3$ you used in making the five mixtures of reagents, you have probably already found the number of moles of iodate ion **originally** in each mixture. From the concentration of the IO$_3^-$ ion that we measured for each of the **equilibrium** solutions we calculate the number of moles of that ion in those solutions, with its 12-mL volume. The difference between the original number of moles and the equilibrium number of moles must be the number of moles of iodate ion in the precipitate at the bottom of the tube. The number of moles of barium in the precipitate must equal **half** of that value, given the formula of barium iodate, Ba(IO$_3$)$_2$. Why?

The number of moles of barium ion that remain in the solution will equal the difference between the number in the original mixture and the number you found were in the precipitate in each of the mixtures. Knowing the volume of the equilibrium solution, 12.00 mL, you can calculate the concentration of the Ba^{2+} ion in each mixture. Record the results of these calculations on the Data and Calculations page or on your Excel sheet. Finally, using the concentrations you have recorded, find the values of K$_{sp}$ for each of the five mixtures. Calculate the mean value of K$_{sp}$ and the standard deviation. (See Appendix VIII.)

*The actual iodine species is I$_3^-$, but I$_2$ is simpler so we use it.

Name _____   Section _____

Experiment 26

Data and Calculations: The Solubility Product of $Ba(IO_3)_2$

The calculations are similar to those in Experiment 23, and are outlined in the Experimental Procedure section. To use Excel, set up a table like the one below.

Test Tube No.	1	2	3	4	5
mL 0.0350 M KIO_3	_____	_____	_____	_____	_____
mL 0.0200 M $Ba(NO_3)_2$	_____	_____	_____	_____	_____
Total volume in mL	_____	_____	_____	_____	_____
Absorbance of soln.	_____	_____	_____	_____	_____
$[IO_3^-]$ in soln.	_____	_____	_____	_____	_____

Processing the Data (all data $\times 10^{-5}$ except for K_{sp} values)

Initial no. moles Ba^{2+}	_____	_____	_____	_____	_____
Initial no. moles IO_3^-	_____	_____	_____	_____	_____
No. moles IO_3^- in equil. soln. (12 mL)	_____	_____	_____	_____	_____
No. moles IO_3^- precipitated	_____	_____	_____	_____	_____
No. moles Ba^{2+} precipitated	_____	_____	_____	_____	_____
No. moles Ba^{2+} in equil. soln. (12 mL)	_____	_____	_____	_____	_____
$[Ba^{2+}]$ in equil. soln.	_____	_____	_____	_____	_____
K_{sp} for $Ba(IO_3)_2$	_____	_____	_____	_____	_____

Mean value of K_{sp} _____   Standard deviation _____

How many mg of solid barium iodate were there in the precipitate in Test Tube 1? MM $Ba(IO_3)_2 = 487.14$

_____ mg

Does $[Ba^{2+}]$ vary as expected in the experiment? Explain.

What is your best value of the solubility of $Ba(IO_3)_2$ in water?

Experiment 26

Advance Study Assignment: Determination of the Solubility Product
of $Ba(IO_3)_2$

1. State in words what is meant by the solubility product equation for $Ba(IO_3)_2$:

$$K_{sp} = [Ba^{2+}] [IO_3^-]^2$$

2. If 10 mL 0.10 M $Ba(NO_3)_2$ is mixed with 10 mL 0.10 M KIO_3, a precipitate forms. Which ion will still be present at appreciable concentration in the equilibrium mixture if K_{sp} for barium iodate is very small? Indicate your reasoning. What would that concentration be?

_____ _____ moles/L

3. Lead chloride, $PbCl_2$, is moderately soluble, and has a K_{sp} equal to about 1.7×10^{-5}.

   a. What would be the solubility of lead chloride in water?

   _____ moles/L

   b. What would be its solubility in 0.10 M NaCl?

   _____ moles/L

   c. The difference is caused by what is called the common ion effect. State in words what the common ion effect predicts.

Experiment 27

Relative Stabilities of Complex Ions and Precipitates Prepared from Solutions of Copper(II)

In aqueous solution, typical cations, particularly those produced from atoms of the transition metals, do not exist as free ions but rather consist of the metal ion in combination with some water molecules. Such cations are called complex ions. The water molecules, usually two, four, or six in number, are bound chemically to the metallic cation, but often rather loosely, with the electrons in the chemical bonds being furnished by one of the unshared electron pairs from the oxygen atoms in the H_2O molecules. Copper ion in aqueous solution may exist as $Cu(H_2O)_4^{2+}$, with the water molecules arranged in a square around the metal ion at the center.

If a hydrated cation such as $Cu(H_2O)_4^{2+}$ is mixed with other species that can, like water, form coordinate covalent bonds with Cu^{2+}, those species, called ligands, may displace one or more H_2O molecules and form other complex ions containing the new ligands. For instance, NH_3, a reasonably good coordinating species, may replace H_2O from the hydrated copper ion, $Cu(H_2O)_4^{2+}$, to form $Cu(H_2O)_3NH_3^{2+}$, $Cu(H_2O)_2(NH_3)_2^{2+}$, $Cu(H_2O)(NH_3)_3^{2+}$, or $Cu(NH_3)_4^{2+}$. At moderate concentrations of NH_3, essentially all the H_2O molecules around the copper ion are replaced by NH_3 molecules, forming the copper ammonia complex ion.

Coordinating ligands differ in their tendencies to form bonds with metallic cations, so that in a solution containing a given cation and several possible ligands, an equilibrium will develop in which most of the cations are coordinated with those ligands with which they form the most stable bonds. There are many kinds of ligands, but they all share the common property that they possess an unshared pair of electrons that they can donate to form a coordinate covalent bond with a metal ion. In addition to H_2O and NH_3, other uncharged coordinating species include CO and ethylenediamine; some common anions that can form complexes include OH^-, Cl^-, CN^-, SCN^-, and $S_2O_3^{2-}$.

As you know, when solutions containing metallic cations are mixed with other solutions containing ions, precipitates are sometimes formed. When a solution of 0.1 M copper nitrate is mixed with a little 1 M NH_3 solution, a precipitate forms and then dissolves in excess ammonia. The formation of the precipitate helps us to understand what is occurring as NH_3 is added. The precipitate is hydrous copper hydroxide, formed by reaction of the hydrated copper ion with the small amount of hydroxide ion present in the NH_3 solution. The fact that this reaction occurs means that even at very low OH^- ion concentration $Cu(OH)_2(H_2O)_2(s)$ is a more stable species than $Cu(H_2O)_4^{2+}$ ion.

Addition of more NH_3 causes the solid to redissolve. The copper species then in solution cannot be the hydrated copper ion. (Why?) It must be some other complex ion and is, indeed, the $Cu(NH_3)_4^{2+}$ ion. The implication of this reaction is that the $Cu(NH_3)_4^{2+}$ ion is also more stable in NH_3 solution than is the hydrated copper ion. To deduce in addition that the copper ammonia complex ion is also more stable in general than $Cu(OH)_2(H_2O)_2(s)$ is not warranted, since under the conditions in the solution $[NH_3]$ is much larger than $[OH^-]$, and given a higher concentration of hydroxide ion, the solid hydrous copper hydroxide might possibly precipitate even in the presence of substantial concentrations of NH_3.

To resolve this question, you might proceed to add a little 1 M NaOH solution to the solution containing the $Cu(NH_3)_4^{2+}$ ion. If you do this you will find that $Cu(OH)_2(H_2O)_2(s)$ does indeed precipitate. We can conclude from these observations that $Cu(OH)_2(H_2O)_2(s)$ is more stable than $Cu(NH_3)_4^{2+}$ in solutions in which the ligand concentrations (OH^- and NH_3) are roughly equal.

The copper species that will be present in a system depends, as we have just seen, on the conditions in the system. We cannot say in general that one species will be more stable than another; the stability of a given species depends in large measure on the kinds and concentrations of other species that are also present with it.

Another way of looking at the matter of stability is through equilibrium theory. Each of the copper species we have mentioned can be formed in a reaction between the hydrated copper ion and a complexing or precipitating ligand; each reaction will have an associated equilibrium constant, which we might call a formation constant for that species. The pertinent formation reactions and their constants for the copper species we have been considering are listed here:

$$Cu(H_2O)_4^{2+} (aq) + 4\ NH_3(aq) \rightleftharpoons\ Cu(NH_3)_4^{2+}\ (aq) + 4\ H_2O \qquad K_1 = 5 \times 10^{12} \tag{1}$$

$$Cu(H_2O)_4^{2+} (aq) + 2\ OH^-(aq) \rightleftharpoons Cu(OH)_2(H_2O)_2(s) + 2\ H_2O \qquad K_2 = 2 \times 10^{19} \tag{2}$$

The formation constants for these reactions do not involve $[H_2O]$ terms, which are essentially constant in aqueous systems and are included in the magnitude of K in each case. The large size of each formation constant indicates that the tendency for the hydrated copper ion to react with the ligands listed is very high.

In terms of these data, let us compare the stability of the $Cu(NH_3)_4^{2+}$ complex ion with that of solid $Cu(OH)_2(H_2O)_2$. This is most readily done by considering the reaction:

$$Cu(NH_3)_4^{2+} (aq) + 2\ OH^-(aq) + 2\ H_2O \rightleftharpoons\ Cu(OH)_2(H_2O)_2(s) + 4\ NH_3(aq) \tag{3}$$

We can find the value of the equilibrium constant for this reaction by noting that it is the sum of Reaction 2 and the reverse of Reaction 1. By the Law of Multiple Equilibrium, K for Reaction 3 is given by the equation

$$K = \frac{K_2}{K_1} = \frac{2 \times 10^{19}}{5 \times 10^{12}} = 4 \times 10^6 = \frac{[NH_3]^4}{[Cu(NH_3)_4^{2+}][OH^-]^2} \tag{4}$$

From the expression in Equation 4 we can calculate that in a solution in which the NH_3 and OH^- ion concentrations are both about 1 M.

$$[Cu(NH_3)_4^{2+}] = \frac{1}{4 \times 10^6} = 2.5 \times 10^{-7}\ M \tag{5}$$

Since the concentration of the copper ammonia complex ion is very, very low, any copper(II) in the system will exist as the solid hydroxide. In other words, the solid hydroxide is more stable under such conditions than the ammonia complex ion. But that is exactly what we observed when we treated the hydrated copper ion with ammonia and then with an equivalent amount of hydroxide ion.

Starting now from the experimental behavior of the copper ion, we can conclude that since the solid hydroxide is the species that exists when copper ion is exposed to equal concentrations of ammonia and hydroxide ion, the hydroxide is more stable under those conditions, *and* the equilibrium constant for the formation of the hydroxide is larger than the constant for the formation of the ammonia complex. By determining, then, which species is present when a cation is in the presence of equal ligand concentrations, we can speak meaningfully of stability under such conditions and can rank the formation constants for the possible complex ions, and indeed for precipitates, in order of their increasing magnitudes.

In this experiment you will carry out formation reactions for a group of complex ions and precipitates involving the Cu^{2+} ion. You can make these species by mixing a solution of $Cu(NO_3)_2$ with solutions containing NH_3 or anions, which may form either precipitates or complex ions by reaction with $Cu(H_2O)_4^{2+}$, the cation present in aqueous solutions of copper(II) nitrate. By examining whether the precipitates or complex ions formed by the reaction of hydrated copper(II) ion with a given species can, on addition of a second ligand, be dissolved or transformed to another species, you will be able to rank the relative stabilities of the precipitates and complex ions made from Cu^{2+} with respect to one another, and thus rank the equilibrium formation constants for each species in order of increasing magnitude. The species to be reacted with Cu^{2+} ion in aqueous solution are NH_3, Cl^-, OH^-, $C_2O_4^{2-}$, S^{2-}, NO_2^-, and PO_4^{3-}. In each case the test for relative stability will be made in the presence of essentially equal concentrations of the two ligands. When you have completed your ranking of the known species you will test an unknown species and incorporate it into your list.

Experimental Procedure

Obtain from the stockroom an unknown and eight small test tubes.

Add about 1 mL 0.1 M $Cu(NO_3)_2$ solution to each of the test tubes ($\frac{1}{2}$-inch depth).

To one of the test tubes add about 1 mL 1 M NH_3, drop by drop. Note whether a precipitate forms initially, and if it dissolves in excess NH_3. Shake the tube sideways to mix the reagents. In the NH_3-NH_3 space in the table (Data page), write a P in the upper left-hand corner if a precipitate is formed initially. In the rest of the space, write the formula and the color of the species that was present with an excess of NH_3. If a solution is present, the copper ion will be in a complex. Cu(II) will always have a coordination number of 4, so the formula with NH_3 would be $Cu(NH_3)_4^{2+}$. If a precipitate is present, it will be neutral, and in the case of NH_3 it would be a hydroxide with the formula $Cu(OH)_2$. (There should in principle be two H_2O molecules in the formula, but they are usually omitted.) Add 1 mL 1 M NH_3 to the rest of the test tubes.

Now you will test the stability of the species present in excess NH_3 relative to those which might be present with other precipitating or coordinating species. Add, drop by drop, 1 mL of a 1 M solution of each of the anions in the horizontal row of the table to the test tubes you have just prepared, one solution to a test tube. Note any changes that occur in the appropriate spaces in the table. A change in color or the formation of a precipitate implies that a reaction has occurred between the added ligand or precipitating anion and the species originally present. As before, put a P in the upper left-hand corner if a precipitate initially forms on addition of the anion solution. In the rest of the space, write the formula and color of the species present when the anion is in excess. That is the species that is stable in the presence of equal concentrations of NH_3 and the added anion. Again, in complex ions, Cu(II) will normally be attached to four ligands; copper(II) precipitates will be neutral. If a new species forms on addition of the second reagent, its formula should be given. If no change occurs, the original species is more stable, so put its formula in that space.

Repeat the above series of experiments, using 1 M Cl^- as the species originally added to the $Cu(NO_3)_2$ solution. In each case record the color of any precipitates or solutions formed on addition of the reagents in the horizontal row, and the formula of the species stable when an excess of both Cl^- ion and the added species is present in the solution. Since these reactions are reversible, it is not necessary to retest Cl^- with NH_3, since the same results would be obtained as when the NH_3 solution was tested with Cl^- solution.

Repeat the series of experiments for each of the anions in the vertical row in the table, omitting those tests where decisions as to relative stabilities are already clear. Where both ligands produce precipitates it will be helpful to check the effect of the addition of the other ligand to those precipitates. When complete, your table should have at least 28 entries.

Examine your table and decide on the relative stabilities of all species you observed to be present in all the spaces of the table. There should be seven such species; rank them as best you can in order of increasing stability. There is only one correct ranking, and you should be prepared to defend your choices. Although we did not in general prepare the species by direct reaction of $Cu(H_2O)_4^{2+}$ with the added ligand or precipitating anion, the equilibrium formation constants for those species for the direct reactions will have magnitudes that increase in the same order as the relative stabilities of the species you have established.

When you are satisfied that your ranking order is correct, carry out the necessary tests on your unknown to determine its proper position in the list. Your unknown may be one of the Cu(II) species you have already observed, or it may be a different species, present in excess of its ligand or precipitating anion. If your unknown contains a precipitate, shake it well before using a portion of it to make a test.

Dispose of all reaction products in the waste crock provided or as otherwise directed by your instructor.

Alternate Procedure Using Microscale Approach   Optional

From the stockroom obtain an unknown and two plastic well plates (4 × 6).

Align the plates into a 6 × 8 well configuration, with six wells across the top. To each of the six wells in the top row of the plate add six drops of 0.1 M $Cu(NO_3)_2$. Use a Beral pipet for this and all other additions of reagents. To the six wells add six drops of 1 M solutions containing the species in the horizontal row at the top of the Table of Observations; one species to a well, starting with 1 M NH_3 in the first well and ending with 1 M $NaNO_2$ in the sixth (we will omit S^{2-} and come back to it later). Mix the reagents by moving the well plate back and forth on the top of the lab bench. You will notice a difference in the appearance of the mixtures in each of the six wells. Each contains the species that is stable when Cu(II) is mixed with one of the reagents. Report the color and the formula of those six species along the **diagonal** blocks in the Table, using the directions in the first two paragraphs in the ordinary procedure. The contents of these wells will serve as a reference during the rest of the experiment.

To establish the relative stabilities of the six species, we need to mix them with each of the other species in some systematic way and see what happens. We will first test them against NH_3. To do this, in the second row of wells make the same mixtures as you did in the first row. Then, add six drops of 1 M NH_3 to the second through sixth wells in that row (we do not need to test NH_3 with NH_3, since we really have already done that). Mix the reagents, and note any changes that occurred on addition of NH_3. In some of the wells, changes will occur, indicating a reaction to form a more stable species has occurred. Use the wells in the top row for comparison. Record your observations in the blocks in the first row of the Table, noting the color and formula of the stable Cu(II) species in each case.

Now make stability tests for Cl^-, using wells in the third row of the plate. This time start with the third well, since we already know how Cl^- and NH_3 stack up from the last set of tests. (It doesn't make any difference if we change the order in which the reagents are added.) The four wells on the right end of the third row should be prepared in the same way as the corresponding wells in the first row. To each of those wells add six drops of 1 M NaCl, mix, note any changes, and report your observations as before in the blocks in the second row of the Table.

Continue along these lines, making stability tests with 1 M NaOH in the bottom row of the top plate. Report your observations as before. Then, in the top three rows of the lower well plate, make the same sort of tests with 1 M $K_2C_2O_4$, 1 M Na_2HPO_4, and 1 M KNO_2, respectively. When you are done, you should have entries in all of the blocks in the upper right half of the Table except for those involving S^{2-}.

To establish the stability of the Cu(II) species with respect to the sulfide ion, S^{2-}, add a drop or two of 0.1 M $(NH_4)_2S$ to each of the wells in the top row of the top plate. Although the sulfide solution is dilute, its effect on the contents of the wells, and on the nose, should be clear. Fill in the last column of the Table.

Given your entries in the Table, you should be able to decide on the relative stabilities of the seven Cu(II) species studied in this experiment. Repeat any tests that appear ambiguous, changing the order in which reagents are added if necessary. List the seven species, in order of increasing stability. Then carry out the necessary tests on your unknown to establish its position in the list. The unknown may be one of the Cu(II) species you have studied, or it may be different. If the unknown contains a precipitate, stir it up before making a test.

DISPOSAL OF REACTION PRODUCTS.   Pour the contents of all of the wells into the waste crock unless directed otherwise by your instructor.

Experiment 27

Observations and Conclusions: Relative Stabilities of Complex Ions
and Precipitates Prepared from
Solutions of Copper(II)

Table of Observations

	NH_3	Cl^-	OH^-	$C_2O_4^{2-}$	PO_4^{3-}	$Cu(OH)_2$	S^{2-}	Unknown
S^{2-}								
$Cu(OH)_2$								
PO_4^{3-}								
$C_2O_4^{2-}$								
OH^-								
Cl^-								
NH_3								

(continued on following page)

Determination of Relative Stabilities

In each row of the table you can compare the stabilities of species involving the reagent in the horizontal row with those of the species containing the reagent initially added. In the first row of the table, the copper(II)-NH_3 species can be seen to be more stable than some of the species obtained by addition of the other reagents, and less stable than others. Examining each row, make a list of all the complex ions and precipitates you have in the table in order of increasing stability and formation constant.

Reasons

Lowest _____ _____

_____ _____

_____ _____

_____ _____

_____ _____

_____ _____

Highest _____ _____

Stability of Unknown

Indicate the position your unknown would occupy in the above list.

Reasons:

Unknown no._____

Experiment 27

Advance Study Assignment: Relative Stabilities of Complex Ions
and Precipitates Prepared from
Solutions of Copper(II)

1. In testing the relative stabilities of Cu(II) species using a well plate, a student adds 6 drops 1 M NH_3 to 6 drops 0.1 M $Cu(NO_3)_2$. He observes that a blue precipitate initially forms, but that in excess NH_3 the precipitate dissolves and the solution turns blue. Addition of 6 drops 1 M NaOH to the dark-blue solution results in the formation of a blue precipitate.

   a. What is the formula of Cu(II) species in the dark-blue solution?

   b. What is the formula of the blue precipitate present after addition of 1 M NaOH?

   c. Which species is more stable in equal concentrations of NH_3 and OH^-, the one in Part a or the one in Part b?

   _____

2. Given the following two reactions and their equilibrium constants:

   $$Cu(H_2O)_4^{2+} (aq) + 4\ NH_3(aq) \rightleftharpoons Cu(NH_3)_4^{2+} (aq) + 4\ H_2O \quad K_1 = 5 \times 10^{12}$$

   $$Cu(H_2O)_4^{2+} (aq) + CO_3^{2-} \rightleftharpoons CuCO_3(s) + 4\ H_2O \qquad\qquad K_2 = 7 \times 10^9$$

   a. Evaluate the equilibrium constant for the reaction (see discussion):

   $$CuCO_3(s) + 4\ NH_3(aq) \rightleftharpoons Cu(NH_3)_4^{2+}(aq) + CO_3^{2-} (aq)$$

   _____

   b. If 1 M NH_3 were added to some solid $CuCO_3$ in a test tube containing 1 M Na_2CO_3, what, if anything, would happen? Explain your reasoning.

Experiment 28

Determination of the Hardness of Water

One of the factors that establishes the quality of a water supply is its degree of hardness. The hardness of water is defined in terms of its content of calcium and magnesium ions. Since the analysis does not distinguish between Ca^{2+} and Mg^{2+}, and since most hardness is caused by carbonate deposits in the earth, hardness is usually reported as total parts per million calcium carbonate by weight. A water supply with a hardness of 100 parts per million would contain the equivalent of 100 grams of $CaCO_3$ in 1 million grams of water or 0.1 gram in one liter of water. In the days when soap was more commonly used for washing clothes, and when people bathed in tubs instead of using showers, water hardness was more often directly observed than it is now, since Ca^{2+} and Mg^{2+} form insoluble salts with soaps and make a scum that sticks to clothes or to the bathtub. Detergents have the distinct advantage of being effective in hard water, and this is really what allowed them to displace soaps for laundry purposes.

Water hardness can be readily determined by titration with the chelating agent EDTA (ethylenediaminetetraacetic acid). This reagent is a weak acid that can lose four protons on complete neutralization; its structural formula is

$$HOOC-CH_2 \qquad\qquad\qquad CH_2-COOH$$
$$N-CH_2-CH_2-N$$
$$HOOC-CH_2 \qquad\qquad\qquad CH_2-COOH$$

The four acid sites and the two nitrogen atoms all contain unshared electron pairs, so that a single EDTA ion can form a complex with up to six sites on a given cation. The complex is typically quite stable, and the conditions of its formation can ordinarily be controlled so that it contains EDTA and the metal ion in a 1:1 mole ratio. In a titration to establish the concentration of a metal ion, the EDTA which is added combines quantitatively with the cation to form the complex. The end point occurs when essentially all of the cation has reacted.

In this experiment we will standardize a solution of EDTA by titration against a standard solution made from calcium carbonate, $CaCO_3$. We will then use the EDTA solution to determine the hardness of an unknown water sample. Since both EDTA and Ca^{2+} are colorless, it is necessary to use a rather special indicator to detect the end point of the titration. The indicator we will use is called Eriochrome Black T, which forms a rather stable wine-red complex, $MgIn^-$, with the magnesium ion. A tiny amount of this complex will be present in the solution during the titration. As EDTA is added, it will complex free Ca^{2+} and Mg^{2+} ions, leaving the $MgIn^-$ complex alone until essentially all of the calcium and magnesium have been converted to chelates. At this point EDTA concentration will increase sufficiently to displace Mg^{2+} from the indicator complex; the indicator reverts to an acid form, which is sky blue, and this establishes the end point of the titration.

The titration is carried out at a pH of 10, in an $NH_3-NH_4^+$ buffer, which keeps the EDTA (H_4Y) mainly in the form, HY^{3-}, where it complexes the Group 2 ions very well but does not tend to react as readily with other cations such as Fe^{3+} that might be present as impurities in the water. Taking H_4Y and H_3In as the formulas for EDTA and Eriochrome Black T, respectively, the equations for the reactions which occur during the titration are:

$$\text{(main reaction) } HY^{3-}(aq) + Ca^{2+}(aq) \rightarrow CaY^{2-}(aq) + H^+(aq) \text{ (same for } Mg^{2+})$$

$$\text{(at end point) } HY^{3-}(aq) + MgIn^-(aq) \rightarrow MgY^{2-}(aq) + HIn^{2-}(aq)$$
$$\text{wine red} \qquad\qquad\qquad \text{sky blue}$$

Since the indicator requires a trace of Mg^{2+} to operate properly, we will add a little magnesium ion to each solution and titrate it as a blank.

Experimental Procedure

Obtain a 50-mL buret, a 250-mL volumetric flask, and 25- and 50-mL pipets from the stockroom.

Put about a half gram of calcium carbonate in a small 50-mL beaker and weigh the beaker and contents on the analytical balance. Using a spatula, transfer about 0.4 g of the carbonate to a 250-mL beaker and weigh again, determining the mass of the $CaCO_3$ sample by difference.

Add 25 mL of distilled water to the large beaker and then, *slowly*, about 40 drops of 6 M HCl. Cover the beaker with a watch glass and allow the reaction to proceed until all of the solid carbonate has dissolved. Rinse the walls of the beaker down with distilled water from your wash bottle and heat the solution until it just begins to boil. (Be sure not to be confused by the evolution of CO_2 that occurs with the boiling.) Add 50 mL of distilled water to the beaker and carefully transfer the solution, using a stirring rod as a pathway, to the volumetric flask. Rinse the beaker several times with small portions of distilled water and transfer each portion to the flask. All of the Ca^{2+} originally in the beaker should then be in the volumetric flask. Fill the volumetric flask to the horizontal mark with distilled water, adding the last few mL a drop at a time with your wash bottle. Stopper the flask and mix the solution thoroughly by inverting the flask at least 20 times over a period of several minutes.

Clean your buret thoroughly and draw about 200 mL of the stock EDTA solution from the carboy into a dry 250-mL Erlenmeyer flask. Rinse the buret with a few mL of the solution at least three times. Drain through the stopcock and then fill the buret with the EDTA solution.

Determine a blank by adding 25 mL distilled water and 5 mL of the pH 10 buffer to a 250-mL Erlenmeyer flask. Add a small amount of solid Eriochrome Black T indicator mixture from the stock bottle. You need only a small portion, about 25 mg, just enough to cover the end of a small spatula. The solution should turn blue; if the color is weak, add a bit more indicator. Add 15 drops 0.03 M $MgCl_2$, which should contain enough Mg^{2+} to turn the solution wine red. Read the buret to 0.02 mL and add EDTA to the solution until the last tinge of purple just disappears. The color change is rather slow, so titrate slowly near the end point. Only a few mL will be needed to titrate the blank. Read the buret again to determine the volume required for the blank. This volume must be subtracted from the total EDTA volume used in each titration. Save the solution as a reference for the end point in all your titrations.

Pipet three 25-mL portions of the Ca^{2+} solution in the volumetric flask into clean 250-mL Erlenmeyer flasks. To each flask add 5 mL of the pH 10 buffer, a small amount of indicator, and 15 drops of 0.03 M $MgCl_2$. Titrate the solution in one of the flasks until its color matches that of your reference solution; the end point is a reasonably good one, and you should be able to hit it within a few drops if you are careful. Read the buret. Refill the buret, read it, and titrate the second solution, then the third.

Your instructor will furnish you a sample of water for hardness analysis. Since the concentration of Ca^{2+} is probably lower than that in the standard calcium solution you prepared, pipet 50 mL of the water sample for each titration. As before, add some indicator, 5 mL of pH 10 buffer, and 15 drops of 0.03 M $MgCl_2$ before titrating. Carry out as many titrations as necessary to obtain two volumes of EDTA that agree within about 3%. If the volume of EDTA required in the first titration is low due to the fact that the water is not very hard, increase the volume of the water sample so that in succeeding titrations, it takes at least 20 mL of EDTA to reach the end point.

Optional Experiment: Find the mass percent calcium carbonate in a Tums tablet or a sample of limestone.

DISPOSAL OF REACTION PRODUCTS. The chemical waste from this experiment may be diluted with water and poured down the sink.

Experiment 28

Data and Calculations: Determination of the Hardness of Water

Mass of beaker plus $CaCO_3$	_____ g	Volume Ca^{2+} solution prepared	_____ mL
Mass of beaker less sample	_____ g	Molarity of Ca^{2+}	_____ M
Mass of $CaCO_3$ sample	_____ g	Moles Ca^{2+} in each aliquot titrated	_____ moles
Number of moles $CaCO_3$ in sample (Formula mass = 100.1)	_____ moles		

Standardization of EDTA Solution

Determination of blank:

Initial buret reading	_____ mL	Final buret reading	_____ mL	Volume of blank	_____ mL

Titration	I	II	III
Initial buret reading	_____ mL	_____ mL	_____ mL
Final buret reading	_____ mL	_____ mL	_____ mL
Volume of EDTA	_____ mL	_____ mL	_____ mL
Volume EDTA used to titrate blank	_____ mL	_____ mL	_____ mL
Volume EDTA used to titrate Ca^{2+}	_____ mL	_____ mL	_____ mL

Average volume of EDTA required to titrate Ca^{2+} _____ mL

$$\text{Molarity of EDTA} = \frac{\text{no. moles } Ca^{2+} \text{ in aliquot} \times 1000}{\text{average volume EDTA required (mL)}} = \underline{\qquad} \text{ M}$$

(continued on following page)

Determination of Water Hardness

Titration	I	II	III
Volume of water used	_____ mL	_____ mL	_____ mL
Initial buret reading	_____ mL	_____ mL	_____ mL
Final buret reading	_____ mL	_____ mL	_____ mL
Volume of EDTA	_____ mL	_____ mL	_____ mL
Volume of EDTA used to titrate blank	_____ mL	_____ mL	_____ mL
Volume of EDTA required to titrate water	_____ mL	_____ mL	_____ mL
Average volume EDTA per liter of water	_____ mL		

No. moles EDTA per liter water _____ = No. moles $CaCO_3$ per liter water

No. grams $CaCO_3$
per liter water _____ g

Water hardness
(1 ppm = 1 mg/liter) _____ ppm $CaCO_3$

Unknown no. _____

Experiment 28

Advance Study Assignment: Determination of the Hardness of Water

1. A 0.4243-g sample of $CaCO_3$ is dissolved in 12 M HCl and the resulting solution is diluted to 250.0 mL in a volumetric flask.

    a. How many moles of $CaCO_3$ are used (formula mass = 100.1)?

    _____ moles

    b. What is the molarity of the Ca^{2+} in the 250 mL of solution?

    _____ M

    c. How many moles of Ca^{2+} are in a 25.00-mL aliquot of the solution in 1b?

    _____ moles

2. 25.00-mL aliquots of the solution from Problem 1 are titrated with EDTA to the Eriochrome Black T end point. A blank containing a small measured amount of Mg^{2+} requires 3.21 mL of the EDTA to reach the end point. An aliquot to which the same amount of Mg^{2+} is added requires 24.95 mL of the EDTA to reach the end point.

    a. How many milliliters of EDTA are needed to titrate the Ca^{2+} ion in the aliquot?

    _____ mL

    b. How many moles of EDTA are there in the volume obtained in Part a?

    _____ moles

    c. What is the molarity of the EDTA solution?

    _____ M

(continued on following page)

3. A 100-mL sample of hard water is titrated with the EDTA solution in Problem 2. The same amount of Mg^{2+} is added as previously, and the volume of EDTA required is 31.84 mL.

a. What volume of EDTA is used in titrating the Ca^{2+} in the hard water?

_____ mL

b. How many moles of EDTA are there in that volume?

_____ moles

c. How many moles of Ca^{2+} are there in the 100 mL of water?

_____ moles

d. If the Ca^{2+} comes from $CaCO_3$, how many moles of $CaCO_3$ are there in one liter of the water? How many grams of $CaCO_3$ are present per liter of the water?

_____ moles

_____ g

e. If 1 ppm $CaCO_3$ = 1 mg per liter, what is the water hardness in ppm $CaCO_3$?

_____ ppm $CaCO_3$

Experiment 29

Synthesis and Analysis of a Coordination Compound

Some of the most interesting research in inorganic chemistry has involved the preparation and properties of coordination compounds. These compounds, sometimes called *complexes*, are typically salts that contain complex ions. A complex ion contains a central metal atom to which are bonded small polar molecules or anions, called *ligands*. The bonding in the complex is through coordinate covalent bonds, bonds in which the electrons are all furnished by the ligands.

Some coordination compounds can be prepared stoichiometrically pure. That is, the atoms in the compound are present in the exact ratio given by the formula. Many supposedly pure compounds do not meet this criterion, since they contain variable amounts of water, either in hydrates or adsorbed. In this experiment we will be making a compound that is indeed stoichiometrically pure. It is not a hydrate nor does it tend to adsorb water.

The compound we will be preparing contains cobalt ion, ammonia, and chloride ions. Its formula is of the form $Co_x(NH_3)_yCl_z$. Once you have made the compound, you may, depending on the time available, analyze it for its content of one or more of these species and so determine x, y, and/or z.

In many reactions involving complex ion formation the rate of reaction is very fast, so that the thermodynamically stable form of the complex is the one produced. By changing ligand concentrations, one can quickly convert from one complex to another. In Experiment 27, in which many Cu(II) complexes are prepared, all of the complex ions undergo rapid reaction, exchanging ligands very readily. Such complexes are called *labile*. With Cu(II) ion and NH_3 in solution, one can, depending on the NH_3 concentration, have one, two, three, or four NH_3 molecules in the complex, in an equilibrium mixture.

Most, but by no means all, complexes are labile. Some, including the one you will be making, exchange ligands only very slowly. For such species, the preparation reaction takes time, but once formed, the complex ion in solution may resist change rather dramatically. Such complexes are called *inert*.

In the complex ion you will be making, NH_3 and Cl^- may be ligands. Since the overall charge of the compound must be zero, the number of moles of Cl^- ion in a mole of compound will be determined by the charge on the cobalt ion. Some of the Cl^- ions may be in the complex and some may simply be anions in the crystal. When the compound is dissolved in water, the free anions will go into solution, but the ones in the complex will remain firmly attached to cobalt. They will not react with a precipitating agent such as Ag^+ ion, which will tend to form AgCl with any free Cl^- ions. All of the NH_3 molecules will be in the complex ion, and will not react with added H^+ ion, as they would if they were free. To ensure that we find all the Cl^- and NH_3 that is present in the compound, we will destroy the complex before attempting any analyses.

The complete analysis involves a gravimetric procedure for chloride ion, a colorimetric method for cobalt ion, and a volumetric procedure for ammonia. In some of the earlier experiments we have used these techniques (Exps. 4, 7, 23, 24, and 28), so you may be somewhat familiar with them. From the analyses we can determine the number of moles of each species present in a mole of compound. Because the mole ratio must equal the atom or molecule ratio, we can determine the formula of the compound. Although we describe the procedures for analysis of all three species in the compound, the formula can be determined from only two analyses, since the amount of the third species can be found by difference.

The synthesis of the coordination compound will take one lab period. The analysis for both chloride and cobalt content can be done in one period. Determining the amount of ammonia present in the complex will take most of one period.

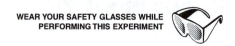

Experimental Procedure

Carry out the procedure in either A.1 or A.2, as directed by your instructor.

A.1 Preparation of $Co_x(NH_3)_yCl_z$ (Procedure I)

Using a piece of weighing paper and a top-loading balance, weigh out 10 ± 0.2 g of ammonium chloride, NH_4Cl. Pour the NH_4Cl into a 250-mL beaker and add 40 mL distilled water. To this add 8 ± 0.2 g of cobalt(II) chloride hexahydrate, $CoCl_2 \cdot H_2O \cdot$ and 0.8 g activated charcoal (Norite). Heat this mixture to the boiling point, stirring to dissolve the soluble components.

Cool the beaker in water, then in an ice bath. Slowly, in a hood, add 40 mL 15 M NH_3, ammonia. **CAUTION:**   *This is a concentrated basic reagent, with a strong pungent odor that can knock you unconscious.* After stirring, add 50 mL 10% H_2O_2, hydrogen peroxide, slowly, a few mL at a time, while continuing to stir. **CAUTION:**   *Be very careful not to spill this reagent on your skin. Use gloves.*

When the bubbling has stopped, place the 250-mL beaker in a 600-mL beaker containing 100 mL of water at about 60°C. Leave the beaker in the water bath for 30 to 40 minutes, holding the temperature of the bath at about 60°C as long as the liquid has a pink color. Stir occasionally.

While the mixture is reacting, prepare a fritted glass filter crucible. Fit the crucible on a filter flask; clean the crucible by pulling, under suction, about 15 mL 1 M HNO_3 through the disk, followed by 50 to 75 mL of distilled water. Put the crucible in an oven at 150°C for about an hour to dry. Prepare a second crucible the same way, and put it in the oven.

When the 40 minutes are up, remove the 250-mL beaker from the water bath and cool it, first in a water bath and then in an ice bath, until the liquid in the beaker has a temperature below 5°C. Hold the mixture at this temperature for at least 5 minutes, stirring occasionally to promote crystallization of the crude product. Set up a Büchner funnel and, with suction, filter the mixture through the filter paper in the funnel. Discard the filtrate in the waste crock.

Scrape the solid from the filter paper into the 250-mL beaker. Add 100 mL distilled water and, in a hood, 4 mL 12 M HCl. **CAUTION:**   *This reagent is a concentrated acid with a choking odor.* Heat the mixture to boiling, with stirring, to dissolve the crystals. Rinse out the suction flask with distilled water, and reassemble the Büchner funnel. Holding the beaker with a pair of tongs or a folded paper towel, filter the hot mixture through the paper in the funnel, under suction. In this operation, the charcoal is removed and should be on the filter paper. The filtrate should be golden yellow and contains the product we seek.

Transfer the filtrate to the (rinsed-out) 250-mL beaker and, in the hood, add 15 mL 12 M HCl. Place the beaker in an ice bath and stir for 5 minutes or so to promote formation of the golden crystals of the product, $Co_x(NH_3)_yCl_z$. Check to see that the temperature is below 5°C.

Pour about 50 mL of ice-cold distilled water on to a piece of filter paper in the Büchner funnel. After 2 or 3 minutes, turn on the suction and pull the water into the suction flask. Empty the flask. Then filter the cold mixture containing the product through the funnel, using suction. Turn the suction off, and pour about 20 mL 95% ethanol onto the crystals. Wait about 10 seconds, and then turn on the suction to pull through the ethanol, which should carry with it most of the water and HCl remaining on the crystals. Draw air through the crystals for several minutes. Weigh a 100-mL beaker to 0.1 g. Transfer the crystals from the filter paper to the beaker. Put the beaker in your locker. Take the fritted glass crucibles out of the oven, using tongs or a folded paper towel, and put the crucibles on a paper towel in your locker. Discard the liquid in the suction flask in the waste crock or as directed by your instructor.

A.2 Alternative Procedure for Preparation of $Co_x(NH_3)_yCl_z$ (Procedure II)

There are several possible $Co_x(NH_3)_yCl_z$ compounds. The following procedure will allow you to make a different one from that produced in Part A.1 of this experiment.

Using a piece of weighing paper and a top-loading balance, weigh out 4.0 ± 0.1 g of ammonium chloride, NH_4Cl. Transfer the NH_4Cl into a 250-mL beaker, and **in the hood,** slowly, add 25 mL of 15 M NH_3. **CAUTION:** *This is a concentrated basic reagent, with a strong pungent odor that can knock you unconscious.* Cover the beaker with a watch glass. Add to the solution 8.0 ± 0.2 g of $CoCl_2.6H_2O$, cobalt(II) chloride hexahydrate, while stirring the mixture with a stirring rod. Continue stirring until most of the solid has dissolved and a slurry of fine tan solid in a deep red solution results.

Measure out 20 mL of 10% hydrogen peroxide, H_2O_2. **CAUTION:** *This reagent can cause severe skin burns; use gloves when handling it.* **In the hood,** add the H_2O_2 to the mixture in 1-2 mL portions while stirring. There will be some effervescence (bubbling) as oxygen gas is evolved. Stir until bubbles of oxygen decrease significantly and all of the solid material dissolves. The initial Co^{2+} has now been oxidized to Co^{3+} and is present in the form of a complex ion. **In the hood,** measure out 25 mL of 12 M HCl. **CAUTION:** *This is a concentrated acidic reagent with a choking odor.* Add the HCl in approximately 5 mL increments while stirring the mixture. The white cloud that forms is solid ammonium chloride, formed by a gaseous acid-base neutralization reaction, dispersed in the air.

Preferably in the hood, or at your lab bench, heat the mixture in the beaker to 75 to 85°C for about 30 minutes, stirring occasionally. Keep the beaker covered with your watch glass. Control the temperature by judicious use of your Bunsen burner. A deep purple solid should gradually form in the beaker. This is your product, which may need to be recrystallized.

While the mixture is being heated, prepare two fritted glass crucibles, according to the directions in the 4th paragraph of Part A.1 of this experiment.

Remove heat from the beaker after 30 minutes and allow it to cool in a cool water bath—a 400 mL beaker containing 150 mL of cold tap water—for two minutes. Set up a vacuum filter apparatus, using a 2-part Büchner funnel. Transfer the smaller beaker to an ice bath. Keep the reaction beaker in the bath until the solution is colder than 10°C.

Using #1 filter paper that has been moistened while in the Büchner funnel, filter the solution, as demonstrated by your instructor. It is not necessary to transfer all of the solid. Knock or scrape the contents of the filter cup, including the paper, back into the 250 mL beaker. Rinse the cup with a few mL of water from your wash bottle. Similarly, using tweezers, rinse any remaining solid from the filter paper into the beaker, and dispose of the paper as directed by your instructor. The filtrate solution should be transferred to a designated waste container.

It may be desirable to recrystallize your product to ensure its purity. If so advised by your lab instructor, and time is available, proceed as follows:

Add water to the beaker from your wash bottle to bring the volume up to 125 mL, using the markings on the beaker. Stir the solution and take the beaker to the hood. **In the hood,** add 5 mL of 15 M NH_3 to the beaker, in small portions, while stirring. Heat gently, if necessary, to dissolve all of the solid. Add 15 mL of 12 M HCl, in small portions, to the reaction beaker, while stirring. Preferably in the hood, or at your lab bench, heat the solution for 30 minutes as before but now **at approximately 60°C,** stirring occasionally and using a watch glass to cover the beaker. At the end of the heating period, place the beaker in a cool water bath (as above) for two minutes. Cool further in an ice bath with occasional stirring, until the temperature falls below 10°C, and for an additional five minutes. You have now recrystallized the desired cobalt-containing product, which should be a violet color.

Reassemble your Büchner filter apparatus. To transfer as much of the product as possible, stir the mixture with your stirring rod just before quickly pouring out small portions of the suspension. Use a minimal amount of water from your wash bottle to transfer the remaining solid. Wash the product on the funnel with two five mL portions of water to aid in removing ammonium chloride. Wash with 10 mL of ethanol. Draw air through your product for five minutes, while weighing a watch glass to ± 0.01 g on a top-loading balance.

Transfer the crystals to the watch glass and scrape any residual product from the filter paper onto the watch glass. Weigh the watch glass again and calculate your yield. Place the watch glass in your drawer until next lab period, at which time you will be analyzing your product. Take the fritted glass crucibles out of the oven, using tongs or a folded paper towel. Put the crucibles, after cooling, in your drawer on a paper towel.

DISPOSAL OF REACTION PRODUCTS.   Discard the liquid in the suction flask into the waste container or as otherwise directed by your instructor.

B. Gravimetric Determination of Chloride Content

Weigh your prepared compound in the 100-mL beaker to 0.0001 g on an analytical balance. (If a week has gone by, the compound should be well dried. If you are attempting to proceed on the same day as you prepared the compound, you must dry it thoroughly. This could be done by putting the sample under vacuum, while keeping the sample warm. If the sample does not lose 0.0001 g in 2 minutes while on the analytical balance, you may presume that it is dry.)

Transfer 0.3 ± 0.05 g of the compound from the beaker to a labeled 250-mL beaker, and reweigh the 100-mL beaker to 0.0001 g. Transfer a second sample of about the same size to another labeled 250-mL beaker, and again weigh the 100-mL beaker and the remaining compound accurately. Record all masses on the Data page. Cover the 100-mL beaker with a watch glass and put it in your locker.

Add 25 mL of distilled water to the sample in one of the 250-mL beakers. Add 5 mL 6 M NaOH and stir the mixture until all of the solid has dissolved. Heat the beaker and its contents to boiling, and simmer for 3 minutes, stirring occasionally. This step will destroy the complex; the mixture will turn black because of the formation of Co_3O_4.

Once the black Co_3O_4 is formed, turn off the heat and let the beaker cool for a few minutes. Then add 6 M HNO_3 until the mixture is acidic to litmus (it will take about 5 mL). Add 1 mL more of the HNO_3. Add a small scoop of solid Na_2SO_3, sodium sulfite (approximately 0.2 g) and stir; this will reduce the Co_3O_4 and produce the Co^{2+} ion. The solution should clarify and turn pink. Wash down the walls of the beaker with water to bring any unreacted dark material into contact with the solution. Reheat to boiling and simmer gently for a minute or so. Then add 75 mL of distilled water.

Slowly, with stirring, add 50 mL 0.1 M $AgNO_3$, silver nitrate. This will precipitate all of the chloride ion in the solution as silver chloride, AgCl. Heat the mixture, with occasional stirring, to the boiling point. This will help coagulate the AgCl crystals and so facilitate filtration. Remove the beaker from the wire gauze and cover with a watch glass. Let it cool for 10 minutes or so.

While the beaker is cooling, take one of the fritted glass crucibles you prepared in the previous session from your locker and weigh it to 0.0001 g on the analytical balance. Then set up the filter crucible apparatus.

Carefully, with suction on, transfer *all* of the AgCl precipitate to the filter crucible. Use your wash bottle and rubber policeman to complete the transfer. Wash the AgCl, with suction on, with 10 mL H_2O, followed by 5 mL 6 M HNO_3, followed by 20 mL H_2O. Keep the suction on for a minute after the last liquid has been pulled through the filter.

Put the crucible and its contents in an oven at 150°C for at least an hour. Then let it cool to room temperature in a covered beaker. When it is cool, weigh it to 0.0001 g on the analytical balance. Put the solid AgCl in the AgCl crock.

Repeat the entire procedure, using the second sample of compound.

C. Colorimetric Determination of Cobalt Content

Transfer 0.5 ± 0.05 g of your compound into a clean, dry 50-mL beaker, using the same procedure as in the chloride determination. Weighings should be made to 0.0001 g. Make another transfer of about 0.5 g into another 50-mL beaker, again weighing the 100-mL beaker and its contents accurately.

Cover each beaker with a watch glass and heat one of them gently with the Bunsen burner until the solid melts, foams, and turns blue. In this process the complex is destroyed and cobalt ion freed from the ligands. Allow the beaker and its contents to cool. Repeat the procedure with the sample in the other beaker.

To the solid in the first beaker add 5 mL 6 M H_2SO_4 and 5 mL distilled water. Heat the beaker until the liquid begins to boil and the solid is all dissolved. Wash any deposits on the watch glass into the beaker with a few milliliters of water from your wash bottle. Repeat these steps with the sample in the other beaker.

Transfer the solution from the first beaker to a clean 25-mL volumetric flask. Rinse the beaker with small portions of distilled water, adding the washings to the volumetric flask, so that all of the solution goes into the flask. Fill the flask to the mark with distilled water from your wash bottle. Stopper the flask and invert it at least twenty times to ensure that the solution inside becomes thoroughly mixed.

Pour some of the solution in the flask into a spectrophotometric test tube (3/4 full). Measure and record the absorbance of the solution at 510 nm. From the absorbance and the calibration curve furnished by your instructor, determine the molarity of cobalt ion in the solution.

DISPOSAL OF REACTION PRODUCTS.   Pour the rest of the solution in the volumetric flask in the waste crock.

Rinse the flask, and use it with the solution you prepared in the other beaker. Treat that solution as you did the first one, recording its absorbance and molarity of cobalt ion on the Data page.

D. Volumetric Determination of Ammonia Content

In this part of the experiment we will first decompose the complex, releasing NH_3 to the solution, as we did in the chloride analysis. We then distill off the NH_3 into another container, and titrate it against an acid. The procedure is called a Kjeldahl analysis, and is used for determining nitrogen content of many organic and inorganic substances.

Assemble the distillation apparatus shown in Figure 29.1. The details of the apparatus will vary from school to school but you will need a 250-mL distilling flask, a distilling head, a condenser, and a receiver adapter, in addition to a 125-mL Erlenmeyer flask, which will serve as a receiving flask. The 250-mL flask should be on a wire gauze on an iron ring. The flask and condenser should be held with clamps, which are to be adjusted so that all joints are tight. On the end of the receiver adapter attach a 4-inch length of latex tubing, which should reach to the bottom of the receiving flask.

Put 50 mL saturated boric acid solution in the 125-mL receiving flask; add five drops bromcresol green indicator, and adjust the level of the flask, if necessary, so that the latex tubing is well under the liquid surface.

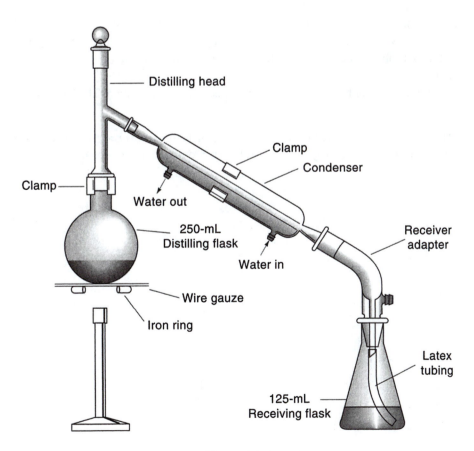

Figure 29.1

Weigh out 1 ± 0.1 g of your compound into a beaker, as you did earlier, making the weighing to 0.0001 g. Disconnect the distilling flask from the head. Transfer the sample to the distilling flask, using a funnel. Rinse the beaker several times with small portions of distilled water, and add the washings to the flask. All the sample must end up in the flask. Add distilled water as necessary to give a final volume of about 50 mL. Swirl the flask to dissolve the sample. Add a few boiling chips and several pieces of granulated zinc. Start water flowing through the condenser, slowly. If you are working with standard taper glassware, lightly grease the lower joint of the distilling head.

Pour 40 mL 6 M NaOH and 50 mL distilled water into the flask and reconnect it promptly to the distilling head. Make sure that all joints are tight, and that the top of the distilling head is stoppered. Turn on the Bunsen burner and bring the liquid in the flask to a boil. The complex will be destroyed, the liquid will turn dark as Co_3O_4 forms, and NH_3 will be driven from the solution into the receiving flask.

Adjust the burner as necessary to maintain smooth boiling and minimize bumping. The color of the indicator will change as the NH_3 is absorbed. There may be a tendency for liquid to rise in the latex tubing during the distillation; this is not serious, but if the level goes up more than a few inches, you can let in a little air by momentarily loosening one of the joints. When the volume of liquid in the receiving flask reaches 100 mL, 50 mL of the solution will have been distilled and essentially all of the NH_3 driven into the boric acid.

With the burner still heating the flask, disconnect the receiver adapter from the condenser. (If the heat is turned off first, liquid will tend to back up into the condenser.) Rinse the adapter with the tubing still in the flask, washing any liquid on the inner or outer surface into the distillate. Then turn the burner off. Pour the distillate into a 250-mL volumetric flask. Rinse the receiving flask with a few small portions of distilled water, adding the washings to the volumetric flask. Fill to the mark with distilled water. Stopper the flask and invert it at least 20 times to ensure that the liquid is thoroughly mixed.

Clean two burets. Rinse one of them with several small portions of the NH_3 solution in the volumetric flask, draining some of the solution through the stopcock. Fill the buret with the NH_3 solution and read and record the level. Rinse the other buret with small portions of the standardized HCl solution from the stock supply, and then fill that buret with that solution. Record the level.

Draw about 40 mL of the NH_3 solution into a 250-mL Erlenmeyer flask, and add five drops of bromcresol green indicator. Titrate with HCl to the end point, where the indicator changes from blue to yellow. The actual end point is green and can be reached by back titrating as necessary with NH_3. Record the final levels in the NH_3 and HCl burets.

Repeat the titration once or twice with 40-mL portions of the NH_3 solution. The NH_3-HCl volume ratios should check within 1% if all goes well.

DISPOSAL OF REACTION PRODUCTS.   The titrated solutions may be poured down the sink. The liquid remaining in the distilling flask should be discarded in the waste crock.

Experiment 29

Data and Calculations: Synthesis and Analysis of a
Coordination Compound

A.1 or A.2 Preparation of $Co_x(NH_3)_yCl_z$

Procedure used _____

Mass of $CoCl_2 \cdot 6 H_2O$ _____ g

Mass of 100-mL beaker _____ g

Approximate mass of product (see Part B, line 1) _____ g

B. Determination of Chloride Ion

Mass of 100-mL beaker plus product _____ g

Mass after first transfer _____ g Mass of sample I _____ g

Mass after second transfer _____ g Mass of sample II _____ g

Mass of first filter crucible _____ g

Mass of second filter crucible _____ g

Mass of first crucible plus AgCl _____ g Mass of AgCl I _____ g

Mass of second crucible plus AgCl _____ g Mass of AgCl II _____ g

C. Determination of Cobalt Ion

Mass of 100-mL beaker plus product _____ g

Mass after first transfer _____ g Mass of sample I _____ g

Mass after second transfer _____ g Mass of sample II _____ g

Absorbance of first solution _____

Molarity of Co ion _____ M

Absorbance of second solution _____

Molarity of Co ion _____ M

D. Determination of Ammonia

Mass of 100-mL beaker plus product _____ g

Mass after transfer _____ g **2nd Trial**

Initial reading NH_3 buret _____ mL _____ mL

Final reading NH_3 buret _____ mL _____ mL

Initial reading HCl buret _____ mL _____ mL

Final reading HCl buret _____ mL _____ mL

Molarity of standardized HCl _____ M

(continued on following page)

E. Calculation of Chloride Content

	Sample I	Sample II
Mass of AgCl	_____ g	_____ g
Mass of Cl^- in AgCl	_____ g	_____ g
Mass of sample	_____ g	_____ g
Mass of Cl^- per 100-g sample	_____ g	_____ g
Moles Cl^- per 100-g sample	_____	_____

F. Calculation of Cobalt Content

	Sample I	Sample II
Molarity of cobalt ion	_____ M	_____ M
Moles cobalt in 25 mL		
= moles cobalt in sample	_____	_____
Mass of sample	_____ g	_____ g
Moles cobalt per 100-g sample	_____	_____

G. Calculation of Ammonia Content

Mass of sample _____ g   Molarity of HCl _____ M

	Trial I	Trial II
Volume NH_3 used	_____ mL	_____ mL
Volume HCl used	_____ mL	_____ mL
Moles HCl = moles NH_3	_____	_____
Moles NH_3 per 250-mL NH_3 solution		
= moles NH_3 in sample	_____	_____
Moles NH_3 per 100-g sample	_____	_____

H. Determination of the Formula of $Co_x(NH_3)_yCl_z$

For a 100-g sample: Moles Cl^- _____   Moles Co ion _____   Moles NH_3 _____

Whole number ratio: z = _____   x = _____   y = _____

Formula of compound:

Experiment 29

Advance Study Assignment: Synthesis and Analysis of a
Coordination Compound

A student prepared 7.2 g of $Co_x(NH_3)_yCl_z$. She then analyzed the compound by the procedure in this experiment.

A. In the gravimetric determination of chloride, she weighed out 0.2988 g of the compound. The following data were obtained:

Mass of crucible plus AgCl	19.0020 g	
Mass of crucible	18.4628 g	
Mass of AgCl	_____ g	MM AgCl = 143.323 g
Mass of Cl⁻ in AgCl	_____ g	MM Cl⁻ = 35.453 g
Moles Cl⁻ in sample	_____	
Moles Cl⁻ in 100-g sample	_____	

B. In the colorimetric determination of cobalt, she used a sample weighing 0.5120 g. The molarity of cobalt ion in the solution from the volumetric flask was 0.085 M.

Moles cobalt ion in 25 mL solution = moles cobalt in sample   _____

Moles cobalt per 100-g sample   _____

C. In the volumetric determination of ammonia, the sample weighed 0.9985 g. In the titration, 0.1000 M HCl was used. She found that 40.00 mL of the NH_3 solution required 39.62 mL of the HCl to reach the end point.

No. moles HCl used   _____ = no. moles NH_3 in 40 mL NH_3 solution

No. moles NH_3 in 250 mL NH_3 _____ = no. moles NH_3 in sample

No. moles NH_3 per 100-g sample   _____

D. Calculation of the formula of $Co_x(NH_3)_yCl_z$:

In 100 g of sample,   no. moles Co ion = _____

no. moles NH_3 = _____

no. moles Cl⁻ = _____

Dividing by smallest number,   no. moles Co ion _____

no. moles NH_3 _____

no. moles Cl⁻ _____

Formula of compound _____

(This may or may not be the formula of the compound you will be making!)

Experiment 30

Determination of Iron by Reaction with Permanganate—A Redox Titration

Potassium permanganate, $KMnO_4$, is widely used as an oxidizing agent in volumetric analysis. In acid solution, MnO_4^- ion undergoes reduction to Mn^{2+} as shown in the following equation:

$$8\ H^+(aq) + MnO_4^-(aq) + 5\ e^- \rightarrow Mn^{2+}(aq) + 4\ H_2O$$

Since the MnO_4^- ion is violet and the Mn^{2+} ion is nearly colorless, the end point in titrations using $KMnO_4$ as the titrant can be taken as the first permanent pink color that appears in the solution.

$KMnO_4$ will be employed in this experiment to determine the percentage of iron in an unknown containing iron(II) ammonium sulfate, $Fe(NH_4)_2(SO_4)_2 \cdot 6\ H_2O$. The titration, which involves the oxidation of Fe^{2+} ion to Fe^{3+} by permanganate ion, is carried out in sulfuric acid solution to prevent the air oxidation of Fe^{2+}. The end point of the titration is sharpened markedly if phosphoric acid is present, because the Fe^{3+} ion produced in the titration forms an essentially colorless complex with the acid.

The number of moles of potassium permanganate used in the titration is equal to the product of the molarity of the $KMnO_4$ and the volume used. The number of moles of iron present in the sample is obtained from the balanced equation for the reaction and the amount of MnO_4^- ion reacted. The percentage by weight of iron in the solid sample follows directly. You will calculate the mean percentage of iron for your values, along with the standard deviation. (See Appendix VIII.)

WEAR YOUR SAFETY GLASSES WHILE
PERFORMING THIS EXPERIMENT

Experimental Procedure

Obtain from the stockroom a buret and an unknown iron(II) sample.

Weigh out accurately on the analytical balance three samples of about 1.0 g of your unknown into clean 250-mL Erlenmeyer flasks.

Clean your buret thoroughly. Draw about 100 mL of the standard $KMnO_4$ solution from the carboy in the laboratory. Rinse the buret with a few milliliters of the $KMnO_4$ three times. Drain and then fill the buret with the $KMnO_4$ solution.

Prepare 150 mL of 1 M H_2SO_4 by pouring 25 mL of 6 M H_2SO_4 into 125 mL of H_2O, while stirring. Add 50 mL of this 1 M H_2SO_4 to *one* of the iron samples. The sample should dissolve completely. Without delay, titrate this iron solution with the $KMnO_4$ solution. When a light-yellow color develops in the iron solution during the titration, add 3 mL of 85% H_3PO_4. **CAUTION:** *Caustic reagent.*

Continue the titration until you obtain the first pink color that persists for 15 to 30 seconds. Repeat the titration with the other two samples.

Optional The $KMnO_4$ solution can be standardized by the method you used in this experiment. $Fe(NH_4)_2(SO_4)_2 \cdot 6\ H_2O$ is a primary standard with a molar mass equal to 392.2 g. For standardization, use 0.7 ± 0.05 g samples of the primary standard, weighed to 0.0001 g. Draw 100 mL of the stock $KMnO_4$ solution and titrate the samples as you did the unknown.

Optional Experiment: Determine the mass percent of iron in iron supplement tablets that contain iron(II), ferrous ion.

DISPOSAL OF REACTION PRODUCTS. Dispose of your titrated solutions as directed by your laboratory supervisor.

Experiment 30

Data and Calculations: Determination of Iron by Reaction with
Permanganate—A Redox Titration

Mass of sample tube plus contents _____ g

Mass after removing Sample I _____ g

Mass after removing Sample II _____ g

Mass after removing Sample III _____ g

Molarity of standard $KMnO_4$ solution _____ M

Sample	I	II	III
Initial buret reading	_____ mL	_____ mL	_____ mL
Final buret reading	_____ mL	_____ mL	_____ mL
Volume $KMnO_4$ required	_____ mL	_____ mL	_____ mL
Moles $KMnO_4$ required	_____	_____	_____
Moles Fe^{2+} in sample	_____	_____	_____
Mass of Fe in sample	_____ g	_____ g	_____ g
Mass of sample	_____ g	_____ g	_____ g
Percentage of Fe in sample	_____ %	_____ %	_____ %

Unknown no. _____   Average % Fe _____ %   Standard deviation _____ %

Standardization of KMnO₄ Optional

Sample	IV	V	VI
Mass of primary standard	_____ g	_____ g	_____ g
Moles Fe in standard	_____	_____	_____

(continued on following page)

Moles $KMnO_4$ required   _____   _____   _____

Initial buret reading   _____ mL   _____ mL   _____ mL

Final buret reading   _____ mL   _____ mL   _____ mL

Volume $KMnO_4$ used   _____ mL   _____ mL   _____ mL

Molarity $KMnO_4$ _____ M   _____ M   _____ M

Average molarity   _____ M

Experiment 30

Advance Study Assignment: Determination of Iron by Reaction with
Permanganate—A Redox Titration

1. Write the balanced net ionic equation for the reaction between MnO_4^- ion and Fe^{2+} ion in acid solution.

2. How many moles of Fe^{2+} ion can be oxidized by 1.3×10^{-2} moles MnO_4^- ion in the reaction in Question 1?

    _____ moles

3. A solid sample containing some Fe^{2+} ion weighs 1.264 g. It requires 38.67 mL 0.02487 M $KMnO_4$ to titrate the Fe^{2+} in the dissolved sample to a pink end point.

    a. How many moles of MnO_4^- ion are required?

    _____ moles

    b. How many moles of Fe^{2+} are there in the sample?

    _____ moles

    c. How many grams of iron are there in the sample?

    _____ g

    d. What is the mass percent of Fe in the sample?

    _____ %

4. What is the mass percent of Fe in iron(II) ammonium sulfate hexahydrate, $Fe(NH_4)_2(SO_4)_2 \cdot 6\,H_2O$?

    _____ %

Experiment 31

Determination of an Equivalent Mass by Electrolysis

In Experiment 24 we determined the molar mass of an acid by titration of a known mass of the acid with a standardized solution of NaOH. We noted that the acid could contain only one acidic hydrogen atom per molecule if we were to be able to find the molar mass. If there are two or three such acidic hydrogen atoms we can only determine the mass of acid that will furnish one mole of H$^+$ ion. That is called the equivalent mass of the acid.

Experimentally we find that the equivalent mass of an element can be related in a fundamental way to the chemical effects observed in that phenomenon known as *electrolysis*. As you know, some liquids, because they contain ions, will conduct an electric current. If the two terminals on a storage battery, or any other source of DC voltage, are connected through metal electrodes to a conducting liquid, an electric current will pass through the liquid and chemical reactions will occur at the two metal electrodes; in this process electrolysis is said to occur, and the liquid is said to be electrolyzed.

At the electrode connected to the *negative* pole of the battery, a *reduction* reaction will invariably be observed. In this reaction electrons will usually be accepted by one of the species present in the liquid, which, in the experiment we shall be doing, will be an aqueous solution. The species reduced will ordinarily be a metallic cation or the H$^+$ ion or possibly water itself; the reaction that is actually observed will be the one that occurs with the least expenditure of electrical energy, and will depend on the composition of the solution. In the electrolysis cell we shall study, the reduction reaction of interest will occur in a slightly acidic medium; hydrogen gas will be produced by the reduction of hydrogen ion:

$$2 \text{ H}^+(\text{aq}) + 2 \text{ e}^- \rightarrow \text{H}_2(\text{g}) \tag{1}$$

In this reduction reaction, which will occur at the negative pole, or *cathode*, of the cell, for every H$^+$ ion reduced *one* electron will be required, and for every molecule of H$_2$ formed, *two* electrons will be needed.

Ordinarily in chemistry we deal not with individual ions or molecules but rather with moles of substances. In terms of moles, we can say that, by Equation 1,

The reduction of one mole of H$^+$ ion requires one mole of electrons

The production of one mole of H$_2$(g) requires two moles of electrons

A mole of electrons is a fundamental amount of electricity in the same way that a mole of pure substance is a fundamental unit of matter, at least from a chemical point of view. A mole of electrons is called a *faraday*, after Michael Faraday, who discovered the basic laws of electrolysis. The amount of a species that will react with a *mole of electrons*, or *one faraday*, is equal to the *equivalent mass* of that species. Since one faraday will reduce one mole of H$^+$ ion, we say that the equivalent mass of hydrogen is 1.008 grams, equal to the mass of one mole of H$^+$ ion (or 1/2 mole of H$_2$(g)). To form one mole of H$_2$(g) one would have to pass two faradays through the electrolysis cell.

In the electrolysis experiment we will perform we will measure the volume of hydrogen gas produced under known conditions of temperature and pressure. By using the Ideal Gas Law we will be able to calculate how many moles of H$_2$ were formed, and hence how many faradays of electricity passed through the cell.

At the positive pole of an electrolysis cell (the metal electrode that is connected to the + terminal of the battery), an *oxidation* reaction will occur, in which some species will give up electrons. This reaction, which takes place at the *anode* in the cell, may involve again an ionic or neutral species in the solution or the metallic electrode itself. In the cell that you will be studying, the pertinent oxidation reaction will be that in which a metal under study will participate:

$$M(s) \rightarrow M^{n+}(aq) + ne^- \tag{2}$$

During the course of the electrolysis the atoms in the metal electrode will be converted to metallic cations and will go into the solution. The mass of the metal electrode will decrease, depending on the amount of electricity passing through the cell and the nature of the metal. To oxidize one mole, or one molar mass, of the metal, it would take *n* faradays, where *n* is the charge on the cation that is formed. By definition, one faraday of electricity would cause one equivalent mass, *EM,* of metal to go into solution. The molar mass, *MM,* and the equivalent mass of the metal are related by the equation:

$$MM = EM \times n \tag{3}$$

In an electrolysis experiment, since *n* is not determined independently, it is not possible to find the molar mass of a metal. It is possible, however, to find equivalent masses of many metals, and that will be our main purpose.

The general method we will use is implied by the discussion. We will oxidize a sample of an unknown metal at the positive pole of an electrolysis cell, weighing the metal before and after the electrolysis and so determining its loss in mass. We will use the same amount of electricity, the same number of electrons, to reduce hydrogen ion at the negative pole of the electrolysis cell. From the volume of H_2 gas that is produced under known conditions we can calculate the number of moles of H_2 formed, and hence the number of faradays that passed through the cell. The equivalent mass of the metal is then calculated as the amount of metal that would be oxidized if one faraday were used. In an optional part of the experiment, your instructor may tell you the nature of the metal you used. Using Equation 3, it will be possible to determine the charge on the metallic cations that were produced during electrolysis.

WEAR YOUR SAFETY GLASSES WHILE PERFORMING THIS EXPERIMENT

Experimental Procedure

Obtain from the stockroom a buret and a sample of metal unknown. Lightly sand the metal to clean it. Rinse the metal with water and then in acetone. Let the acetone evaporate. When the sample is dry, weigh it on the analytical balance to 0.0001 g.

Set up the electrolysis apparatus as indicated in Figure 31.1. There should be about 100 mL 0.5 M $HC_2H_3O_2$ in 0.5 M Na_2SO_4 in the beaker with the gas buret. This will serve as the conducting solution. Immerse the end of the buret in the solution and attach a length of rubber tubing to its upper end. Open the stopcock on the buret and, with suction, carefully draw the acid up to the top of the graduations. Close the stopcock. Insert the bare coiled end of the heavy copper wire up into the end of the buret; all but the coil end of the wire should be covered with watertight insulation. Check the solution level after a few minutes to make sure the stopcock does not leak. Record the level.

The metal unknown will serve as the anode in the electrolysis cell. Connect the metal to the + pole of the power source with an alligator clip and immerse the metal but not the clip in the conducting solution. The copper electrode will be the cathode in the cell. Connect that electrode to the − pole of the power source. Hydrogen gas should immediately begin to bubble from the copper cathode. Collect the gas until about 50 mL have been produced. At that point, stop the electrolysis by disconnecting the copper electrode from the power source. Record the level of the liquid in the buret. Measure and record the temperature and the barometric pressure in the laboratory. (In some cases a cloudiness may develop in the solution during the electrolysis. This is caused by the formation of a metal hydroxide, and will have no adverse effect on the experiment.)

Raise the buret, and discard the conducting solution in the beaker in the waste crock. Rinse the beaker with water, and pour in 100 mL of fresh conducting solution. Repeat the electrolysis, generating about 50 mL of H_2 and recording the initial and final liquid levels in the buret. Take the alligator clip off the metal anode

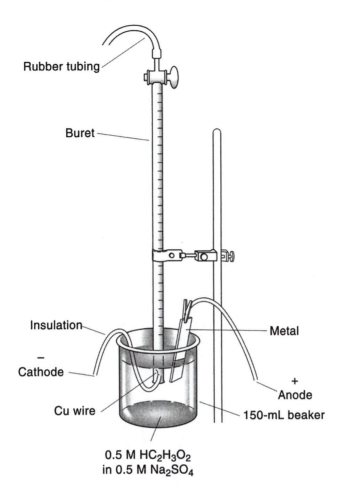

Rubber tubing

Buret

Insulation

– Cathode

Cu wire

Metal

+ Anode

150-mL beaker

0.5 M HC$_2$H$_3$O$_2$
in 0.5 M Na$_2$SO$_4$

Figure 31.1

and wash the anode with 0.1 M HC$_2$H$_3$O$_2$, acetic acid. Rub off any loose adhering coating with your fingers, and then rinse off your hands. Rinse the electrode in water and then in acetone. Let the acetone evaporate. Weigh the dry metal electrode to the nearest 0.0001 g.

DISPOSAL OF REACTION PRODUCTS. When you are finished with the experiment, return the metal electrode to the stockroom and discard the conducting solution in the waste crock unless directed otherwise by your instructor.

Experiment 31

Data and Calculations: Determination of an Equivalent Mass by Electrolysis

Mass of metal anode   _____ g

Mass of anode after electrolysis   _____ g

Initial buret reading   _____ mL

Buret reading after first electrolysis   _____ mL

Buret reading after refilling   _____ mL

Buret reading after second electrolysis   _____ mL

Barometric pressure   _____ mm Hg

Temperature t _____ °C

Vapor pressure of H_2O at t _____ mm Hg

Total volume of H_2 produced, V _____ mL

Temperature T _____ K

Pressure exerted by dry H_2: $P = P_{bar} - VP_{H_2O}$
(ignore any pressure effect due to liquid levels in buret)   _____ mm Hg

No. moles H_2 produced, n
(use Ideal Gas Law, $PV = nRT$)   _____ moles

No. of faradays passed (no. of moles of electrons)   _____

Loss in mass by anode   _____ g

Equivalent mass of metal $\left(EM = \dfrac{\text{no. g lost}}{\text{no. faradays passed}} \right)$ _____ g

Unknown metal number   _____

Optional   Nature of metal _____ MM _____ g

Charge n on cation _____ (Eq. 3)

Experiment 31

Advance Study Assignment: Determination of an Equivalent
Mass by Electrolysis

1. In an electrolysis cell similar to the one employed in this experiment, a student observed that his unknown metal anode lost 0.208 g while a total volume of 96.30 mL of H_2 was being produced. The temperature in the laboratory was 25°C and the barometric pressure was 748 mm Hg. At 25°C the vapor pressure of water is 23.8 mm Hg. To find the equivalent mass of his metal, he filled in the blanks below. Fill in the blanks as he did.

$P_{H_2} = P_{bar} - VP_{H_2O} = $ _____ mm Hg = _____ atm

$V_{H_2} = $ _____ mL = _____ L

$T = $ _____ K

$n_{H_2} = $ _____ moles $\quad n_{H_2} = \dfrac{PV}{RT} \quad$ (where $P = P_{H_2}$)

1 mole H_2 requires passage of _____ faradays

No. of faradays passed = _____

Loss of mass of metal anode = _____ g

No. grams of metal lost per faraday passed $= \dfrac{\text{no. grams lost}}{\text{no. faradays passed}} = $ _____ g = *EM*

The student was told that his metal anode was made of iron.

MM Fe = _____ g. The charge n on the Fe ion is therefore _____. (See Eq. 3.)

2. In standard, IUPAC units, the faraday is equal to 96,480 coulombs. A coulomb is the amount of electricity passed when a current of one ampere flows for one second. Given the charge on an electron, 1.6022×10^{-19} coulombs, calculate a value for Avogadro's number.

_____

Experiment 32

Voltaic Cell Measurements

Many chemical reactions can be classified as oxidation-reduction reactions, because they involve the oxidation of one species and the reduction of another. Such reactions can conveniently be considered as the result of two half-reactions, one of oxidation and the other reduction. In the case of the oxidation-reduction reaction

$$Zn(s) + Pb^{2+}(aq) \rightarrow Zn^{2+}(aq) + Pb(s)$$

that would occur if a piece of metallic zinc were put into a solution of lead nitrate, the two reactions would be

$$Zn(s) \rightarrow Zn^{2+}(aq) + 2\ e^{-} \quad \text{oxidation}$$

$$2\ e^{-} + Pb^{2+}(aq) \rightarrow Pb(s) \quad \text{reduction}$$

The tendency for an oxidation-reduction reaction to occur can be measured if the two reactions are made to occur in separate regions connected by a barrier that is porous to ion movement. An apparatus, called a voltaic cell, in which this reaction might be carried out under this condition is shown in Figure 32.1.

If we connect a voltmeter between the two electrodes, we will find that there is a voltage, or potential, between them. The magnitude of the potential is a direct measure of the driving force or thermodynamic tendency of the spontaneous oxidation-reduction reaction to occur.

If we study several oxidation-reduction reactions, we find that the voltage of each associated voltaic cell can be considered to be the sum of a potential for the oxidation reaction and a potential for the reduction reaction. In the $Zn,Zn^{2+} \parallel Pb^{2+},Pb$ cell we have been discussing, for example,

$$E_{cell} = E_{Zn.Zn^{2+} \text{ oxidation reaction}} + E_{Pb^{2+}.Pb \text{ reduction reaction}} \tag{1}$$

By convention, the negative electrode in a voltaic cell is taken to be the one from which electrons are emitted (i.e., where oxidation occurs). The negative electrode is the one that is connected to the minus pole of the voltmeter when the voltage is measured.

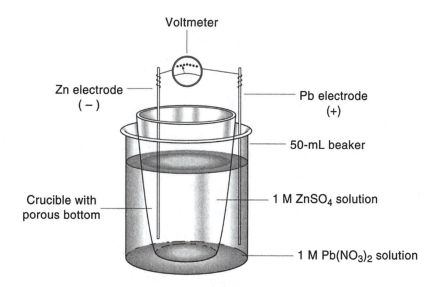

Figure 32.1

Since any cell potential is the sum of two electrode potentials, it is not possible, by measuring cell potentials, to determine individual absolute electrode potentials. However, if a value of potential is arbitrarily assigned to one electrode reaction, other electrode potentials can be given definite values, based on the assigned value. The usual procedure is to assign a value of 0.0000 volts to the standard potential for the electrode reaction

$$2\ H^+(aq) + 2\ e^- \rightarrow H_2(g); \qquad E^0_{H^+,H_2\ red} = 0.0000\ V$$

For the Zn,Zn^{2+} || H$^+$,H$_2$ cell, the measured potential is 0.76 V, and the zinc electrode is negative. Zinc metal is therefore oxidized, and the cell reaction must be

$$Zn(s) + 2\ H^+(aq) \rightarrow Zn^{2+}(aq) + H_2(g); \qquad E^0_{cell} = 0.76\ V$$

Given this information, one can readily find the potential for the oxidation of Zn to Zn^{2+}.

$$E^0_{cell} = E^0_{Zn,Zn^{2+}\ oxid} + E^0_{H^+,H_2\ red}$$

$$0.76\ V = E^0_{Zn,Zn^{2+}\ oxid} + 0.00\ V; \qquad E^0_{Zn,Zn^{2+}\ oxid} = +0.76\ V$$

If the potential for a half-reaction is known, the potential for the reverse reaction can be obtained by changing the sign. For example:

$$if \quad E^0_{Zn,Zn^{2+}\ oxid} = +0.76\ V, \quad then \quad E^0_{Zn^{2+},Zn\ red} = -0.76\ V$$

$$if \quad E^0_{Pb^{2+},Pb\ red} = +Y\ V, \quad then \quad E^0_{Pb,Pb^{2+}\ oxid} = -Y\ V$$

In the first part of this experiment you will measure the voltages of several different cells. By arbitrarily assigning the potential of a particular half-reaction to be 0.00 V, you will be able to calculate the potentials corresponding to all of the various half-reactions that occurred in your cells.

In our discussion so far we have not considered the possible effects of such system variables as temperature, potential at the liquid-liquid junction, size of metal electrodes, and concentrations of solute species. Although temperature and liquid junctions do have a definite effect on cell potentials, taking account of their influence involves rather complex thermodynamic concepts and is usually not of concern in an elementary course. The size of a metal electrode has no appreciable effect on electrode potential, although it does relate directly to the capacity of the cell to produce useful electrical energy. In this experiment we will operate the cells so that they deliver essentially no energy but exert their maximum potentials.

The effect of solute ion concentrations is important and can be described relatively easily. For the cell reaction at 25°C:

$$aA(s) + bB^+(aq) \rightarrow cC(s) + dD^{2+}(aq)$$

$$E_{cell} = E^0_{cell} - \frac{0.0592}{n} \log \frac{[D^{2+}]^d}{[B^+]^b} \tag{2}$$

where E^0_{cell} is a constant for a given reaction and is called the standard cell potential, and n is the number of electrons in either electrode reaction.

By Equation 2 you can see that the measured cell potential, E_{cell}, will equal the standard cell potential if the molarities of D^{2+} and B$^+$ are both unity, or, if d equals b, if the molarities are simply equal to each other. We will carry out experiments under such conditions that the cell potentials you observe will be very close to the standard potentials given in the tables in your chemistry text.

Considering the Cu,Cu^{2+} || Ag$^+$,Ag cell as a specific example, the observed cell reaction would be

$$Cu(s) + 2\ Ag^+(aq) \rightarrow Cu^{2+}(aq) + 2\ Ag(s)$$

For this cell, Equation 2 takes the form

$$E_{cell} = E^0_{cell} - \frac{0.0592}{2} \log \frac{[Cu^{2+}]}{[Ag^+]^2} \tag{3}$$

In the equation n is 2 because in the cell reaction two electrons are transferred in each of the two half-reactions. E^0 would be the cell potential when the copper and silver salt solutions are both 1 M, since then the logarithm term is equal to zero.

If we decrease the Cu^{2+} concentration, keeping that of Ag^+ at 1 M, the potential of the cell will go up by about 0.03 volts for every factor of ten by which we decrease $[Cu^{2+}]$. Ordinarily it is not convenient to change concentrations of an ion by several orders of magnitude, so in general, concentration effects in cells are relatively small. However, if we should add a complexing or precipitating species to the copper salt solution, the value of $[Cu^{2+}]$ would drop drastically, and the voltage change would be appreciable. In the experiment we will illustrate this effect by using NH_3 to complex the Cu^{2+}. Using Equation 3, we can actually calculate $[Cu^{2+}]$ in the solution of its complex ion. It is very low.

In an analogous experiment we will determine the solubility product of AgCl. In this case we will surround the Ag electrode in a $Cu,Cu^{2+} \parallel Ag^+,Ag$ cell, with a solution of known Cl^- ion concentration that is saturated with AgCl. From the measured cell potential, we can use Equation 3 to calculate the very small value of $[Ag^+]$ in the chloride-containing solution.

In the two experiments in which we change the cation concentrations, first Cu^{2+} and then Ag^+, we end up with systems in equilibrium. In the first case Cu^{2+} ion is in equilibrium with $Cu(NH_3)_4^{2+}$ and NH_3. Using the cell potential, we can calculate $[Cu^{2+}]$. From the way we made up the system we can calculate $[Cu(NH_3)_4^{2+}]$ and $[NH_3]$. If we wish to do so, we can use these data to find the dissociation constant for the $Cu(NH_3)_4^{2+}$ ion. Many equilibrium constants are found by experiments of this sort.

In the last experiment we have an equilibrium system containing Ag^+, Cl^-, and AgCl(s). Here we can use the cell potential to find $[Ag^+]$. There isn't much Ag^+ present, but there is a tiny bit, and the cell potential lets us find it. We know $[Cl^-]$ from the way we made up the mixture in the crucible. From these data we can easily calculate K_{sp} for AgCl:

$$AgCl(s) \rightleftharpoons Ag^+(aq) + Cl^-(aq) \qquad K_{sp} = [Ag^+][Cl^-] \tag{4}$$

WEAR YOUR SAFETY GLASSES WHILE PERFORMING THIS EXPERIMENT

Experimental Procedure

You may work in pairs in this experiment.

A. Cell Potentials

In this experiment you will be working with these seven electrode systems:

$Ag^+,Ag(s)$	$Br_2(l),Br^-,Pt$
$Cu^{2+},Cu(s)$	$Cl_2(g, 1 \text{ atm}),Cl^-,Pt$
Fe^{3+},Fe^{2+},Pt	$I_2(s),I^-,Pt$
$Zn^{2+},Zn(s)$	

Your purpose will be to measure enough voltaic cell potentials to allow you to determine the electrode potentials of each electrode by comparing it with an arbitrarily chosen electrode potential.

Using the apparatus shown in Figure 32.1, set up a voltaic cell involving any two of the electrodes in the list. About 10 mL of each solution should be enough for making the cell. The solute ion concentrations may be assumed to be one molar and all other species may be assumed to be at unit activity, so that the potentials of the cells you set up will be essentially the standard potentials. Measure the cell potential and record it along with which electrode has negative polarity. Pour both solutions into a beaker and rinse the crucible and 50-mL beaker with distilled water.

In a similar manner set up and measure the cell voltage and polarities of other cells, sufficient in number to include all of the electrode systems on the list at least once. Do not combine the silver electrode system with any of the halogen electrode systems because a precipitate will form; for the same reason, do not use the copper and iodine electrodes in the same cell. Any other combinations may be used. The data from this part of the experiment should be entered in the first three columns of the table in Part A.1 of your report.

B. Effect of Concentration on Cell Potentials

1. Complex Ion Formation.   Set up the $Cu,Cu^{2+} \parallel Ag^+,Ag$ cell, using 10 mL of the $CuSO_4$ solution in the crucible and 10 mL of $AgNO_3$ in the beaker. Measure the potential of the cell. While the potential is being measured, add 10 mL of 6 M NH_3 to the $CuSO_4$ solution, stirring carefully with your stirring rod. Measure the potential when it becomes steady.

2. Determination of the Solubility Product of AgCl.   Remove the crucible from the cell you have just studied and discard the $Cu(NH_3)_4^{2+}$. Clean the crucible by drawing a little 6 M NH_3 through it, using the adapter and suction flask. Then draw some distilled water through it. Reassemble the Cu-Ag cell, this time using the beaker for the $Cu-CuSO_4$ electrode system. Put 10 mL 1 M KCl in the crucible; immerse the Ag electrode in that solution. Add a drop of $AgNO_3$ solution to form a little AgCl, so that an equilibrium between Ag^+ and Cl^- can be established. Measure the potential of this cell, noting which electrode is negative. In this case $[Ag^+]$ will be very low, which will decrease the potential of the cell to such an extent that its polarity may change from that observed previously.

> **DISPOSAL OF REACTION PRODUCTS.**   As you finish each voltage measurement, pour the two solutions into a beaker. At the end of the experiment, pour the contents of the beaker into the waste crock, unless directed otherwise by your instructor.

Experiment 32

Data and Calculations: Voltaic Cell Measurements

A.1. Cell Potentials

Electrode Systems Used in Cell	Cell Potential, E^0_{cell} (volts)	Negative Electrode	Oxidation Reaction	$E^0_{oxidation}$ in volts	Reduction Reaction	$E^0_{reduction}$ in volts
1.						
2.						
3.						
4.						
5.						
6.						
7.						

Calculations

A. Noting that oxidation occurs at the *negative* pole in a cell, write the oxidation reaction in each of the cells. The other electrode system must undergo reduction; write the reduction reaction that occurs in each cell.

B. Assume that $E^0_{Ag^+,Ag} = 0.00$ volts (whether in reduction or oxidation). Enter that value in the table for all of the silver electrode systems you used in your cells. Since $E^0_{cell} = E^0_{oxidation} + E^0_{reduction}$, you can calculate E^0 values for all the electrode systems in which the Ag,Ag^+ system was involved. Enter those values in the table.

C. Using the values and relations in Part B and taking advantage of the fact that for any given electrode system, $E^0_{oxidation} = -E^0_{reduction}$, complete the table of E^0 values. The best way to do this is to use one of the E^0 values you found in Part B in another cell with that electrode system. That potential, along with E^0_{cell}, will allow you to find the potential of the other electrode. Continue this process with other cells until all the electrode potentials have been determined.

(continued on following page)

A.2. Table of Electrode Potentials

In Table A.1, you should have a value for E_{red}^0 or E_{oxid}^0 for each of the electrode systems you have studied. Remembering that for any electrode system, $E_{red}^0 = -E_{oxid}^0$, you can find the value for E_{red}^0 for each system. List those potentials in the left column of the table below in order of decreasing value.

$E_{reduction}^0$ $(E_{Ag^+,Ag}^0 = 0.00 \text{ V})$	Electrode Reaction in Reduction	$E_{reduction}^0$ $(E_{H^+,H_2}^0 = 0.00 \text{ V})$
_____	_____	_____
_____	_____	_____
_____	_____	_____
_____	_____	_____
_____	_____	_____
_____	_____	_____

The electrode potentials you have determined are based on $E_{Ag^+,Ag}^0 = 0.00$ V. The usual assumption is that $E_{H^+,H_2}^0 = 0.00$ volts, under which conditions $E_{Ag^+,Ag \text{ red}}^0 = 0.80$ V. Convert from one base to the other by adding 0.80 volts to each of the electrode potentials, and enter these values in the third column of the table.

Why are the values of E_{red}^0 on the two bases related to each other in such a simple way?

B. Effect of Concentration on Cell Potentials

1. Complex ion formation:

Potential, E_{cell}^0, before addition of 6 M NH_3 _____ V

Potential, E_{cell}, after $Cu(NH_3)_4^{2+}$ formed _____ V

Given Equation 3

$$E_{cell} = E_{cell}^0 - \frac{0.0592}{2} \log \frac{[Cu^{2+}]}{[Ag^+]^2} \tag{3}$$

calculate the residual concentration of free Cu^{2+} ion in equilibrium with $Cu(NH_3)_4^{2+}$ in the solution in the crucible. Take $[Ag^+]$ to be 1 M.

$[Cu^{2+}] =$ _____ M

(continued on following page)

2. **Solubility product of AgCl:**

Potential, E_{cell}^0 , of the Cu, Cu^{2+} ǁ Ag^+, Ag cell (from B.1)

_____V   Negative electrode_____

Potential, E_{cell}, with 1 M KCl present   _____V   Negative electrode_____

Using Equation 3, calculate [Ag^+] in the cell, where it is in equilibrium with 1 M Cl^- ion. (E_{cell} in Equation 3 is the *negative* of the measured value if the polarity is not the same as in the standard cell.) Take [Cu^{2+}] to be 1 M.

[Ag^+] = _____ M

Since Ag^+ and Cl^- in the crucible are in equilibrium with AgCl, we can find K_{sp} for AgCl from the concentration of Ag^+ and Cl^-, which we now know. Formulate the expression for K_{sp} for AgCl, and determine its value.

K_{sp} = _____

Experiment 32

Advance Study Assignment: Voltaic Cell Measurements

1. A student measures the potential of a cell made up with 1 M $CuSO_4$ in one solution and 1 M $AgNO_3$ in the other. There is a Cu electrode in the $CuSO_4$ and an Ag electrode in the $AgNO_3$, and the cell is set up as in Figure 32.1. She finds that the potential, or voltage, of the cell, E^0_{cell} , is 0.45 V, and that the Cu electrode is negative.

   a. At which electrode is oxidation occurring?   _____

   b. Write the equation for the oxidation reaction.

   c. Write the equation for the reduction reaction.

   d. If the potential of the silver, silver ion electrode, $E^0_{Ag^+,Ag}$ is taken to be 0.000 V in oxidation or reduction, what is the value of the potential for the oxidation reaction, $E^0_{Cu,Cu^{2+} oxid}$? $E^0_{cell} = E^0_{oxid} + E^0_{red}$.

                                                                           _____ volts

   e. If $E^0_{Ag^+,Ag\ red}$ equals 0.80 V, as in standard tables of electrode potentials, what is the value of the potential of the oxidation reaction of copper, $E^0_{Cu,Cu^{2+} oxid}$?

                                                                           _____ volts

   f. Write the net ionic equation for the spontaneous reaction that occurs in the cell that the student studied.

   g. The student adds 6 M NH_3 to the $CuSO_4$ solution until the Cu^{2+} ion is essentially all converted to $Cu(NH_3)_4^{2+}$ ion. The voltage of the cell, E_{cell}, goes up to 0.92 V and the Cu electrode is still negative. Find the residual concentration of Cu^{2+} ion in the cell. (Use Eq. 3.)

                                                                           _____ M

   h. In Part g, $[Cu(NH_3)_4^{2+}]$ is about 0.05 M, and $[NH_3]$ is about 3 M. Given those values and the result in Part 1g for $[Cu^{2+}]$, calculate K for the reaction:

   $$Cu(NH_3)_4^{2+}(aq) \rightleftarrows Cu^{2+}(aq) + 4\ NH_3(aq)$$

                                                                           _____

Experiment 33

Preparation of Copper(I) Chloride

O xidation-reduction reactions are, like precipitation reactions, often used in the preparation of inorganic substances. In this experiment we will employ a series of such reactions to prepare one of the less commonly encountered salts of copper, copper(I) chloride. Most copper compounds contain copper(II), but copper(I) is present in a few slightly soluble or complex copper salts.

The process of synthesis of CuCl we will use begins by dissolving copper metal in nitric acid:

$$Cu(s) + 4\ H^+\ (aq) + 2\ NO_3^-\ (aq) \rightarrow Cu^{2+}\ (aq) + 2\ NO_2(g) + 2\ H_2O \tag{1}$$

The solution obtained is treated with sodium carbonate in excess, which neutralizes the remaining acid with evolution of CO_2 and precipitates Cu(II) as the carbonate:

$$2\ H^+\ (aq) + CO_3^{2-}\ (aq) \rightleftharpoons (H_2CO_3)(aq) \rightleftharpoons CO_2(g) + H_2O \tag{2}$$

$$Cu^{2+}\ (aq) + CO_3^{2-}\ (aq) \rightleftharpoons CuCO_3(s) \tag{3}$$

The $CuCO_3$ will be purified by filtration and washing and then dissolved in hydrochloric acid. Copper metal added to the highly acidic solution reduces the Cu(II) to Cu(I) and is itself oxidized to Cu(I) in a disproportionation reaction. In the presence of excess chloride, the copper will be present as a $CuCl_4^{3-}$ complex ion. Addition of this solution to water destroys the complex, and white CuCl precipitates.

$$CuCO_3(s) + 2\ H^+\ (aq) + 4\ Cl^-\ (aq) \rightarrow CuCl_4^{2-}\ (aq) + CO_2(g) + H_2O \tag{4}$$

$$CuCl_4^{2-}\ (aq) + Cu(s) + 4\ Cl^-\ (aq) \rightarrow 2\ CuCl_4^{3-}\ (aq) \tag{5}$$

$$CuCl_4^{3-}(aq) \xrightarrow{H_2O} CuCl(s) + 3\ Cl^-\ (aq) \tag{6}$$

Because CuCl is readily oxidized, due care must be taken to minimize its exposure to air during its preparation and while it is being dried.

In an optional part of the experiment, we prove that the formula of the compound synthesized is CuCl.

WEAR YOUR SAFETY GLASSES WHILE
PERFORMING THIS EXPERIMENT

Experimental Procedure

Obtain a 1-gram sample of copper metal turnings, a Buchner funnel, and a filter flask from the stockroom. Weigh the copper metal on the top-loading balance to 0.1 g.

Put the metal in a 150-mL beaker and **under a hood** add 5 mL 15 M HNO_3. **CAUTION:** *Caustic reagent.* Brown NO_2 gas will be evolved and an acidic blue solution of $Cu(NO_3)_2$ produced. If it is necessary, you may warm the beaker gently with a Bunsen burner to dissolve all of the copper. When all of the copper is in solution, add 50 mL water to the solution and allow it to cool.

Weigh out about 5 grams of sodium carbonate in a small beaker on the top-loading balance. Add small amounts of the Na_2CO_3 to the solution with your spatula, adding the solid as necessary when the evolution of CO_2 subsides. Stir the solution to expose it to the solid. When the acid is neutralized, a blue-green precipitate of $CuCO_3$ will begin to form. At that point, add the rest of the Na_2CO_3, stirring the mixture well to ensure complete precipitation of the copper carbonate.

Transfer the precipitate to the Buchner funnel and use suction to remove the excess liquid. Use your rubber policeman and a spray from your wash bottle to make a complete transfer of the solid. Wash the precipitate well with distilled water with suction on; then let it remain on the filter paper with suction on for a minute or two.

Remove the filter paper from the funnel and transfer the solid $CuCO_3$ to the 150-mL beaker. Add 10 mL water and then 30 mL 6 M HCl slowly to the solid, stirring continuously. When the $CuCO_3$ has all dissolved, add 1.5 g Cu turnings to the beaker and cover it with a watch glass.

Heat the mixture in the beaker to the boiling point and keep it at that temperature, just simmering, for about 10 minutes. It may be that the dark-colored solution that forms will clear to a yellow color before that time is up, and if it does, you may stop heating and proceed to the next step.

While the mixture is heating, put 150 mL distilled water in a 400-mL beaker and put the beaker in an ice bath. Cover the beaker with a watch glass. After you have heated the acidic Cu-$CuCl_2$ mixture for 10 minutes or as soon as it turns light colored, carefully decant the hot liquid into the beaker of water, taking care not to transfer any of the excess Cu metal to the beaker. White crystals of CuCl should form. Continue to cool the beaker in the ice bath to promote crystallization and to increase the yield of solid.

Cool 25 mL distilled water, to which you have added five drops 6 M HCl, in an ice bath. Put 20 mL acetone into a small beaker. Filter the crystals of CuCl in the Buchner funnel using suction. Swirl the beaker to aid in transferring the solid to the funnel. Just as the last of the liquid is being pulled through, wash the CuCl with 1/3 of the acidified cold water. Rinse the last of the CuCl into the funnel with another portion of the water and use the final 1/3 to rewash the solid. Turn off suction and add 1/2 of the acetone to the funnel; wait about 10 seconds and turn on the suction. Repeat this operation with the other half of the acetone. Draw air through the sample for a few minutes to dry it. If you have properly washed the solid, it will be pure white: if the moist compound is allowed to come into contact with air, it will tend to turn pale green, due to oxidation of Cu(I) to Cu(II). Weigh the CuCl in a previously weighed beaker to 0.1 g. Show your sample to your instructor for evaluation.

DISPOSAL OF REACTION PRODUCTS. The contents of the suction flask after the first filtration (of $CuCO_3$) can be discarded down the sink. The contents after the second filtration include an appreciable amount of copper, so they should be put in the waste crock.

Determination of the Formula of Copper(I) Chloride Optional

There are many ways to prove that the formula of the copper compound you have prepared is indeed CuCl. The following procedure is easy to carry out and gives good results.

Dry the copper compound by putting it under a heat lamp or in a drying oven at 110°C for 10 minutes. Weigh out about a 0.1 g sample of the compound accurately on the analytical balance, using a weighed 50-mL beaker as a container. Dissolve the sample in 5 mL of 6 M HNO_3. Add 10 mL of distilled water, mix, and transfer the solution to a clean 100-mL volumetric flask. Use several 5-mL portions of distilled water to rinse the remaining solution from the beaker into the flask. Fill the flask to the mark with distilled water. Stopper the flask and invert it about 20 times to ensure that the solution in the flask is thoroughly mixed.

Use a clean 10-mL pipet to transfer a 10-mL aliquot of the solution to a clean, dry 50-mL beaker. Add 10 mL 6 M NH_3 to the beaker, using the 10-mL pipet, after you have rinsed it with 6 M NH_3. This will convert any Cu^{2+} in the solution to deep blue $Cu(NH_3)_4^{2+}$. After mixing, measure the absorbance of the solution of $Cu(NH_3)_4^{2+}$ at a wavelength of 575 nm. Determine $[Cu(NH_3)_4^{2+}]$ from a graph that is furnished to you or by comparison with a standard solution. (You can prepare the standard by accurately weighing about 0.06 g of copper turnings and putting them into a 50-mL beaker. Dissolve the copper in 5 mL of 6 M HNO_3, and proceed as you did with the solution of the copper compound in the previous paragraph. Measure the absorbance of the standard at 575 nm. Calculate $[Cu(NH_3)_4^{2+}]$ in the solution of the compound by assuming that Beer's Law holds.) Knowing that concentration, and the fact that the original sample was effectively diluted to a volume of 200 mL, calculate the number of moles of Cu in the sample. Then calculate the number of grams of Cu in the sample, and the number of grams of Cl by difference. Use that value to find the number of moles of Cl, and from the mole ratio Cu:Cl obtain the formula of the compound.

Experiment 33

Data and Calculations: Preparation of Copper(I) Chloride

Mass of Cu sample   _____ g

Mass of beaker   _____ g

Mass of beaker plus CuCl   _____ g

Mass of CuCl prepared   _____ g

Theoretical yield   _____

Percentage yield   _____ %

Determination of the Formula of Copper(I) Chloride   Optional

Mass of Cu(I) chloride sample   _____ g

Absorbance of solution of $Cu(NH_3)_4^{2+}$ _____

Mass of Cu turnings   _____ g

Absorbance of standard solution   _____

No. moles Cu in turnings   _____ moles

All of the copper in the turnings is converted to $Cu(NH_3)_4^{2+}$
in a solution whose total volume would be 200 mL.

$[Cu(NH_3)_4^{2+}]$ in standard solution   _____ M

If Beer's Law holds,

$$[Cu(NH_3)_4^{2+}] \text{ in solution of sample} = [Cu(NH_3)_4^{2+}] \text{ in standard} \times \frac{\text{Abs of sample}}{\text{Abs of standard}}$$

$[Cu(NH_3)_4^{2+}]$ in solution of sample   _____ M

No. moles Cu in sample (volume = 0.200 L)   _____ moles

No. grams Cu in sample   _____ g

No. grams Cl in sample (sample mass = mass Cu + mass Cl)   _____ g

No. moles Cl in sample   _____ g

Mole ratio Cl:Cu   _____

Formula of prepared compound (rounded to nearest integers)   _____

Experiment 33

Advance Study Assignment: Preparation of Copper(I) Chloride

1. The Cu^{2+} ions in this experiment are produced by the reaction of 1.0 g of copper turnings with excess nitric acid. How many moles of Cu^{2+} are produced?

2. Why isn't hydrochloric acid used in a direct reaction with copper to prepare the $CuCl_2$ solution?

3. How many grams of metallic copper are required to react with the number of moles of Cu^{2+} calculated in Problem 1 to form the CuCl? The overall reaction can be taken to be:

$$Cu^{2+}(aq) + 2\ Cl^-(aq) + Cu(s) \rightarrow 2\ CuCl(s)$$

_____ g

4. What is the maximum mass of CuCl that can be prepared from the reaction sequence of this experiment, using 1.0 g of Cu turnings to prepare the Cu^{2+} solution?

_____ g

5. A sample of the compound prepared in this experiment, weighing 0.1021 g, is dissolved in HNO_3, and diluted to a volume of 100 mL. A 10-mL aliquot of that solution is mixed with 10 mL 6 M NH_3. The $[Cu(NH_3)_4^{2+}]$ in the resulting solution is found to be 5.16×10^{-3} M.

   a. How many moles of Cu were in the original sample, which had been effectively diluted to a volume of 200 mL?

_____ moles

   b. How many grams of Cu were in the sample?

_____ g

   c. How many grams of Cl were in the sample? How many moles?

_____ g _____ moles

   d. What is the formula of the copper chloride compound?

Experiment 34

Development of a Scheme for Qualitative Analysis

In many of the previous experiments in this book you were asked to find out how much of a given species is present in a sample. You may have determined the concentration of chloride in an unknown solution, the molarity of an NaOH solution, and the amount of calcium ion in a sample of hard water. All of these experiments fall into that part of chemistry called quantitative analysis.

Sometimes chemists are interested more in the nature of the species in a sample than the amount of those species. In that sort of problem we find out what the sample contains but not how much. For example, in Experiment 12 students are asked to determine which alkaline earth halide is present in a solution. That experiment is one involving qualitative analysis. This and the next four experiments deal with the qualitative analysis of solutions containing various anions and metallic cations.

The procedures in qualitative analyses of this sort involve using precipitation reactions to remove the cations sequentially from a mixture. If the precipitate can contain only one cation under the conditions that prevail, that precipitate serves to prove the presence of that cation. If the precipitate may contain several cations, it can be dissolved and further resolved in a series of steps that may include acid-base, complex ion formation, redox, or other precipitation reactions. The ultimate result is a resolution of the sample into fractions that can contain only one cation, whose presence is established by formation of a characteristic precipitate or a colored complex ion.

In this experiment you will be asked to develop a scheme for the qualitative analysis of four cations, using this systematic approach. The behavior of the cations toward a set of common test reagents differs from one cation to another and furnishes the basis for their separation.

The cations we will study are Ba^{2+}, Mg^{2+}, Cd^{2+}, and Al^{3+}. The test reagents we will use are 1 M Na_2SO_4, 1 M Na_2CO_3, 6 M NaOH, 6 M NH_3, and 6 M HNO_3. These reagents furnish anions or molecules that will precipitate or form complex ions with the cations. So you may observe formation of insoluble sulfates, carbonates, and hydroxides (the last with either NaOH or NH_3). In addition you may form complexes with OH^- and NH_3 when 6 M NaOH or 6 M NH_3 is added to the cation-containing solutions. The complexes may be quite stable, stable enough to prevent precipitation of an otherwise insoluble salt on addition of a particular anion. The complexes, however, are all unstable in excess acid and can be broken down on addition of 6 M HNO_3, releasing the cation for further reactions.

In the first part of the experiment you will observe the behavior of the four cations in the presence of one or more of the reagents we have listed. On the basis of your observations you can set up the scheme for identifying the cations in a mixture. After testing the scheme with a known containing all of the cations, you will be given an unknown to analyze.

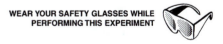

WEAR YOUR SAFETY GLASSES WHILE
PERFORMING THIS EXPERIMENT

Experimental Procedure

To four small test tubes add 1 mL of 0.1 M solutions of the nitrates or chlorides of Ba^{2+}, Mg^{2+}, Cd^{2+}, and Al^{3+}, one solution to a test tube (depth of 1 cm $\cong$ 1 mL). To each solution add one drop of 1 M Na_2SO_4, and stir with your stirring rod. Keep the rod in a 400-mL beaker of distilled water between tests. If a precipitate forms, write the formula of the precipitate in the box on the Data page corresponding to the cation-SO_4^{2-} pair, to indicate that the sulfate of that cation is insoluble in water. Then add 1 mL more of the Na_2SO_4 to each test tube. If the precipitate dissolves on stirring, the cation forms a complex ion with sulfate ion. Under such conditions, put the formula of the complex ion in the box. In this experiment you can assume that any cations that form complex ions have a coordination number of four. To any precipitates or complex ions, add 6 M HNO_3, drop by drop, until the solution is acidic to litmus (blue to red). If the precipitate dissolves, note that with an A, meaning that the precipitate dissolves in acid. In the case of complexes, the precipitate that originally formed may reprecipitate if the ligands react with acid, and then dissolve again when the solution becomes acidic. If it reprecipitates, indicate that with a P, and as before use an A if the precipitate dissolves when the solution becomes acidic.

Pour the contents of the four tubes into a beaker and rinse the tubes with distilled water. Repeat the tests, first with 1 M Na_2CO_3, then with 6 M NaOH, and finally with 6 M NH_3. Although in most cases you will not observe formation of complex ions, you will see a few, and it is important not to miss them. One drop of reagent typically will produce a precipitate if the cation-anion compound is insoluble in water. The complex may be very stable and form readily in excess reagent. You may not get a precipitate reforming when you add HNO_3. If the solution becomes acidic, and you saw no precipitate or a faint one, slowly add 1 M Na_2CO_3. If the carbonate comes down, write its formula in the box.

When your table is complete you should be able to use it to state whether the sulfate, carbonate, and hydroxide of each of the cations is insoluble in water, and whether it dissolves in acid. You should also know whether the cation forms a complex ion with SO_4^{2-}, CO_3^{2-}, OH^-, or NH_3, and whether the complex is destroyed by acid.

Now, given the solubility and reaction data you have obtained, your problem is to devise a step-by-step procedure for establishing whether each cation is present in a mixture. If you think about it for a while, several possible approaches should occur to you. As you construct your scheme, there are a few things to keep in mind besides the solubility and reaction data.

1. To separate a precipitate from a solution, you can use a centrifuge. Decant the solution into a test tube for use in further steps. See Appendix IV for some suggestions regarding procedures in qualitative analysis.
2. If a precipitate can contain only one cation, its presence serves to prove the presence of that cation. If it may involve more than one cation, it must be further resolved. In that case, the precipitate must be washed free of any cations that did not precipitate in that step because those cations would possibly interfere with later steps. To clean a precipitate, wash it twice with a 1:1 mixture of water and the precipitating reagent, stirring before centrifuging out the wash liquid.
3. pH is important. Your original tests were made with the cations in a water solution. If you want to obtain a precipitate, the solution must have a pH where that precipitate can form. 6 M HNO_3 or 6 M NaOH can be used to bring a mixture to a pH of about 7.

When you have your separation scheme in mind, describe it by constructing a flow diagram. The design of a flow diagram is discussed in the Advance Study Assignment. On the flow diagram the formulas of all species involving the cations should be given at the beginning and end of each step. Reagents are shown alongside the line connecting reactants and products.

When you have completed your flow diagram, test your scheme with a known containing all four cations. If your scheme works, show your flow diagram to your instructor, who will give you an unknown to analyze.

DISPOSAL OF REACTION PRODUCTS. On completing the experiment, pour the contents of the beaker used for reaction products into the waste crock unless directed otherwise by your instructor.

Experiment 34

Observations and Report Sheet: Development of a Scheme
for Qualitative Analysis

Table of Solubility Properties

	Ba^{2+}	Mg^{2+}	Cd^{2+}	Al^{3+}
1 M Na_2SO_4				
1 M Na_2CO_3				
6 M NaOH				
6 M NH_3				

Key: No entry = soluble on mixing reagents
    Top formula = precipitate insoluble in water
    Second formula = complex ion that forms in excess reagent
    Bottom formula = carbonate that precipitates from acidified complex
    A = precipitate dissolves in acid
    P = precipitate reforms when complex is slowly acidified

Flow Diagram for Separation Scheme:

Observations on known:

Observations on unknown:

Cations present in unknown: _____ _____ _____ _____

Unknown no. _____

Experiment 34

Advance Study Assignment: Development of a Scheme
for Qualitative Analysis

Qualitative analysis schemes can be summarized by flow diagrams. The flow diagram for a scheme that might be used to analyze a mixture that could contain Cu^{2+}, Pb^{2+}, and Sn^{2+} is shown below:

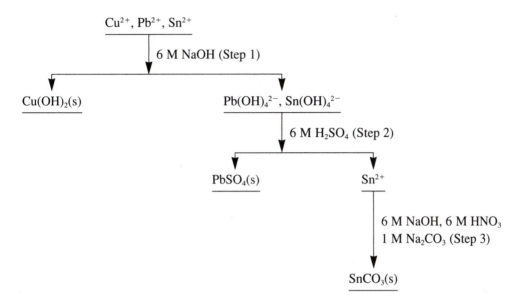

The meaning of the flow diagram is almost obvious. In the first step, 6 M NaOH is added in excess. $Cu(OH)_2$ precipitates, and Pb^{2+} and Sn^{2+} remain in solution because of the formation of hydroxo-complex ions. $Cu(OH)_2$ is removed by centrifuging. In Step 2 the solution is made acidic with H_2SO_4, and $PbSO_4$ precipitates after the complex is destroyed by the H^+ ion in the acid. $PbSO_4$ is removed by centrifuging. In Step 3 pH control is used to bring the pH to about 7. Then addition of Na_2CO_3 precipitates $SnCO_3$.

1. Construct the flow diagram for the separation scheme for a solution containing Ag^+, Ni^{2+}, and Zn^{2+}. The steps in the procedure are as follows:

    Step 1. Add 6 M HCl to precipitate Ag^+ as AgCl. Ni^{2+} and Zn^{2+} are not affected. Centrifuge out the AgCl.

    Step 2. Add 6 M NaOH in excess, precipitating $Ni(OH)_2(s)$ and converting Zn^{2+} to the $Zn(OH)_4^{2-}$ complex ion. Centrifuge out the $Ni(OH)_2$.

    Step 3. Neutralize the solution with 6 M HCl to pH = 7. Add 1 M Na_2CO_3, precipitating $ZnCO_3$.

    Use the following page for your flow diagram.

Experiment 35

Spot Tests for Some Common Anions

There are two broad categories of problems in analytical chemistry. Quantitative analysis deals with the determination of the amounts of certain species present in a sample; there are several experiments in this manual involving quantitative analysis, and you probably have performed some of them. The other area of analysis, called qualitative analysis, has a more limited purpose, establishing whether given species are or are not present in detectable amounts in a sample. Several of the experiments in this manual have dealt with problems in qualitative analysis.

One can carry out the qualitative analysis of a sample in various ways. Probably the simplest approach, which we will use in this experiment, is to test for the presence of each possible component by adding a reagent that will cause the component, if it is in the sample, to react in a characteristic way. This method involves a series of "spot" tests, one for each component, carried out on separate samples of the unknown. The difficulty with this way of doing qualitative analysis is that frequently, particularly in complex mixtures, one species may interfere with the analytical test for another. Although interferences are common, there are many ions that can, under optimum conditions at least, be identified in mixtures by simple spot tests.

In this experiment we will use spot tests for the analysis of a mixture that may contain the following commonly encountered anions in solution:

$$CO_3^{2-} \qquad PO_4^{3-} \qquad Cl^- \qquad SCN^-$$

$$SO_4^{2-} \qquad SO_3^{2-} \qquad C_2H_3O_2^- \qquad NO_3^-$$

The procedures we will use involve simple acid-base, precipitation, complex ion formation, or oxidation-reduction reactions. In each case you should try to recognize the kind of reaction that occurs, so that you can write the net ionic equation that describes it.

WEAR YOUR SAFETY GLASSES WHILE
PERFORMING THIS EXPERIMENT

Experimental Procedure

Carry out the test for each of the anions as directed. Repeat each test using a solution made by diluting the anion solution 9:1 with distilled water; use your 10-mL graduated cylinder to make the dilution and make sure you mix well before taking the sample for analysis. In some of the tests, a boiling-water bath containing about 100 mL water in a 150-mL beaker will be needed, so set that up before proceeding. When performing a test, if no reaction is immediately apparent, stir the mixture with your stirring rod to mix the reagents. These tests can easily be used to detect the anions at concentrations of 0.02 M or greater, but in dilute solutions careful observation may be required.

Test for the Presence of Carbonate Ion, CO_3^{2-}

Cautiously add 1 mL of 6 M HCl to 1 mL of 1 M Na_2CO_3 in a small test tube. With concentrated solutions, bubbles of carbon dioxide gas are immediately evolved. With dilute solutions, the effervescence will be much less obvious. Warming in the water bath, with stirring, will increase the amount of bubble formation. Carbon dioxide is colorless and odorless.

Test for the Presence of Sulfate Ion, SO_4^{2-}

Add 1 mL of 6 M HCl to 1 mL of 0.5 M Na_2SO_4. Add a few drops of 1 M $BaCl_2$. A white, finely divided precipitate of $BaSO_4$ indicates the presence of SO_4^{2-} ion.

Test for the Presence of Phosphate Ion, PO_4^{3-}

Add 1 mL 6 M HNO_3 to 1 mL of 0.5 M Na_2HPO_4. Then add 1 mL of 0.5 M $(NH_4)_2MoO_4$ and stir thoroughly. A yellow precipitate of ammonium phosphomolybdate, $(NH_4)_3PO_4 \cdot 12\ MoO_3$, establishes the presence of phosphate. The precipitate may form slowly, particularly in the more dilute solution; if it does not appear promptly, put the test tube in the boiling-water bath for a few minutes.

Test for the Presence of Sulfite Ion, SO_3^{2-}

Sulfite ion in acid solution tends to evolve SO_2, which can be detected by its odor even at low concentrations. Sulfite ion is slowly oxidized to sulfate in moist air, so sulfite-containing solutions will usually test positive for sulfate ion.

To 1 mL 0.5 M Na_2SO_3 add 1 mL 6 M HCl, and mix with your stirring rod. Cautiously, sniff the rod to try to detect the acrid odor of SO_2, which is a good test for sulfite. If the odor is too faint to detect, put the test tube in the hot-water bath for 10 seconds and sniff again.

To proceed with a chemical test, add 1 mL 1 M $BaCl_2$ to the solution in the tube. Stir, and centrifuge out any precipitate of $BaSO_4$. Decant the clear solution into a test tube and add 1 mL 3% H_2O_2, hydrogen peroxide. Stir the solution and let it stand for a few seconds. If sulfite is present, its oxidation to sulfate will cause a new precipitate of $BaSO_4$ to form.

If thiocyanate is present, it may interfere with the chemical test. In that case, add 1 mL 1 M $BaCl_2$ to 1 mL of the sample. Centrifuge out any precipitate, which will contain $BaSO_3$ if sulfite is present, but will not contain thiocyanate. Wash the solid with 2 mL water, stir, centrifuge, and discard the wash. To the precipitate add 1 mL 6 M HCl and 2 mL water, and stir. Centrifuge out any $BaSO_4$, and decant the clear liquid into a test tube. To the liquid add 1 mL 3% H_2O_2. If you have sulfite present, you will observe a new precipitate of $BaSO_4$ within a few seconds.

Test for the Presence of Thiocyanate Ion, SCN^-

Add 1 mL 6 M acetic acid, $HC_2H_3O_2$, to 0.5 M KSCN and stir. Add one or two drops 0.1 M $Fe(NO_3)_3$. A deep red coloration as a result of formation of $FeSCN^{2+}$ ion is proof for the presence of SCN^- ion.

Test for the Presence of Chloride Ion, Cl^-

Add 1 mL 6 M HNO_3 to 1 mL 0.5 M NaCl. Add two or three drops 0.1 M $AgNO_3$. A white, curdy precipitate of AgCl will form if chloride ion is present.

Several anions interfere with this test, because they too form white precipitates with $AgNO_3$ under these conditions. In this experiment only SCN^- ion will interfere. If the sample contains SCN^- ion, put 1 mL of the solution into a 30- or 50-mL beaker and add 1 mL 6 M HNO_3. Boil the solution gently until its volume is decreased by half. This will destroy most of the thiocyanate. To the solution in a small test tube add 1 mL 6 M HNO_3 and a few drops of $AgNO_3$ solution. If Cl^- is present you will get a curdy precipitate. If Cl^- is absent, you may see some cloudiness due to residual amounts of SCN^-.

Test for the Presence of Acetate Ion, $C_2H_3O_2^-$

To 1 mL 0.5 M $NaC_2H_3O_2$ add 6 M NH_3 or 6 M HNO_3 until the solution is just basic to litmus. Add one drop of 1 M $BaCl_2$. If a precipitate forms, add 1 mL of the $BaCl_2$ solution to precipitate interfering anions. Stir, centrifuge, and decant the clear liquid into a test tube. Add one drop $BaCl_2$ to make sure that precipitation was complete.

To 1 mL of the liquid add 0.1 M KI_3, drop by drop, until the solution takes on a fairly strong rust color. Add 0.5 mL 0.1 M $La(NO_3)_3$ and six drops 6 M NH_3. Stir, and put the test tube in the water bath. If acetate is present, the mixture will darken to nearly black in a few minutes. The color is due to iodine adsorbed on the basic lanthanum acetate precipitate.

Test for the Presence of Nitrate Ion, NO_3^-

To 1 mL 0.5 M $NaNO_3$ add 1 mL 6 M NaOH. Then add a few granules of Al metal, using your spatula, and put the test tube in the hot-water bath. In a few seconds, the Al-NaOH reaction will produce H_2 gas, which will reduce the NO_3^- ion to NH_3, which will come off as a gas. To detect the NH_3, hold a piece of moistened red litmus paper just above the end of the test tube. If the sample contains nitrate ion, the litmus paper will gradually turn blue, within a minute or two. Blue spots caused by effervescence are not to be confused with the blue color over all of the litmus exposed to NH_3 vapors. Cautiously sniff the vapors at the top of the tube; you may be able to detect the odor of ammonia.

If SCN^- is present, it will interfere with the test. In that case, first add 1 mL 1 M $CuSO_4$ to 1 mL of the sample and put the test tube in the water bath for a minute or two. Centrifuge out the precipitate, and decant the solution into a test tube. Add 1 mL 1 Na_2CO_3 to remove excess Cu^{2+} ion. Centrifuge out the precipitate, and decant the solution into a test tube. To 1 mL of the solution add 1 mL 6 M NaOH, and proceed, starting with the second sentence of this procedure.

When you have completed all of the tests, obtain an unknown from your laboratory supervisor, and analyze it by applying the tests to separate 1-mL portions. The unknown will contain three or four ions on the list, so your test for a given ion may be affected by the presence of others. When you think you have properly analyzed your unknown, you may, if you wish, make a "known" with the composition you found and test it to see if it behaves as your unknown did.

DISPOSAL OF REACTION PRODUCTS. As you complete each test, pour the products into a beaker. When you have finished the experiment, pour the contents of the beaker into the waste crock, unless directed otherwise by your instructor.

Name _____   **Section** _____

Experiment 35

Observations and Report Sheet: Spot Tests for Some Common Anions

Observations and Comments on Spot Tests

Ion	Stock Solution	9:1 Dilution	Unknown
CO_3^{2-}			
SO_4^{2-}			
PO_4^{3-}			
SO_3^{2-}			
Cl^-			
$C_2H_3O_2^-$			
SCN^-			
NO_3^-			

Unknown no. _____ contains _____

Experiment 35

Advance Study Assignment: Spot Tests for Some Common Anions

1. Each of the observations listed was made on a different solution. Given the observations, state which ion studied in this experiment is present. If the test is not definitive, indicate that with a question mark.

   A. Addition of 6 M NaOH and Al to the solution produces a vapor that turns red litmus blue.
   Ion present:

   B. Addition of 6 M HCl produces a vapor with an acrid odor.
   Ion present:

   C. Addition of 6 M HCl produces an effervescence.
   Ion present:

   D. Addition of 6 M HNO_3 plus 0.1 M $AgNO_3$ produces a precipitate.
   Ion present:

   E. Addition of 6 M HNO_3 plus 1 M $BaCl_2$ produces a precipitate.
   Ion present:

   F. Addition of 6 M HNO_3 plus 0.5 M $(NH_4)_2MoO_4$ produces a precipitate.
   Ion present:

2. An unknown containing one or more of the ions studied in this experiment has the following properties:

   A. No effect on addition of 6 M HNO_3.

   B. No effect on addition of 0.1 M $AgNO_3$ to solution in Part A.

   C. White precipitate on addition of 1 M $BaCl_2$ to solution in Part A.

   D. Yellow precipitate on addition of $(NH_4)_2MoO_4$, to solution in Part A.

   On the basis of this information which ions are present, which are absent, and which are in doubt?

Present	Absent	In doubt

(continued on following page)

3. The chemical reactions that are used in the anion spot tests in this experiment are for the most part simple precipitation or acid-base reactions. Given the information in each test procedure, try to write the net ionic equation for the key reaction in each test.

A. CO_3^{2-}

B. SO_4^{2-}

C. PO_4^{3-} (Reactants are HPO_4^{2-}, NH_4^+, MoO_4^{2-}, and H^+; products are $(NH_4)_3PO_4 \cdot 12\ MoO_3$ and H_2O; no oxidation or reduction occurs.)

D. SO_3^{2-}

E. SCN^-

F. Cl^-

G. NO_3^- (Take as reactants Al and NO_3^-; as products NH_3 and AlO_2^-; final equation also contains OH^- and H_2O as reactants.)

Experiment 36

Qualitative Analysis of Group I Cations

Precipitation and Separation of Group I Ions

The chlorides of Pb^{2+}, Hg_2^{2+}, and Ag^+ are all insoluble in cold water. They can be removed as a group from solution by the addition of HCl. The reactions that occur are simple precipitations and can be represented by the equations:

$$Ag^+ (aq) + Cl^- (aq) \rightarrow AgCl(s) \tag{1}$$

$$Pb^{2+} (aq) + 2\ Cl^- (aq) \rightarrow PbCl_2(s) \tag{2}$$

$$Hg_2^{2+} (aq) + 2\ Cl^- (aq) \rightarrow Hg_2Cl_2(s) \tag{3}$$

It is important to add enough HCl to ensure complete precipitation, but not too large an excess. In concentrated HCl solution these chlorides tend to dissolve, producing chloro-complexes such as $AgCl_2^-$.

Lead chloride is separated from the other two chlorides by heating with water. The $PbCl_2$ dissolves in hot water by the reverse of Reaction 2:

$$PbCl_2(s) \rightarrow Pb^{2+} (aq) + 2\ Cl^- (aq) \tag{4}$$

Once Pb^{2+} has been put into solution, we can check for its presence by adding a solution of K_2CrO_4. The chromate ion, CrO_4^{2-}, gives a yellow precipitate with Pb^{2+}:

$$Pb^{2+} (aq) + CrO_4^{2-} (aq) \rightarrow PbCrO_4(s) \tag{5}$$
$$\text{yellow}$$

The other two insoluble chlorides, AgCl and Hg_2Cl_2, can be separated by adding aqueous ammonia. Silver chloride dissolves, forming the complex ion $Ag(NH_3)_2^+$:

$$AgCl(s) + 2\ NH_3 (aq) \rightarrow Ag(NH_3)_2^+ (aq) + Cl^- (aq) \tag{6}$$

Ammonia also reacts with Hg_2Cl_2 via a rather unusual oxidation-reduction reaction. The products include finely divided metallic mercury, which is black, and a compound of formula $HgNH_2Cl$, which is white:

$$Hg_2Cl_2(s) + 2\ NH_3 (aq) \rightarrow Hg(l) + HgNH_2Cl(s) + NH_4^+ (aq) + Cl^- (aq) \tag{7}$$
$$\text{white} \qquad\qquad \text{black} \quad\ \text{white}$$

As this reaction occurs, the solid appears to change color, from white to black or gray.

The solution containing $Ag(NH_3)_2^+$ needs to be further tested to establish the presence of silver. The addition of a strong acid (HNO_3) to the solution destroys the complex ion and reprecipitates silver chloride. We may consider that this reaction occurs in two steps:

$$Ag(NH_3)_2^+ (aq) + 2\ H^+ (aq) \rightarrow Ag^+ (aq) + 2\ NH_4^+ (aq)$$

$$\frac{Ag^+ (aq) + Cl^- (aq) \rightarrow AgCl(s)}{Ag(NH_3)_2^+ (aq) + 2\ H^+ (aq) + Cl^- (aq) \rightarrow AgCl(s) + 2\ NH_4^+ (aq)}$$
white

(8)

Experimental Procedure

WEAR YOUR SAFETY GLASSES WHILE
PERFORMING THIS EXPERIMENT

See Appendix IV for some suggestions regarding procedures in qualitative analysis.

Step 1. Precipitation of the Group I Cations. To gain familiarity with the procedures used in qualitative analysis we will first analyze a known Group I solution, made by mixing equal volumes of 0.1 M $AgNO_3$, 0.2 M $Pb(NO_3)_2$, and 0.1 M $Hg_2(NO_3)_2$.

Add two drops of 6 M HCl to 1 mL of the known solution in a small test tube. (1 mL $\cong$ 1 cm depth in the tube.) Mix with your stirring rod. Centrifuge the mixture, making sure there is a blank test tube containing about the same amount of water in the opposite opening in the centrifuge. Add one more drop of the 6 M HCl to test for completeness of precipitation. Centrifuge again if necessary. Decant the supernatant liquid into another test tube and save it for further tests if cations from other groups may be present. The precipitate will be white and will contain the chlorides of the Group I cations.

Step 2. Separation of Pb^{2+}. Wash the precipitate with 1 or 2 mL of water. Stir, centrifuge, and decant the liquid, which may be discarded.

Add 2 mL distilled water from your wash bottle to the precipitate in the test tube, and place the test tube in a 250-mL beaker that is about half full of boiling water. Leave the test tube in the bath for a minute or two, stirring occasionally with a glass rod. This will dissolve most of the $PbCl_2$, but not the other two chlorides. Centrifuge the hot mixture, and decant the hot liquid into a test tube. Save the remaining precipitate for further tests.

Step 3. Identification of Pb^{2+}. Add one drop of 6 M acetic acid and a few drops of 0.1 M K_2CrO_4 to the solution from Step 2. If Pb^{2+} is present, a bright yellow precipitate of $PbCrO_4$ will form.

Step 4. Separation and Identification of Hg_2^{2+}. To the precipitate from Step 2 add 1 mL 6 M NH_3 and stir thoroughly. Centrifuge the mixture and decant the liquid into a test tube. A gray or black precipitate, produced by reaction of Hg_2Cl_2 with ammonia, proves the presence of Hg_2^{2+}.

Step 5. Identification of Ag^+. Add 6 M HNO_3 to the solution from Step 4 until it is acidic to litmus paper. It will take about 1 mL. Test for acidity by dipping the end of your stirring rod in the solution, then touching it to a piece of blue litmus paper (red in acidic solution). If Ag^+ is present, in the acidified solution it will precipitate as white AgCl.

Step 6. When you have completed the tests on the known solution, obtain an unknown and analyze it for the possible presence of Ag^+, Pb^{2+}, and Hg_2^{2+}.

DISPOSAL OF REACTION PRODUCTS. All reaction products in this experiment should be dealt with as directed by your instructor.

Flow Diagrams

It is possible to summarize the directions for analysis of the Group I cations in a flow diagram. In the diagram, successive steps in the procedure are linked with arrows. Reactant cations or reactant substances containing the ions are at one end of each arrow and products formed are at the other end. Reagents and conditions used to carry out each step are placed alongside the arrows. A partially completed flow diagram for the Group I ions follows:

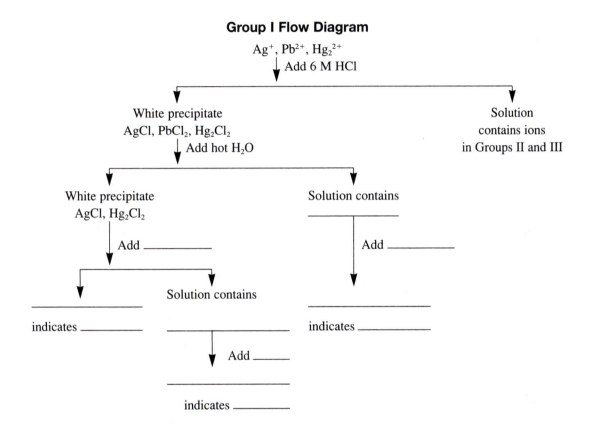

Group I Flow Diagram

Ag^+, Pb^{2+}, Hg_2^{2+}
↓ Add 6 M HCl

White precipitate
$AgCl$, $PbCl_2$, Hg_2Cl_2
↓ Add hot H_2O

Solution
contains ions
in Groups II and III

White precipitate
$AgCl$, Hg_2Cl_2
↓ Add _____

Solution contains
_____

_____   Solution contains

indicates _____   _____
↓ Add _____

Add _____

_____

indicates _____

_____

indicates _____

You will find it useful to construct flow diagrams for each of the cation groups. You can use such diagrams in the laboratory as brief guides to procedure, and you can use them to record your observations on your known and unknown solutions.

Name _____   Section _____

Experiment 36

Observations and Report Sheet: Qualitative Analysis of Group I Cations

Flow Diagram for Group I

Observations on known (record on diagram if different from those on prepared diagram).

Observations on unknown (record on diagram in colored pencil to distinguish from observations on known).

Unknown no. _____

Ions reported present _____ _____ _____

Experiment 36

Advance Study Assignment: Qualitative Analysis of Group I Cations

1. On the report sheet on page 289, complete the flow diagram for the separation and identification of the ions in Group I.

2. Write balanced net ionic equations for the following reactions:

   a. The precipitation of the chloride of Hg_2^{2+}.

   b. The dissolving of $PbCl_2$ in hot water.

   c. The dissolving of $AgCl$ in aqueous ammonia.

3. You are given an unknown solution that contains only one of the Group I cations and no other metallic cations. Develop the simplest procedure you can think of to determine which cation is present. Draw a flow chart showing the procedure and the observations to be expected at each step with each of the possible cations. The information in Appendix IIA should be helpful.

4. A solution may contain Ag^+, Pb^{2+}, and Hg_2^{2+}. A white precipitate forms on addition of 6 M HCl. The precipitate is insoluble in hot water. The residue turns black on addition of ammonia. Which of the ions are present, which are absent, and which remain undetermined? State your reasoning. *Note:* On paper unknowns such as this one, confirmatory tests are usually not included.

Present _____

Absent _____

In doubt _____

Experiment 37

Qualitative Analysis of Group II Cations

Precipitation and Separation of Group II Ions

The sulfides of the four Group II ions, Bi^{3+}, Sn^{4+}, Sb^{3+}, and Cu^{2+}, are insoluble at a pH of 0.5. The solution is adjusted to this pH and then saturated with H_2S, which precipitates Bi_2S_3, SnS_2, Sb_2S_3, and CuS. The reaction with Bi^{3+} is typical:

$$2\ Bi^{3+}\,(aq) + 3\ H_2S\,(aq) \rightarrow Bi_2S_3(s) + 6\ H^+\,(aq) \qquad (1)$$
$$\text{black}$$

Saturation with H_2S could be achieved by simply bubbling the gas from a generator through the solution. A more convenient method, however, is to heat the acid solution after adding a small amount of thioacetamide. This compound, CH_3CSNH_2, hydrolyzes when heated in aqueous solution to liberate H_2S:

$$CH_3CSNH_2\,(aq) + 2\ H_2O \rightarrow H_2S\,(aq) + CH_3COO^-\,(aq) + NH_4^+\,(aq) \qquad (2)$$

Using thioacetamide as the precipitating reagent has the advantage of minimizing odor problems and giving denser precipitates.

The four insoluble sulfides can be separated into two subgroups by extracting with a solution of sodium hydroxide. The sulfides of tin and antimony dissolve, forming hydroxo-complexes:

$$SnS_2\,(s) + 6\ OH^-\,(aq) \rightarrow Sn(OH)_6^{2-}\,(aq) + 2\ S^{2-}\,(aq) \qquad (3)$$

$$Sb_2S_3\,(s) + 8\ OH^-\,(aq) \rightarrow 2\ Sb(OH)_4^-\,(aq) + 3\ S^{2-}\,(aq) \qquad (4)$$

Since Cu^{2+} and Bi^{3+} do not readily form hydroxo-complexes, CuS and Bi_2S_3 do not dissolve in solutions of NaOH.

The solution containing the $Sb(OH)_4^-$ and $Sn(OH)_6^{2-}$ complex ions is treated with HCl and thioacetamide. The H^+ ions of the strong acid HCl destroy the hydroxo-complexes; the free cations then reprecipitate as the sulfides. The reaction with $Sn(OH)_6^{2-}$ may be written as:

$$Sn(OH)_6^{2-}\,(aq) + 6\ H^+\,(aq) \rightarrow Sn^{4+}\,(aq) + 6\ H_2O$$

$$\frac{Sn^{4+}\,(aq) + 2\ H_2S(aq) \rightarrow SnS_2(s) + 4\ H^+\,(aq)}{Sn(OH)_6^{2-}\,(aq) + 2\ H^+\,(aq) + 2\ H_2S(aq) \rightarrow SnS_2(s) + 6\ H_2O} \qquad (5)$$
$$\text{tan}$$

The $Sb(OH)_4^-$ ion behaves in a very similar manner, being converted first to Sb^{3+} and then to Sb_2S_3. The Sb_2S_3 and SnS_2 are then dissolved as chloro-complexes in hydrochloric acid and their presence is confirmed by appropriate tests.

The confirmatory test for tin takes advantage of the two oxidation states, +2 and +4, of the metal. Aluminum is added to reduce Sn^{4+} to Sn^{2+}:

$$2\ Al(s) + 3\ Sn^{4+}\,(aq) \rightarrow 2\ Al^{3+}\,(aq) + 3\ Sn^{2+}\,(aq) \qquad (6)$$

The presence of Sn^{2+} in this solution is detected by adding mercury(II) chloride, $HgCl_2$. This brings about another oxidation-reduction reaction:

$$Sn^{2+}\,(aq) + 2\ Hg^{2+}\,(aq) + 2\ Cl^-\,(aq) \rightarrow Sn^{4+}\,(aq) + Hg_2Cl_2\,(s) \qquad (7)$$
$$\text{white}$$

Formation of a white precipitate of insoluble Hg_2Cl_2 confirms the presence of tin. (The precipitate may be grayish because of the formation of finely divided Hg by an oxidation-reduction reaction similar to Reaction 7.)

The Sb^{3+} ion is difficult to confirm in the presence of Sn^{4+}; the colors of the sulfides of these two ions are similar. To prevent interference by Sn^{4+}, the solution to be tested for Sb^{3+} is first treated with oxalic acid. This forms a very stable oxalato-complex with Sn^{4+}, $Sn(C_2O_4)_3^{2-}$. Treatment with H_2S then gives a bright-orange precipitate of Sb_2S_3 if antimony is present:

$$2\ Sb^{3+}(aq) + 3\ H_2S(aq) \rightarrow Sb_2S_3(s) + 6\ H^+(aq) \qquad (8)$$
$$\text{orange}$$

As pointed out earlier, CuS and Bi_2S_3 are insoluble in NaOH solutions, and they do not dissolve in hydrochloric acid. However, these two sulfides can be brought into solution by treatment with the oxidizing acid, HNO_3. The reactions that occur are of the oxidation-reduction type. The NO_3^- ion is reduced, usually to NO_2; S^{2-} ions are oxidized to elementary sulfur and the cation, Cu^{2+} or Bi^{3+}, is brought into solution. The equations are:

$$CuS(s) + 4\ H^+(aq) + 2\ NO_3^-(aq) \rightarrow Cu^{2+}(aq) + S(s) + 2\ NO_2(g) + 2\ H_2O \qquad (9)$$

$$Bi_2S_3(s) + 12\ H^+(aq) + 6\ NO_3^-(aq) \rightarrow 2\ Bi^{3+}(aq) + 3\ S(s) + 6\ NO_2(g) + 6\ H_2O \qquad (10)$$

The two ions, Cu^{2+} and Bi^{3+}, are easily separated by the addition of aqueous ammonia. The Cu^{2+} ion is converted to the deep-blue complex, $Cu(NH_3)_4^{2+}$:

$$Cu^{2+}(aq) + 4\ NH_3(aq) \rightarrow Cu(NH_3)_4^{2+}(aq) \qquad (11)$$
$$\text{deep blue}$$

The reaction of ammonia with Bi^{3+} is quite different. The OH^- ions produced by the reaction of NH_3 with water precipitate Bi^{3+} as $Bi(OH)_3$. We can consider that the reaction occurs in two steps:

$$3\ NH_3(aq) + 3\ H_2O \rightleftarrows 3\ NH_4^+(aq) + 3\ OH^-(aq)$$

$$\frac{Bi^{3+}(aq) + 3\ OH^-(aq) \rightarrow Bi(OH)_3(s)}{Bi^{3+}(aq) + 3\ NH_3(aq) + 3\ H_2O \rightarrow Bi(OH)_3(s) + 3\ NH_4^+(aq)} \qquad (12)$$
$$\text{white}$$

To confirm the presence of Bi^{3+}, the precipitate of $Bi(OH)_3$ is dissolved by treating with hydrochloric acid:

$$Bi(OH)_3(s) + 3\ H^+(aq) \rightarrow Bi^{3+}(aq) + 3\ H_2O \qquad (13)$$

The solution formed is poured into distilled water. If Bi^{3+} is present, a white precipitate of bismuth oxychloride, BiOCl, will form:

$$Bi^{3+}(aq) + H_2O + Cl^-(aq) \rightarrow BiOCl(s) + 2\ H^+(aq) \qquad (14)$$
$$\text{white}$$

WEAR YOUR SAFETY GLASSES WHILE PERFORMING THIS EXPERIMENT

Experimental Procedure

Step 1. **Adjustment of pH Prior to Precipitation.** Pour a 1-mL sample of the "known" solution for Group II, containing equal volumes of 0.1 M solutions of the nitrates or chlorides of Sn^{4+}, Sb^{3+}, Cu^{2+}, and Bi^{3+}, into a small test tube. Prepare a boiling-water bath, using a 250-mL beaker about 2/3 full of water. Use another beaker full of water as the storage and rinsing place for your stirring rods. Prepare a pH test paper by making ten spots of methyl violet indicator on a piece of filter paper, about one drop to a spot. Let the paper dry in the air.

The Group II known will be very acidic because of the presence of HCl, which is necessary to keep the salts of Bi^{3+}, Sn^{4+}, and Sb^{3+} in solution. Add 6 M NH_3, drop by drop, until the solution, after stirring, produces a violet spot on the pH test paper; test for pH by dipping your stirring rod in the solution and touching it to the paper. (A precipitate will probably form during this step because of the formation of insoluble salts of the Group II cations.) Then add 1 drop of 6 M HCl for each milliliter of solution. This should bring the pH of the solution to about 0.5. Test the pH, using the test paper. At a pH of 0.5, methyl violet will have a blue-green color. Compare your spot with that made by putting a drop of 0.3 M HCl on the test paper (pH = 0.5). Adjust the pH of your solution as necessary by adding HCl or NH_3, until the pH test gives about the right color on the paper. If you have trouble deciding on the color, centrifuging out the precipitate may help. When you have established the proper pH, add 1 mL 1 M thioacetamide to the solution and stir.

Step 2.   **Precipitation of the Group II Sulfides.** Heat the test tube in the boiling water bath for at least five minutes. **CAUTION:** *Small amounts of H_2S will be liberated; this gas is toxic, so avoid inhaling it unnecessarily.*

In the presence of Group II ions, a precipitate will form: typically, its color will be initially light, gradually darkening, and finally becoming black. Continue to heat the tube for at least two minutes after the color has stopped changing. Cool the test tube under the water tap and let it stand for a minute or so. Centrifuge out the precipitate and decant the solution, which will contain any Group III ions if they are present, into a test tube. Test the solution for completeness of precipitation by adding two drops of thioacetamide and letting it stand for a minute. If a precipitate forms, add a few drops of thioacetamide and heat again in the water bath. Combine the two batches of precipitate. Wash the precipitate with 2 mL 1 M NH_4Cl solution; stir thoroughly. Centrifuge and discard the wash into a waste beaker.

Step 3.   **Separation of the Group II Sulfides into Two Subgroups.** To the precipitate from Step 2, add 2 mL 1 M NaOH. Heat in the water bath, with stirring, for two minutes. Any SnS_2 or Sb_2S_3 should dissolve. The residue will typically be dark and may contain CuS and Bi_2S_3. Centrifuge and decant the yellow liquid into a test tube. Wash the precipitate twice with 2 mL water and a few drops of 1 M NaOH. Stir, centrifuge, and decant, discarding the liquid each time into a waste beaker. To the precipitate add 2 mL 6 M HNO_3 and put the test tube aside (Step 8). At this point the tin and antimony are present in the solution as complex ions, and the copper and bismuth are in the sulfide precipitate.

Step 4.   **Reprecipitation of SnS_2 and Sb_2S_3.** To the yellow liquid from Step 3 add 6 M HCl drop by drop until the mixture is just acidic to litmus. Upon acidification, the tin and antimony will again precipitate as orange sulfides. Add five drops of 1 M thioacetamide and heat in the water bath for two minutes to complete the precipitation. Centrifuge, and decant the liquid, which may be discarded.

Step 5.   **Dissolving SnS_2 and Sb_2S_3 in Acid Solution.** Add 2 mL 6 M HCl to the precipitate from Step 4, and heat in the water bath for a minute or two to dissolve the precipitate. Transfer the solution to a 30-mL beaker, and boil it, gently, for about a minute, to drive out the H_2S. Add 1 mL 6 M HCl and 1 mL water and pour the liquid back into a small test tube. If there is an insoluble residue, centrifuge it out and decant the liquid into a test tube.

Step 6.   **Confirmation of the Presence of Tin.** Pour half of the solution from Step 5 into a test tube and add 2 mL 6 M HCl and a 1-cm length of 24-gauge aluminum wire. Heat the test tube in the water bath to promote reaction of the Al and production of H_2. In this reducing medium, any tin present will be converted to Sn^{2+} and any antimony to the metal, which will appear as black specks. Heat the tube for two minutes after *all* of the wire has reacted. Centrifuge out any solid and decant the liquid into a test tube. To the liquid add a drop or two of 0.1 M $HgCl_2$. A white or gray cloudiness, produced as Hg_2Cl_2 or Hg slowly forms, establishes the presence of tin.

Step 7.   **Confirmation of the Presence of Antimony.** To the other half of the solution from Step 5, add 1 M NaOH to bring the pH to 0.5. Add 2 mL water and about 0.5 g of oxalic acid, and stir until no more crystals dissolve. Oxalic acid forms a very stable complex with the Sn^{4+} ion. Add 1 mL 1 M thioacetamide and put the test tube in the water bath. The formation of a red-orange precipitate of Sb_2S_3 confirms the presence of antimony.

Step 8. **Dissolving the CuS and Bi_2S_3.** Heat the test tube containing the precipitate from Step 3 in the hot water bath. Any sulfides that have not already dissolved should go into solution in a minute or two, possibly leaving some insoluble sulfur residue. Continue heating until no further reaction appears to occur, at least two minutes after the initial changes. Centrifuge and decant the solution, which may contain Cu^{2+} and Bi^{3+}, into a test tube. Discard the residue.

Step 9. **Confirmation of the Presence of Copper.** To the solution from Step 8 add 6 M NH_3 dropwise until the mixture is just basic to litmus. Add 0.5 mL more. Centrifuge out any white precipitate that forms, and decant the liquid into a test tube. If the liquid is deep blue, the color is due to the $Cu(NH_3)_4^{2+}$ ion, and copper is present.

Step 10. **Confirmation of the Presence of Bismuth.** Wash the precipitate from Step 9, which probably contains bismuth, with 1 mL water and 0.5 mL 6 M NH_3. Stir, centrifuge, and discard the wash. To the precipitate add 0.5 mL 6 M HCl and 0.5 mL water. Stir to dissolve any $Bi(OH)_3$ that is present. Add the solution drop by drop, to 400 mL water in a 600-mL beaker. A white cloudiness, caused by slow precipitation of BiOCl, confirms the presence of bismuth.

Step 11. When you have completed the analysis of your known, obtain a Group II unknown and test it for the presence of Sn^{4+}, Sb^{3+}, Cu^{2+}, and Bi^{3+}.

Optional Experiment: Establish the presence of copper ion in a mineral supplement tablet. Estimate the amount of copper present by comparison with a suitable standard.

DISPOSAL OF REACTION PRODUCTS. As you complete each part of the experiment, put the waste products in a beaker. At the end of the experiment, pour the contents of the beaker into a waste crock, unless directed otherwise by your instructor.

Experiment 37

Observations and Report Sheet: Qualitative Analysis of
Group II Cations

Flow Diagram for Group II

Observations on known (record on diagram if different from those on prepared diagram).

Observations on unknown (record on diagram in colored pencil to distinguish from observations on known).

Unknown no. _____

Ions reported present _____ _____ _____ _____

Experiment 37

Advance Study Assignment: Qualitative Analysis of
Group II Cations

1. Prepare a complete flow diagram for the separation and identification of the Group II cations and put it on your report sheet on page 297.

2. Write balanced net ionic equations for the following reactions:

   a. Precipitation of the tin(IV) sulfide with H_2S.

   b. The confirmatory test for antimony.

   c. The confirmatory test for bismuth.

   d. The dissolving of Bi_2S_3 in hot nitric acid.

3. A solution that may contain Cu^{2+}, Bi^{3+}, Sn^{4+}, or Sb^{3+} ions is treated with thioacetamide in an acid medium. The black precipitate that forms is partly soluble in strongly alkaline solution. The precipitate that remains is soluble in 6 M HNO_3 and gives only a blue solution on treatment with excess NH_3. The alkaline solution, when acidified, produces an orange precipitate. On the basis of this information, which ions are present, which are absent, and which are still in doubt? (As with Group I, evidence other than confirmatory tests may show the presence or absence of an ion.)

   Present _____

   Absent _____

   In doubt _____

Experiment 38

Qualitative Analysis of Group III Cations

Precipitation and Separation of the Group III Ions

Four ions in Group III are Cr^{3+}, Al^{3+}, Fe^{3+}, and Ni^{2+}. The first step in the analysis involves treating the solution with sodium hydroxide, NaOH, and sodium hypochlorite, NaOCl. The OCl^- ion oxidizes Cr^{3+} to CrO_4^{2-}:

$$2\,Cr^{3+}(aq) + 3\,OCl^-(aq) + 10\,OH^-(aq) \rightarrow 2\,CrO_4^{2-}(aq) + 3\,Cl^-(aq) + 5\,H_2O \qquad (1)$$
yellow

The chromate ion, CrO_4^{2-}, stays in solution. The same is true of the hydroxo-complex ion $Al(OH)_4^-$, formed by the reaction of Al^{3+} with excess OH^-:

$$Al^{3+}(aq) + 4\,OH^-(aq) \rightarrow Al(OH)_4^-(aq) \qquad (2)$$

In contrast, the other two ions in the group form insoluble hydroxides under these conditions:

$$Ni^{2+}(aq) + 2\,OH^-(aq) \rightarrow Ni(OH)_2(s) \qquad (3)$$
green

$$Fe^{3+}(aq) + 3\,OH^-(aq) \rightarrow Fe(OH)_3(s) \qquad (4)$$
red

The Ni^{2+} and Fe^{3+} ions, unlike Al^{3+}, do not readily form hydroxo-complexes. Unlike Cr^{3+}, they do not have a stable higher oxidation state and so are not oxidized by OCl^-.

To separate aluminum from chromium, the solution containing CrO_4^{2-} and $Al(OH)_4^-$ is first acidified. This destroys the hydroxo-complex of aluminum:

$$Al(OH)_4^-(aq) + 4\,H^+(aq) \rightarrow Al^{3+}(aq) + 4\,H_2O \qquad (5)$$

Treatment with aqueous ammonia then gives a white gelatinous precipitate of aluminum hydroxide. We can think of this reaction as occurring in two steps:

$$3\,NH_3(aq) + 3\,H_2O \rightleftharpoons NH_4^+(aq) + 3\,OH^-(aq)$$

$$\underline{Al^{3+}(aq) + 3\,OH^-(aq) \rightarrow Al(OH)_3(s)}$$
$$Al^{3+}(aq) + 3\,NH_3(aq) + 3\,H_2O \rightarrow Al(OH)_3(s) + 3\,NH_4^+(aq) \qquad (6)$$
white

The concentration of OH^- in dilute NH_3 is too low to form the $Al(OH)_4^-$ complex ion by Reaction 2.

The CrO_4^{2-} ion remains in solution after Al^{3+} has been precipitated. It can be tested for by precipitation as yellow $BaCrO_4$ by the addition of $BaCl_2$ solution:

$$Ba^{2+}(aq) + CrO_4^{2-}(aq) \rightarrow BaCrO_4(s) \qquad (7)$$
light yellow

The precipitate of $BaCrO_4$ is dissolved in acid; the solution formed is then treated with hydrogen peroxide, H_2O_2. A deep-blue color is produced, because of the presence of a peroxo-compound, probably CrO_5. The reaction may be represented by the overall equation:

$$2\,BaCrO_4(s) + 4\,H^+(aq) + 4\,H_2O_2(aq) \rightarrow 2\,Ba^{2+}(aq) + 2\,CrO_5(aq) + 6\,H_2O \qquad (8)$$
blue

The mixed precipitate of $Ni(OH)_2$ and $Fe(OH)_3$ formed by Reactions 3 and 4 is dissolved by adding a strong acid, HNO_3. An acid-base reaction occurs:

$$Ni(OH)_2(s) + 2\ H^+(aq) \rightarrow Ni^{2+}(aq) + 2\ H_2O \tag{9}$$

$$Fe(OH)_3(s) + 3\ H^+(aq) \rightarrow Fe^{3+}(aq) + 3\ H_2O \tag{10}$$

At this point, the Ni^{2+} and Fe^{3+} ions are separated by adding ammonia. The Ni^{2+} ion is converted to the deep-blue complex $Ni(NH_3)_6^{2+}$, which stays in solution:

$$Ni^{2+}(aq) + 6\ NH_3(aq) \rightarrow Ni(NH_3)_6^{2+}(aq) \tag{11}$$
$$\text{blue}$$

while the Fe^{3+} ion, which does not readily form a complex with NH_3, is reprecipitated as $Fe(OH)_3$:

$$3\ NH_3(aq) + 3\ H_2O \rightleftharpoons 3\ NH_4^+(aq) + 3\ OH^-(aq)$$
$$\underline{Fe^{3+}(aq) + 3\ OH^-(aq) \rightarrow Fe(OH)_3(s)}$$
$$Fe^{3+}(aq) + 3\ NH_3(aq) + 3\ H_2O \rightarrow Fe(OH)_3(s) + 3\ NH_4^+(aq) \tag{12}$$
$$\text{red}$$

The confirmatory test for nickel in the solution is made by adding an organic reagent, dimethylglyoxime, $C_4H_8N_2O_2$. This gives a deep rose-colored precipitate with nickel:

$$Ni^{2+}(aq) + 2\ C_4H_8N_2O_2(aq) \rightarrow Ni(C_4H_7N_2O_2)_2(s) + 2\ H^+(aq) \tag{13}$$
$$\text{rose}$$

We can confirm Fe^{3+} by dissolving the precipitate of $Fe(OH)_3$ in HCl (Reaction 10) and adding KSCN solution. If iron(III) is present, the blood-red $FeSCN^{2+}$ complex ion will form:

$$Fe^{3+}(aq) + SCN^-(aq) \rightarrow FeSCN^{2+}(aq) \tag{14}$$
$$\text{red}$$

WEAR YOUR SAFETY GLASSES WHILE PERFORMING THIS EXPERIMENT

Experimental Procedure

Step 1. If you are testing a solution from which the Group II ions have been precipitated (Experiment 37), remove the excess H_2S and excess acid by boiling the solution until the volume is reduced to about 1 mL. Remove any sulfur residue by centrifuging the solution.

   If you are working on the analysis of Group III cations only, prepare a known solution containing Fe^{3+}, Al^{3+}, Cr^{3+}, and Ni^{2+} by mixing together 0.5-mL portions of each of the appropriate 0.1 M solutions containing those cations.

Step 2. Oxidation of Cr(III) to Cr(VI) and Separation of Insoluble Hydroxides. Add 1 mL 6 M NaOH to 1 mL of the known solution in a 30-mL beaker. Boil very gently for a minute, stirring to minimize bumping. Remove heat, and slowly add 1 mL 1 M NaOCl, sodium hypochlorite. Swirl the beaker for 30 seconds, using your tongs if necessary. Then boil the mixture gently for a minute. Add 0.5 mL 6 M NH_3 and let stand for 30 seconds. Then boil for another minute. Transfer the mixture to a test tube and centrifuge out the solid, which contains iron and nickel hydroxides. Decant the solution, which contains chromium and aluminum (CrO_4^{2-} and $Al(OH)_4^-$) ions, into a test tube (Step 3). Wash the solid twice with 2 mL water and 0.5 mL 6 M NaOH; after mixing, centrifuge each time, discarding the wash. Add 1 mL water and 1 mL 6 M H_2SO_4 to the solid and put the test tube aside (Step 6).

Step 3. Separation of Al from Cr. Acidify the solution from Step 2 by adding 6 M acetic acid slowly until, after stirring, the mixture is definitely acidic to litmus. If necessary, transfer the solution to a 50-mL beaker and boil it to reduce its volume to about 3 mL. Pour the solution into a test tube. Add 6 M NH_3, drop by drop, until the solution is basic to litmus, and then add 0.5 mL in excess. Sir the mixture for a minute or so to bring the system to equilibrium. If aluminum is present, a light, translucent, gelatinous white precipitate of $Al(OH)_3$ should be floating in the clear (possibly yellow) solution. Centrifuge out the solid, and transfer the liquid, which may contain CrO_4^{2-}, into a test tube (Step 5).

Step 4. Confirmation of the Presence of Aluminum. Wash the precipitate from Step 3 with 3 mL water once or twice, while warming the test tube in the water bath and stirring well. Centrifuge and discard the wash. Dissolve the precipitate in 2 drops 6 M $HC_2H_3O_2$, acetic acid, no more, no less. Add 3 mL water and 2 drops catechol violet reagent and stir. If Al^{3+} is present, the solution will turn blue.

Step 5. Confirmation of the Presence of Chromium. If the solution from Step 3 is yellow, chromium is probably present; if it is colorless, chromium is absent. To the solution add 0.5 mL 1 M $BaCl_2$. In the presence of chromium you obtain a finely divided yellow precipitate of $BaCrO_4$, which may be mixed with a white precipitate of $BaSO_4$. Put the test tube in the boiling water bath for a few minutes; then centrifuge out the solid and discard the liquid. Wash the solid with 2 mL water; centrifuge and discard the wash. To the solid add 0.5 mL 6 M HNO_3, and stir to dissolve the $BaCrO_4$. Add 1 mL water, stir the orange solution, and add two drops of 3% H_2O_2, hydrogen peroxide. A deep blue solution, which may fade quite rapidly, is confirmatory evidence for the presence of chromium.

Step 6. Separation of Iron and Nickel. Returning to the precipitate from Step 2, stir to dissolve the solid in the H_2SO_4. If necessary, warm the test tube in the water bath to complete the solution process. Then add 6 M NH_3 until the solution is basic to litmus. At that point iron will precipitate as brown $Fe(OH)_3$. Add 1 mL more of the NH_3, and stir to bring the nickel into solution as $Ni(NH_3)_6^{2+}$ ion. Centrifuge and decant the liquid into a test tube. Save the precipitate (Step 8).

Step 7. Confirmation of the Presence of Nickel. If the solution from Step 6 is blue, nickel is probably present. To that solution add 0.5 mL dimethylglyoxime reagent. Formation of a rose-red precipitate proves the presence of nickel.

Step 8. Confirmation of the Presence of Iron. Dissolve the precipitate from Step 6 in 0.5 mL 6 M HCl. Add 2 mL water and stir. Then add 2 drops of 0.5 M KSCN. Formation of a deep red solution of $FeSCN^{2+}$ is a definitive test for iron.

Step 9. When you have completed your analysis of the known solution, obtain a Group III unknown and test it for the possible presence of Fe^{3+}, Al^{3+}, Cr^{3+}, and Ni^{2+}.

DISPOSAL OF REACTION PRODUCTS.   As you complete each step in the procedure, put the waste products into a beaker. When you are finished with the experiment, pour the contents of the beaker into a waste crock, unless otherwise directed by your instructor.

Name _____ Section _____

Experiment 38

Observations and Report Sheet: Qualitative Analysis of
Group III Cations

Flow Diagram for Group III

Observations on known (record on diagram if different from those on prepared diagram).

Observations on unknown (record on diagram in colored pencil to distinguish from observations on known).

Unknown no. _____

Ions reported present _____ _____ _____ _____

Experiment 38

Advance Study Assignment: Qualitative Analysis of Group III Cations

1. Prepare a flow diagram for the separation and identification of the ions in Group III and put it on your report sheet.

2. Write balanced net ionic equations for the following reactions:

   a. Dissolving of $Fe(OH)_3$ in nitric acid.

   b. Oxidation of Cr^{3+} to CrO_4^{2-} by ClO^- in alkaline solution (ClO^- is converted to Cl^-).

   c. The confirmatory test for Ni^{2+}.

   d. The confirmatory test for Fe^{3+}.

3. A solution may contain any of the Group III cations. Treatment of the solution with ClO^- in alkaline medium yields a yellow solution and a colored precipitate. The acidified solution is unaffected by treatment with NH_3. The colored precipitate dissolves in nitric acid; addition of excess NH_3 to this acid solution produces only a blue solution. On the basis of this information, which Group III cations are present, absent, or still in doubt?

                                                    Present _____

                                                    Absent _____

                                                    In doubt _____

Experiment 39

Identification of a Pure Ionic Solid

In the previous set of experiments we have seen how one can identify which anions, or cations, are actually present in an unknown solution. In this experiment you will be asked to determine the cation and anion that are present in a solid sample of a pure ionic salt. The possible cations will include those we studied in Experiments 36, 37, and 38:

$$Ag^+ \quad Pb^{2+} \quad Hg_2^{2+} \quad Sn^{2+} \quad Sb^{3+}$$
$$Cu^{2+} \quad Bi^{3+} \quad Cr^{3+} \quad Al^{3+} \quad Ni^{2+} \quad Fe^{3+}$$

The anions we will consider are those in Experiment 35:

$$CO_3^{2-} \quad PO_4^{3-} \quad Cl^- \quad SCN^- \quad SO_4^{2-} \quad SO_3^{2-} \quad C_2H_3O_2^- \quad NO_3^-$$

There are 88 possible compounds that you might be given to identify. There are, of course, many ionic compounds that are not included, but the set is reasonably representative. At first sight it might seem that the best approach would be to carry out the procedures for analysis of Groups I, II, and III, in succession, and stop when you get to the cation in your unknown. That attack should work, but it would take longer than necessary, and would require that you have a solution of the compound to work with. Given a knowledge of the properties of the possible compounds, an experienced chemist would do some preliminary tests to find out the solubility properties of the sample and try to use those to narrow down very significantly the number of compounds she needed to consider. The solubility in various solvents, plus color, and odor if any, should afford some very useful information.

Most of the possible compounds are not soluble in water, but may dissolve in strong acids, like 6 M HCl, or strong bases, like 6 M NaOH, or solutions containing ligands that form stable complex ions with the cation in the compound. Some important complexing ligands that may dissolve otherwise insoluble solids containing the cations we are considering include:

$$NH_3 \quad OH^- \quad Cl^- \quad C_2H_3O_2^-$$

Nearly every compound in our set of 88 will dissolve in, or react with, at least one of the above species in solution.

Before beginning the analysis, it will be helpful to note some of the properties of the cations that distinguish them from one another. We have given those properties in Appendix IIA. You should read that Appendix and complete the Advance Study Assignment before coming to lab. You will need the information in the Appendix in your analysis.

WEAR YOUR SAFETY GLASSES WHILE PERFORMING THIS EXPERIMENT

DISCARD ALL REACTION PRODUCTS IN A 250-mL BEAKER

Experimental Procedure

In this experiment you will be given two unknowns to identify. The cation in your first sample will be colored, or will produce colored solutions. The second unknown will be colorless and somewhat more difficult to identify. In both cases the analysis will involve determining the solubility of the solid in water and some acidic or basic solutions. You should have about a half-gram of finely divided solid available. Read the section on qualitative analysis at the end of Appendix IV if you haven't already. Prepare a boiling-water bath and a bath for rinsing your stirring rods.

Identifying the Cation in a Colored Ionic Solid

The solvents we will consider are the following:

1. Water	2. 6 M HNO_3	3. 6 M HCl
4. 6 M NaOH	5. 6 M NH_3	6. 6 M H_2SO_4

Determine the solubility behavior of the solid in as many of the solvents as you need to find one in which the solid will dissolve. Use a small sample, about 50 mg, enough to cover a 2-mm circle on the end of a small spatula. Put the solid in a small test tube and add about 1 mL of water. Stir the solid with a stirring rod for a minute or so, noting any color changes of the solid or the solution. If the solid does not dissolve, put the tube in your boiling-water bath for a few minutes. If at this point you have a solution, add 5 mL more water and stir. From the color of the solution you should be able to make a good guess as to which cation is in your unknown. (The information in Appendix II or IIA should be helpful.) Use 1 mL of your solution to carry out the confirmatory test for that cation, as directed in Experiment 37 or 38. Make sure that you set the pH and any other conditions for the test properly before making a decision. If your guess is confirmed, you may proceed to analysis for the anion. If you were wrong, perform the confirmatory test for another cation that seems likely. It should not be difficult to identify the cation that is present.

If your sample does not dissolve in water, try 6 M HNO_3 as a solvent, using the same procedure as with water (about 1 mL of reagent and 50 mg of solid). If the sample goes into solution, in the cold, or in the water bath, add 5 mL of water, stir, and proceed to identify the cation present. With HNO_3 you may get some effervescence. This indicates that the anion is either carbonate or sulfite; with the former, odorless CO_2, is given off, while with the latter, SO_2, with its characteristic sharp odor, is evolved. Nearly all the possible compounds in our set that are colored are also soluble in either hot water or hot HNO_3. If you by chance have one that will not dissolve in those reagents, make the solubility test using 6 M HCl as the solvent. That should work if the first two solvents have not. In this part of the experiment you should not need to use the last three solvents on the list.

Anion Analysis

Before beginning analysis for the anion in your unknown, you should find it helpful to know that some anions are much more common than others in the chemicals that are ordinarily available in the stockroom. It is easiest to work with chemicals that are soluble in water. Nitrates, chlorides, sulfates, and acetates are often soluble, and are the most common salts we find on the shelves. Thiocyanates and sulfites are sometimes soluble, but are usually found as sodium or potassium salts, and so are not likely in the compounds in the set we are using. Carbonates and phosphates are insoluble in water, and occasionally are available. Hydroxides and oxides are commonly found in stockrooms, since they are easy to synthesize. They are insoluble, and will not be in our unknowns unless your instructor tells you otherwise.

Given the above information and the spot tests in Experiment 35, determine which anion is in your solid. For each test, use about 1 or 2 mL of the solution you prepared. You don't need to do all the spot tests, only those that seem reasonable in light of the previous discussion. If you used water as your solvent, any positive test for an anion should be conclusive. If you used HNO_3, you can't test for nitrate ion in that solution. Similarly you can't test for chloride ion in an HCl solution.

Some anions may be difficult to determine by the usual spot test, since the cation present may interfere. That is the case with acetate and nitrate, where the test solution is basic and will cause the cation to precipitate. With acetates, a simple test is to add 1 mL 6 M H_2SO_4 to 50 mg of the solid, as if you were testing for solubility. Heat the solution, or slurry, in the hot-water bath. If the sample contains acetate, acetic acid will be volatilized and can be detected by its characteristic odor. Put your stirring rod in the warm liquid, and, cautiously, sniff the rod. If there are no interfering anions, the odor of vinegar is an excellent test for acetate. With nitrates, begin by adding 1 mL 6 M NaOH to a small sample of solid unknown. If you get a hydroxide precipitate, you can try simply proceeding with the spot test. If the cation forms a hydroxide complex, you can try the spot test, and it may work. In many cases, the aluminum metal will reduce the cation to the metal, and you will get a black solid. With excess Al, you still may be able to detect evolved NH_3 and establish the presence of nitrate. The test for nitrate is perhaps the most difficult, and if worst comes to worst, you can treat your acidic solution of unknown with excess 1 M Na_2CO_3 until effervescence ceases and the solution is definitely basic; add 0.5 mL more. Centrifuge out the solid cation-containing carbonate, which can be discarded. Acidify the decanted liquid with 6 M H_2SO_4 drop by drop until CO_2 evolution ceases, and the solution is acidic. Carefully boil this solution down to about 2 mL and use it for the nitrate test and any other anion tests that seem indicated.

Analysis of a Colorless Ionic Solid

As your second unknown to analyze you will be given a white solid, which may have very limited solubility in water. It may also be insoluble in some of the acids and bases on our original list.

The procedure initially at least is the same as you used with the colored sample. Check the solubility of the solid in water, both at room temperature and in the water bath. Some of the solids do dissolve. With these, and the rest of the unknowns, the solubility in the other solvents is likely to furnish clues about the cation that is present. So, carry out solubility tests with all six of the solvents we listed (again, about 1 mL of reagent added to 50 mg of solid). Before deciding on the solubility in a given solvent, make sure you give the solid time to dissolve if it is going to, with stirring, particularly in the hot-water bath. Some solids have been in storage for years and are slow to get their act together. If the solid does not dissolve, add 1 mL water and stir; that may help. Solids containing cations that form complexes with Cl^- or OH^- are likely to dissolve in 6 M HCl or 6 NaOH, respectively. Those with cations that form ammonia complexes will tend to dissolve in 6 M NH_3. The solubility depends on the K_{sp} of the solid and the stability of the complex, so a complexing ligand may dissolve one solid and not another. Nitrate ion and sulfate ion are not very good complexing ions, so 6 M HNO_3 and 6 M H_2SO_4 will not dissolve solids by complex ion formation. If they dissolve a solid, it is by reacting with an anion in the solid to form a weak acid or by reacting with a basic salt like BiOCl. There are a few compounds in our possible set of unknowns that will not dissolve in any of the six reagents, or may dissolve in only one reagent. Those compounds are all discussed in Appendix IIA.

When you have completed the solubility tests on your solid, compare your results with those you obtained in your Advance Study Assignment from information in Appendix IIA. You should be able to narrow down the list of possible cations to only a few, and in some cases you will be able to decide not only which cation is present but also which anion. If you can dissolve the solid in water or in acid, you can proceed to carry out a confirmatory test for any likely cations, as described in Experiments 36, 37, and 38. With the colorless unknowns the number of likely anions is even less than with the first unknown, with nitrate and chloride being most common. If the sample will dissolve in water, or in HNO_3 or HCl, you can carry out spot tests for any anions that are not present in the solvent. For some solids, you will need to identify the anion on the basis of the solubility behavior of the salt.

On the report page note all your observations and the name and formula of the compound you believe is in your unknown.

DISPOSAL OF REACTION PRODUCTS. When you are finished with the experiment, discard all the reaction products in your 250-mL beaker in the waste crock.

Name _____ **Section** _____

Experiment 39

Observations and Report Sheet: Identification of a Pure Ionic Solid

Color of first unknown _____

Solubility properties of unknown

_____

_____

_____

Cations that are likely components of unknown _____ _____

Confirmatory tests performed to identify cation

    Procedures used

_____

_____

_____

    Results observed

_____

_____

_____

_____ is present

Spot tests performed to identify anion

    Procedures used

_____

_____

_____

    Results observed

_____

_____

_____

_____ is present

    Formula of colored compound _____

Solubility behavior of white unknown

Water _____   6 M HNO$_3$ _____

6 M HCl _____   6 M NaOH _____

6 M NH$_3$ _____   6 M H$_2$SO$_4$ _____

Reasoning and results of confirmatory tests

_____

_____

_____

_____

_____

_____

_____

_____

Formula of white compound _____

Experiment 39

Advance Study Assignment: Identification of a Pure Ionic Solid

Use the information in Appendix II and Appendix IIA to find answers to the questions.

1. What are the formulas and colors of the colored cations used in this experiment?

   _____

2. Which of the colorless cations

   A. will form complex ions with NH_3?   _____

   B. will form complex ions with OH^-?   _____

   C. will form complex ions with Cl^-?   _____

   D. have colored, rather than black, sulfides?   _____

   E. have one or more water-soluble salts?   _____

3. State the formula of a colorless compound that

   A. dissolves in 6 M NH_3.   _____

   B. turns black when NH_3 is added.   _____

   C. is much more soluble in hot water than in cold.   _____

   D. is soluble in 6 M NaOH but not in 6 M H_2SO_4.   _____

   E. is not soluble in any of the solvents we use.   _____

4. A colorless ionic solid is insoluble in water, and in 6 M NH_3, but will dissolve in 6 M HCl and 6 M NaOH. Name two compounds that would have these properties.

   _____ and _____

Experiment 40

The Ten Test Tube Mystery

Having carried out at least some of the previous experiments on qualitative analysis, you should have acquired some familiarity with the properties of the cations and anions you studied. Along with this you should at this point have had some experience with interpretation of the data in Appendix II on the solubilities of ionic substances in different media.

In this experiment we are going to ask you to apply what you have learned to a related but somewhat different kind of problem. You will be furnished ten numbered test tubes in each of which there will be a solution of a single substance. You will be provided, one week in advance, with a list giving the formula and molarity of each of the ten solutes that will be used. Your problem in the laboratory will be to find out which solution is in which test tube, that is, to assign a test tube number to each of the solution compositions. You are to do this by intermixing small volumes of the solutions in the test tubes. No external reagents or acid-base indicators such as litmus are allowed. You are permitted, however, to use the odor and color of the different solute species and to make use of heat effects in reactions in your system for identification. Each student will have a different set of test tube number-solution composition correlations, and there will be several different sets of compositions as well.

Of the ten solutions, four are common laboratory reagents. They are 6 M HCl, 3 M H_2SO_4, 6 M NH_3, and 6 M NaOH. The other solutions will usually be 0.1 M nitrate, chloride, or sulfate solutions of the cations, which have been studied in previous experiments in this manual (Experiments 12, 36, 37, and 38).

To determine which solution is in each test tube, you will need to know what happens when the various solutions are mixed, one with another. In some cases, nothing happens that you can observe. This will often be the case when a solution containing one of the cations is mixed with a solution of another. When one of the reagents is mixed with a cation solution you may get a precipitate, white or colored, and that precipitate may dissolve in excess reagent by complex ion formation. In a few cases a gas may be evolved. When one laboratory reagent is mixed with another, you may find that the resulting solution gets very hot and/or that a visible vapor is produced.

There is no way that you will be able to solve your particular test tube mystery without doing some preliminary work. You will need to know what to expect when any two of your ten solutions are mixed. You can find this out, for any pair, by scrutiny of Appendix II, by consulting your chemistry text, and by referring to various reference works on qualitative analysis.* A convenient way to tabulate the information you obtain is to set up a matrix with ten columns and ten rows, one for each solution. The key information about a mixture of two solutions is put in the space where the row for one solution and the column for the other intersect (as is done in Appendix II for various cations and anions). In that space you might put an NR, to indicate no apparent reaction on mixing of the two solutions. If a precipitate forms, put a P, followed by a D if the precipitate dissolves in excess reagent. If the precipitate or final solution is colored, state the color. If heat is evolved, write an H; if a gas or smoke is formed, put a G or S. Since mixing solution A with B is the same as mixing B with A, not all 100 spaces in the 10-by-10 matrix need to be filled. Actually, there are only 45 possible different pairs, since A with A is not very informative either.

Because you are allowed to use the odor or color of a solution to identify it, the problem is somewhat simpler than it might first appear. In each set of ten solutions you will probably be able to identify at least two solutions by odor and color tests. Knowing those solutions, you can make mixtures with the other solutions in which one of the components is known. From the results obtained with those mixtures, and the information in

---

*For example, the following references may be useful:

[1]*Qualitative Analysis and the Properties of Ions in Aqueous Solutions.* 2nd edition by E. J. Slowinski and W. L. Masterton, Saunders College Publishing, 1990.

[2]*Handbook of Chemistry and Physics.* Chemical Rubber Publishing, 2010.

the matrix, you can identify other solutions. These can be used to identify still others, until finally the entire set of ten is identified unequivocally.

Let us now go through the various steps that would be involved in the solution to a somewhat simpler problem than the one you will solve. There are several steps, including constructing the reaction matrix, identifying solutions by simple observations, and a rationale for efficient mixing tests. Let us assume we have to identify the following six solutions in numbered test tubes:

0.1 M $NiSO_4$	0.1 M $BiCl_3$ (in 3 M HCl)
0.1 M $BaCl_2$	6 M NaOH
0.1 M $Al(NO_3)_3$	3 M H_2SO_4

Construction of Reaction Matrix

In the matrix there will be six rows and six columns, one for each of the solutions. The matrix would be set up as in Table 40.1.

Each solution contains a cation and an anion, both of which need to be considered. We need to include every possible pair of solutions: for six solutions there are 15 possible pairs. So we need 15 pieces of information as to what happens when a pair of solutions is mixed. The first pair we might consider is $Al(NO_3)_3$ and $BaCl_2$, containing Al^{3+}, NO_3^-, Ba^{2+}, and Cl^- ions. Consulting Appendix II, we see that in the Al^{3+}, Cl^- space there is an S, meaning that $AlCl_3$ is soluble and will not precipitate when aluminum and chloride ions are mixed. Similarly, $Ba(NO_3)_2$ is soluble because all nitrates are soluble. This means that there will be no reaction when the solutions of $Al(NO_3)_3$ and $BaCl_2$ are mixed. So, in the space for $Al(NO_3)_3$-$BaCl_2$, we have written NR. The same is true if we mix solutions of $Al(NO_3)_3$ and $NiSO_4$, so we insert NR in that space. Since $BiCl_3$ won't react with $Al(NO_3)_3$ solutions, there is an NR in that space too. However, when $Al(NO_3)_3$ is mixed with NaOH, one of the possible products is $Al(OH)_3$, which we see in Appendix II is insoluble in water, but dissolves in acid and excess OH^- ion (A, B). This means that on adding NaOH to $Al(NO_3)_3$, we would initially get a precipitate of $Al(OH)_3$, P, but that it would dissolve in excess NaOH, D. The precipitate is white and the solution is colorless, so we have just a P and a D in the space for that mixture.

Proceeding now to the $BaCl_2$ row, we don't need to consider the first two columns, because they would give no new information. If $BaCl_2$ and $NiSO_4$ were mixed, $BaSO_4$ is a possible product. In Appendix II we see that $BaSO_4$ is insoluble in all common solvents (I). Hence, on mixing those solutions, we would get a precipitate of $BaSO_4$, P. Using the *Handbook of Chemistry and Physics,* or some other source, we find that it is white, so a P goes into that space. With $BiCl_3$ there would be no reaction; with NaOH we might get a slight precipitate (S$^-$ for $Ba(OH)_2$ in Appendix II), so we put a P in the appropriate space. With H_2SO_4, we would again expect to get a white precipitate of $BaSO_4$, P.

In the row for $NiSO_4$ we note that the color of the solution is green, since in Appendix II we see that Ni^{2+}(aq) ion is green. The first column is that for $BiCl_3$; no precipitate forms on mixing solutions of $NiSO_4$ and $BiCl_3$, so NR is in that space. With NaOH, $Ni(OH)_2$ would precipitate, since the entry in Appendix II for Ni^{2+}, OH^-, is not S. From the *Handbook* or other sources we find that the precipitate is green, so in the space we put a P (green). In Appendix II the $Ni(OH)_2$ entry is A, N, which tells us that the precipitate will dissolve

Table 40.1

Reaction Matrix for Six Solutions						
	$Al(NO_3)_3$	$BaCl_2$	$NiSO_4$	$BiCl_3$	NaOH	H_2SO_4
$Al(NO_3)_3$		NR	NR	NR	P, D	NR
$BaCl_2$			P	NR	sl P	P
$NiSO_4$(green)				NR	P (green)	NR
$BiCl_3$					P, H	NR
NaOH						H
H_2SO_4						

in 6 M NH_3 (but not in 6 M NaOH since there is no B in the space). There would be no reaction between solutions of $NiSO_4$ and H_2SO_4(NR).

With $BiCl_3$ in 3 M HCl we have both a salt and an acid in the solution. Addition of 6 M NaOH should produce a precipitate (white) of $Bi(OH)_3$, P. There would also be a substantial exothermic acid-base reaction between H^+ ions in the HCl and OH^- ions in the NaOH, so there would be a very noticeable rise in temperature on mixing, as denoted by H in the space. There would be no reaction with H_2SO_4.

When 6 M NaOH is mixed with 3 M H_2SO_4 there should be a large heat effect because of the acid-base reaction that occurs, so in the space for NaOH-H_2SO_4 we have an H.

The matrix in Table 40.1 summarizes the reaction information that will be needed for identification of the solutions in the test tubes. Your matrix can be constructed by the reasoning used in making this one.

Identifying Solutions by Simple Observations

When we made the matrix for the six test tube system, we noted that the $NiSO_4$ solution would be green, since Ni^{2+} ion is green. So, as soon as we see the six text tubes, we can pick out the 0.1 M $NiSO_4$. Let us assume that it is in Test Tube 4. We have now identified one solution out of six.

Selection of Efficient Mixing Tests

Knowing that Test Tube 4 contains $NiSO_4$, we mix the solution in Test Tube 4 with all of the other solutions. For each mixture, we record what we observe. The results might be as follows:

4 + 1	no obvious reaction
4 + 2	no obvious reaction
4 + 3	green precipitate forms
4 + 5	no obvious reaction
4 + 6	white precipitate

Referring to the matrix, we would expect precipitates for mixtures of $NiSO_4$ and NaOH, and $NiSO_4$ and $BaCl_2$. The former precipitate would be green, the latter white. Clearly the solution in Test Tube 3 is 6 M NaOH. The solution in Test Tube 6 must be 0.1 M $BaCl_2$. At this point three solutions have been identified.

Because 6 M NaOH reacts with several of the solutions, and because we know now that it is in Test Tube 3, we mix the solution in Test Tube 3 with those that remain unidentified, with the following results:

3 + 1	solution becomes hot
3 + 2	white precipitate, solution gets hot
3 + 5	apparently no reaction

Again, referring to the matrix we would expect heat evolution with mixtures of NaOH and H_2SO_4, and NaOH and the HCl in the $BiCl_3$. In addition, we would get a white precipitate with $BiCl_3$. It is obvious that Solution 1 is 3 M H_2SO_4. Solution 2 must be 0.1 M $BiCl_3$ in 3 M HCl. By elimination Solution 5 must be $Al(NO_3)_3$. We repeat the 3 + 5 mixture, since there should have been an initial precipitate that dissolved in excess NaOH. This time, using one drop of Solution 3 added to 1 mL of Solution 5, we get a precipitate. In excess of Solution 3 the precipitate dissolves as the aluminum hydroxide complex ion is formed.

WEAR YOUR SAFETY GLASSES WHILE
PERFORMING THIS EXPERIMENT

Experimental Procedure

Using the approach we have outlined in our example, carry out tests as necessary for the identification of the solutions in your set. After you have tentatively assigned each solution to its test tube, carry out some confirmatory mixing reactions, so that you have at least two mixtures that give results that support your conclusions.

DISPOSAL OF REACTION PRODUCTS. Dispose of all reaction products in the waste crock, unless directed otherwise by your instructor.

Experiment 40

Observations and Report Sheet: The Ten Test Tube Mystery

Solutions That Were Identified Directly

Solution	Observations	Conclusions	Test tube no.

Mixing Tests: First Series

Test tube nos.		Observations	Conclusions

Mixing Tests: Second Series

Test tube nos.		Observations	Conclusions

Confirming Tests (total of two per solution)

Test tube nos.		Observations	Conclusions

Final identifications: No. 1 _____   No. 2 _____   No. 3 _____

No. 4 _____   No. 5 _____   No. 6 _____   No. 7 _____

No. 8 _____   No. 9 _____   No. 10 _____

Unknown set no. _____

Experiment 40

Advance Study Assignment: The Ten Test Tube Mystery

1. Construct the reaction matrix for your set of ten solutions. This should go on the reverse side of this page. This matrix can be made using information in your chemistry text or in Appendix II or, if your instructor allows it, you can go into the lab prior to the scheduled session and work with known solutions to obtain the needed information.

2. Which solutions should you be able to identify by simple observations?

3. Outline the procedure you will follow in identifying the solutions that will require mixing tests. Be as specific as you can about what you will look for and what conclusions you will be able to draw from your observations.

Experiment 41

Preparation of Aspirin

One of the simpler organic reactions that one can carry out is the formation of an ester from an acid and an alcohol:

$$R-\overset{\overset{\displaystyle O}{\|}}{C}-OH + HO-R' \rightarrow R-\overset{\overset{\displaystyle O}{\|}}{C}-O-R' + H_2O \tag{1}$$

an acid   an alcohol   an ester

In the equation, R and R' are H atoms or organic fragments like CH_3, C_2H_5, or more complex aromatic groups. There are many esters, because there are many organic acids and alcohols, but they all can be formed, in principle at least, by Reaction 1. The driving force for the reaction is, in general, not very great, so that one ends up with an equilibrium mixture of ester, water, and acid, and alcohol.

There are some esters that are solids because of their high molecular weight or other properties. Most of these esters are not soluble in water, so they can be separated from the mixture by crystallization. This experiment deals with an ester of this sort, the substance commonly called aspirin.

Aspirin has been used since the beginning of the 20th century to treat headaches and fever. It is the most used medicinal, with a daily production of about 50 tons, about one pill for one out of every two people in this country. Its effectiveness is actually two-fold: as an inflammatory, which explains its pain-killing properties; and as an anticoagulant, which is important in its use in prevention of heart attacks and strokes. It works by inhibiting an enzyme that catalyzes formation of precursors of prostaglandins, rather large organic molecules that regulate inflammation of tissues, and of thromboxanes, which stimulate platelet aggregation, the first step in blood clotting. Researchers on prostaglandins have received two Nobel Prizes, the first in 1970, to their discoverer, and the second in 1982, for work relating to the effects of aspirin on prostaglandin behavior.

Aspirin can be made by the reaction of the —OH group in the salicylic acid molecule with the carboxyl (—COOH) group in acetic acid:

acetic acid   salicylic acid   aspirin (salicylic acid acetate)

A better preparative method, which we will use in this experiment, employs acetic anhydride in the reaction instead of acetic acid. The anhydride can be considered to be the product of a reaction in which two acetic acid molecules combine, with the elimination of a molecule of water. The anhydride will react with the water produced in the esterification reaction and will tend to drive the reaction to the right. A catalyst, normally sulfuric or phosphoric acid, is also used to speed up the reaction.

acetic anhydride   salicylic acid   aspirin   acetic acid

The aspirin you will prepare in this experiment is relatively impure and should certainly not be taken internally, even if the experiment gives you a bad headache.

There are several ways by which the purity of your aspirin can be estimated. Probably the simplest way is to measure its melting point. If the aspirin is pure, it will melt sharply at the literature value of the melting point. If it is impure, the melting point will be lower than the literature value by an amount that is roughly proportional to the amount of impurity present.

A more quantitative measure of the purity of your aspirin sample can be obtained by determining the percentage of salicylic acid it contains. Salicylic acid is the most likely impurity in the sample because, unlike acetic acid, it is not very soluble in water. Salicylic acid forms a highly colored magenta complex with Fe(III). By measuring the absorption of light by a solution containing a known amount of aspirin in excess Fe^{3+} ion, one can easily determine the percentage of salicylic acid present in the aspirin.

WEAR YOUR SAFETY GLASSES WHILE PERFORMING THIS EXPERIMENT

Experimental Procedure

Weigh a 50-mL Erlenmeyer flask on a triple-beam or top-loading balance and add 2.0 g of salicylic acid. Measure out 5.0 mL of acetic anhydride in your graduated cylinder, and pour it into the flask in such a way as to wash any crystals of salicylic acid on the walls down to the bottom. Add five drops of 85% phosphoric acid to serve as a catalyst. **CAUTION:** *Both acetic anhydride and phosphoric acid are reactive chemicals that can give you a bad chemical burn, so use due caution in handling them. If you get any of either on your hands or clothes, wash thoroughly with soap and water.*

Clamp the flask in place in a beaker of water supported on a wire gauze on a ring stand. Heat the water with a Bunsen burner to about 75°C, stirring the liquid in the flask occasionally with a stirring rod. Maintain this temperature for about 15 minutes, by which time the reaction should be complete. *Cautiously,* add 2 mL of water to the flask to decompose any excess acetic anhydride. There will be some hot acetic acid vapor evolved as a result of the decomposition.

When the liquid has stopped giving off vapors, remove the flask from the water bath and add 20 mL of water. Let the flask cool for a few minutes in air, during which time crystals of aspirin should begin to form. Put the flask in an ice bath to hasten crystallization and increase the yield of product. If crystals are slow to appear, it may be helpful to scratch the inside of the flask with a stirring rod. Leave the flask in the ice bath for at least 5 minutes.

Collect the aspirin by filtering the cold liquid through a Buchner funnel using suction. Turn off the suction and pour about 5 mL of ice-cold distilled water over the crystals; after about 15 seconds turn on the suction to remove the wash liquid along with most of the impurities. Repeat the washing process with another 5-mL sample of ice-cold water. Draw air through the funnel for a few minutes to help dry the crystals and then transfer them to a piece of dry, weighed filter paper. Weigh the sample on the paper to ±0.1 g.

Test the solubility properties of the aspirin by taking samples of the solid the size of a pea on your spatula and putting them in separate 1-mL samples of each of the following solvents and stirring:

1. toluene, $C_6H_5CH_3$, nonpolar aromatic
2. heptane, C_7H_{16}, nonpolar aliphatic
3. ethyl acetate, $C_2H_5OCOCH_3$, aliphatic ester

4. ethyl alcohol, C_2H_5OH, polar aliphatic, hydrogen bonding
5. acetone, CH_3COCH_3, polar aliphatic, nonhydrogen bonding
6. water, highly polar, hydrogen bonding

To determine the melting point of the aspirin, add a small amount of your prepared sample to a melting point tube (made from 5-mm tubing), as directed by your instructor. Shake the solid down by tapping the tube on the bench top, using enough solid to give you a depth of about 5 mm. Set up the apparatus shown in Figure 41.1. Fasten the melting point tube to the thermometer with a small rubber band, which should be above the surface of the oil. The thermometer bulb and sample should be about 2 cm above the bottom of the tube. Heat the oil bath *gently,* especially after the temperature gets above 100°C. As the melting point is approached, the crystals will begin to soften. Report the melting point as the temperature at which the last crystals disappear.

To analyze your aspirin for its salicylic acid impurity, weigh out 0.10 ± 0.01 g of your sample into a weighed 100-mL beaker. Dissolve the solid in 5 mL 95% ethanol. Add 5 mL 0.025 M $Fe(NO_3)_3$ in 0.5 M HCl and 40 mL distilled water. Make all volume measurements with a graduated cylinder. Stir the solution to mix all reagents.

Rinse out a spectrophotometer tube with a few milliliters of the solution and then fill the tube with that solution. Measure the absorbance of the solution at 525 nm. From the calibration curve or equation provided, calculate the percentage of salicylic acid in the aspirin sample.

DISPOSAL OF REACTION PRODUCTS.   The contents of the suction flask may be poured down the drain. The toluene and heptane from the solubility tests should be put in the waste crock.

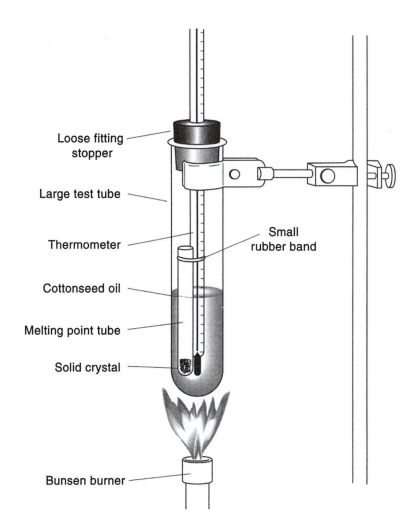

Figure 41.1

Experiment 41

Data and Calculations: Preparation of Aspirin

Mass of salicylic acid used   _____ g

Volume of acetic anhydride used   _____ mL

Mass of acetic anhydride used
(density = 1.08 g/mL)   _____ g

Mass of aspirin obtained   _____ g

Theoretical yield of aspirin   _____ g

Percentage yield of aspirin   _____ %

Melting point of aspirin   _____ °C

Absorbance of aspirin solution   _____

Percentage of salicylic acid impurity   _____ %

Solubility properties of aspirin

    Toluene   _____   Ethyl alcohol   _____

    Heptane   _____   Acetone   _____

    Ethyl acetate _____   Water   _____

S = soluble   I = insoluble   SS = slightly soluble

Underline those characteristics listed which would be likely to be present in a good solvent for aspirin.

      organic   aliphatic   polar   hydrogen bonding

      inorganic   aromatic   nonpolar   nonhydrogen bonding

Experiment 41

Advance Study Assignment: Preparation of Aspirin

1. Calculate the theoretical yield of aspirin to be obtained in this experiment, starting with 2.0 g of salicylic acid and 5.0 mL of acetic anhydride (density = 1.08 g/mL).

_____ g

2. If 1.6 g of aspirin were obtained in this experiment, what would be the percentage yield?

_____ %

3. The name acetic anhydride implies that the compound will react with water to form acetic acid. Write the equation for the reaction.

4. Identify R and R′ in Equation 1 when the ester, aspirin, is made from salicylic acid and acetic acid.

5. Write the equation for the reaction by which aspirin decomposes in an aqueous ethanol solution.

Experiment 42

Rate Studies on the Decomposition of Aspirin

Once aspirin has been prepared, it is relatively stable and in tablet form in a tight container resists decomposition for years. If it is exposed to moist air or heat, it will slowly break down by the reverse of the reaction we used to make aspirin in Experiment 41:

$$\text{Aspirin} + \text{Water} \rightarrow \text{Salicylic Acid} + \text{Acetic Acid} \qquad (1)$$

In this experiment we will study the rate at which aspirin in water solution will decompose at about 80°C. Like many organic reactions, this one is slow at room temperature. Raising the temperature allows us to carry out the reaction in a few minutes rather than several hours. To analyze the reaction mixture we will use the same method as in Experiment 41, where we find the amount of salicylic acid impurity in the aspirin sample. Iron(III) forms a very intensely colored violet complex with salicylic acid, so we can use the amount of light the reaction mixture absorbs to find out how far the reaction has gone. This will allow us to determine the rate of reaction at different concentrations. Our goal will be to use the observed rates to establish the order of the reaction, the rate constant, and the activation energy. The theory and terminology we will use are found in Experiment 20, which you should review now, even though you may have performed the experiment earlier in this course.

WEAR YOUR SAFETY GLASSES WHILE
PERFORMING THIS EXPERIMENT

Experimental Procedure

Obtain a digital thermometer from the stockroom.

A. Measuring the Rate of Reaction

Step 1. Setting the stage. Set up a hot-water bath, using a 600-mL beaker about 3/4 full of water. As a heater, use an electric hot plate with magnetic stirrer if such is available. Otherwise, use a Bunsen burner, supporting the beaker on a wire screen on an iron ring. If you use a burner, you should put a larger iron ring around the beaker, near the top, to prevent its tipping over. Clamp the ring to the ring stand. Prepare an ice bath by filling a 400-mL beaker with crushed ice; add water until the level is at the top of the ice. Clean three regular (18 × 150-mm) test tubes. Rinse them with a few mL of acetone; pour the acetone into one of the tubes, shake it, then pour it into the next, and the next. Dry the tubes in a gentle stream of compressed air. Number the test tubes and put them in a test tube rack. Put a small plastic funnel in Tube #1.

Step 2. Weighing the sample. While the water is heating, weigh out on an analytical balance about 60 mg of aspirin, to 0.0001 g. This takes a bit of doing, since the sample is so small. It is probably easiest to make the weighing on a piece of weighing paper. Tare the paper, and use a small spatula to add a very small amount of aspirin. Do this several times, until you are within about 5 mg of the desired mass. Record the mass. Then, carefully, remove the paper from the balance. Fold it at the edge, and transfer the sample to the funnel. Tap the paper and the funnel to get as much sample as possible into Tube #1.

Step 3. Preparing the reaction mixture. With a rinsed 10-mL graduated cylinder, or a pipet, add 10.0 mL of distilled water to the aspirin in Tube #1. The aspirin will not dissolve in the water; some will stay at the bottom of the tube, and some will float on the water. Put your stirring rod in the tube.

Step 4. Carrying out the reaction. By now the water in the 600-mL beaker should be getting warm. If it is boiling, slowly add tap water or ice to lower the temperature to 80°C, as measured to 0.1°C on your digital thermometer. **CAUTION:** *Boiling water can give you a bad burn, so be careful not to tip over the beaker.* Adjust your heat source as necessary to hold the bath steady at about 80°C, for at least five minutes. When you have managed that, and it may take a little time, you are ready to make your first run. Pick up the tube, and, noting the time to 1 second, put the tube in the water bath. Stir the mixture continuously, to help dissolve the aspirin and bring the reaction mixture up to temperature quickly. When the aspirin is all dissolved, which should happen within a minute, keep the rod in the tube, stirring occasionally. Measure and record the temperature of the bath at one-minute intervals as the reaction proceeds; it will fall initially, but should come back to the original temperature fairly quickly. Leave the tube in the bath for exactly **six** minutes. Then remove the tube and put it in the ice-water bath to stop the reaction. Stir the reaction mixture to hasten the cooling. Within a minute the solution should be cool, and you can put the tube back in the test tube rack.

Step 5. Adding the indicator. Drain the graduated cylinder or pipet, and use it to measure out 5.0 mL of 0.025 M $Fe(NO_3)_3$ into the test tube. The solution should turn violet as the iron-salicylic acid complex forms. Stir the mixture until it is of uniform color.

Step 6. Finding the salicylic acid concentration. Measure the absorbance of the solution at 525 nm on a Spectronic 20 or other spectrophotometer. Use the procedure described in Appendix IV if this is the first time you have measured absorbance. After adjusting the spectrophotometer for 0 and 100% transmittance, rinse the spectrophotometer tube twice with a small amount of the solution from Step 5. Then add a few mL of that solution to the tube and record the absorbance you obtain. Find the concentration of the salicylic acid in the reaction mixture, using the graph that is provided or the equation you are given relating absorbance and concentration. This will be the salicylic acid concentration in the reaction mixture **before** the indicator was added. By Equation 1, that concentration will equal the **change** that occurs in the **aspirin concentration.**

Step 7. Correcting for warmup time. Once the reaction mixture is up to 80°C, the reaction rate is essentially constant, but during the first minute, when the aspirin is dissolving and the mixture is warming up, the reaction is much slower. To correct for this, you might run the same size sample at 80°C, for one minute, and subtract the absorbance for that sample from the total absorbance. You could then use 5 minutes as the time the reaction occurred. The authors actually did this, but found that if they took account of the warmup period by using the total absorbance and 5.5 minutes for the time of reaction, the rates were essentially the same as with the more accurate, but more complex, procedure. So, when calculating the rate, we will adopt the simpler approach and use **5.5 minutes** as the **time,** and the **total absorbance** observed in Step 6.

B. Determining the Order of the Reaction and the Rate Constant

Reset your water bath to the temperature you used in Part A, and prepare a new ice bath.

Repeat Steps 1 through 7, at 80°C, using about 40 mg of sample, weighed to 0.0001 g, in Test Tube #2. Be sure to rinse out your graduated cylinder or pipet with distilled water before measuring out the 10 mL of water. The results you obtain with this sample, along with those from Part A will allow you to find the order of the reaction and the rate constant at 80°C. On the Data page carry out the calculations for the order and the rate constant of the reaction. Calculate the mean value of the rate constant and the standard deviation. (See Appendix VIII.) Then proceed to Part C.

C. Finding the Activation Energy

The last piece of information to be obtained in this experiment is the activation energy for the reaction. That is the energy that must be present in the key step when the reaction proceeds. It is found by measuring the dependence on temperature of the rate constant for the reaction.

We will use nearly the same procedure as in Part A, Steps 1 through 7, except that we will carry out the reaction in Test Tube #3 at about 70°C instead of 80°C. The sample size should be about 60 mg, weighed accurately. Bring the water bath to about 70°C, or a bit higher, and hold it there for five minutes. Make a new ice bath, and make sure that you rinse your pipet or graduated cylinder before measuring out the water.

At 70°C the reaction will go a lot more slowly than in Part A. In order to get easily measured absorbance values, run the reaction for **twelve** minutes, rather than six. The aspirin will be slower to dissolve, so the warmup period will be **two** minutes, rather than one. It will require a considerably longer time to dissolve the aspirin, but, with continuous stirring, it should go into solution in two minutes. Record the temperature at 2-minute intervals. Record the absorbance that you obtain. To correct for warmup time, calculate the reaction rate using the **total absorbance** and **time** equal to **11 minutes.**

With the data you have obtained, you can proceed to evaluate the energy of activation, E_a, for the reaction. This is done using the rate constants at 80°C and 70°C and the Arrhenius equation, as in Experiment 20:

$$\ln k = -E_a/RT + a \text{ constant} \qquad (2)$$

DISPOSAL OF REACTION PRODUCTS.   All of the reaction products may be poured down the sink, with the water running as you pour.

Experiment 42

Data and Calculations: Rate Studies on the Decomposition of Aspirin

Let Asp = aspirin and Sal = salicylic acid (MM Asp = 180 g)

	Part A Tube #1	Part B Tube #2	Part C Tube #3

1. Mass of Asp in tube in g   _____   _____   _____

   No. moles Asp in tube   _____   _____   _____

   [Asp] in moles/L (10 mL)   _____   _____   _____

   (Temperature measurements are on following page)

2. Absorbance, Abs, of solution   _____   _____   _____

   [Sal] in reaction mixture   _____ M _____ M _____ M

   Change in [Asp], Δ[Asp] = $-$[Sal]:   _____ M _____ M _____ M

   Note that Δ[Asp] $\ll$ [Asp]

3. rate of reaction = $-\Delta$[Asp]$/\Delta t$ = k[Asp]^n (3)

   rate in moles/L min Δt = 5.5 min   _____   _____

                        = 11 min   _____

4. (rate in Tube #1)/(rate in Tube #2)   _____

5. ([Asp] in Tube #1)/([Asp] in Tube #2)   _____

   The reaction is _____ order (Use Eqn. 3 for Tubes #1 and 2)
   (Select nearest integer, and give reasoning.)

6. Value of rate constant k   _____   _____   _____

   Average value of k at 80°C _____   Standard deviation _____

7. k_{ave} at ~80°C/k at ~70°C   _____

(continued on following page)

8. Temperature measurements during reactions:

time(min)	0	1	2	3	4	5	6	
or	0	2	4	6	8	10	12	
								Ave temp
t, °C in #1	____	____	____	____	____	____	____	_____
in #2	____	____	____	____	____	____	____	_____
in #3	____	____	____	____	____	____	____	_____

9.

$$\ln k = -E_a/RT + \text{a constant} \qquad (2)$$

Use average temperatures in K, not °C, in Eqn. 2

$$\ln (k_{80}/k_{70}) = \underline{\hspace{2cm}} = \underline{\hspace{2cm}}$$

Solve for E_a

The activation energy, E_a = _____ J = _____ kJ

Experiment 42

Advance Study Assignment: The Rate of Decomposition of Aspirin

1. A student studies the rate at which aspirin decomposes by the reaction:

$$Asp + H_2O \rightarrow Sal + Acetic\ Acid$$

She weighs out 60.8 mg of aspirin and dissolves it, making 10 mL of solution in water. She heats the solution for 5 minutes at 90°C, and finds that about 10% of the aspirin is converted to salicylic acid and acetic acid.

a. How many moles of aspirin are in the initial solution? What is the molarity of the aspirin?
   MM aspirin = 180 g.

   _____ moles _____ M

b. What is the rate of reaction? (See Eqn. 1 on Data page.)

   _____ moles/L min

c. In a similar experiment with a smaller sample, she finds that once again about 10% of the aspirin decomposes in 5 minutes at 100°C. What is the order of the reaction? Why?

   _____

d. What is the rate constant for the reaction?

   _____

e. In another experiment she finds that the energy of activation E_a for the reaction is about 90 kJ. How long would it take for 10% of an aspirin sample to decompose in a person's stomach at body temperature, 37°C?

   _____ minutes

Experiment 43

Analysis for Vitamin C

Vitamin C, known chemically as ascorbic acid, is an important component of a healthy diet. In the mid-eighteenth century the British navy found that the addition of citrus fruit to the sailors' diet prevented the malady called scurvy. Humans are one of the few members of the animal kingdom unable to synthesize vitamin C, resulting in the need for regular ingestion in order to remain healthy. The National Academy of Sciences has established the threshold of 60 mg/day for adults as the Recommended Dietary Allowance (RDA). Linus Pauling, a chemist whose many contributions to chemical bonding theory should be well-known to you, recommended a level of 500 mg/day to help ward off the common cold. He also suggested that large doses of vitamin C are helpful in preventing cancer.

The vitamin C content of foods can easily be determined by oxidizing ascorbic acid, $C_6H_8O_6$ to dehydro-L-ascorbic acid, $C_6H_6O_6$:

$$C_6H_8O_6 \rightarrow C_6H_6O_6 + 2\,H^+ + 2e^-$$

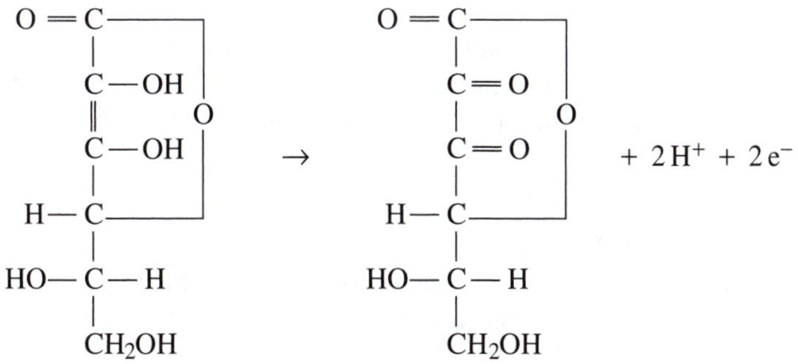

This reaction is very slow for ascorbic acid in the dry state, but occurs readily when in contact with moisture. A reagent that is particularly good for the oxidation is an aqueous solution of iodine, I_2. Since I_2 is not very soluble in water, we dissolve it in a solution of potassium iodide, KI, in which the I_2 exists mainly as I_3^-, a complex ion. The reaction with ascorbic acid involves I_2, which is reduced to I^- ion.

$$2e^- + I_2 \rightarrow 2\,I^-$$

In the overall reaction, one mole of ascorbic acid requires one mole of I_2 for complete oxidation.

When the red-colored I_2 solution is added to the ascorbic solution, the characteristic iodine color disappears because of the above reaction. Although we could use the first permanent appearance of the yellow color of dilute iodine to mark the end point of the titration, better results are obtained when starch is added as an indicator. Starch reacts with I_2 to form an intensely colored blue complex. In the titration I_2 reacts preferentially with ascorbic acid, and so its concentration remains very low until the ascorbic acid is all oxidized. At that point, the I_2 concentration begins to go up and the reaction with the indicator occurs:

$$I_2 + \text{starch} \rightarrow \text{starch-}I_2 \text{ complex}$$

yellow   blue

Because an I_2 solution cannot be prepared accurately by direct weighing, it is necessary to standardize the I_2 against a reference substance of known purity. We will use pure ascorbic acid for this reference, or primary standard. After standardization you can use the iodine solution for the direct determination of vitamin C in any kind of sample.

Experimental Procedure

A. Standardization of the Iodine Solution

Obtain from the storeroom a buret and an unknown vitamin C sample. Weigh out accurately on the analytical balance three ascorbic acid samples of approximately 0.10 g into clean 250-mL Erlenmeyer flasks. Dissolve each sample in approximately 100 mL of water.

Clean your buret thoroughly. Draw about 100 mL of the stock I_2 solution from the carboy in the laboratory into a 400-mL beaker and add approximately 150 mL of water. Stir thoroughly and cover with a piece of aluminum foil. Rinse the buret with a few milliliters of the I_2 solution three times. Drain and then fill the buret with the I_2 solution.

After taking an initial reading of the buret (you may find looking toward a light source will make it easier to see the bottom of the meniscus), add 1 mL of starch indicator to the first ascorbic acid sample and titrate with the iodine solution. Note the change of the I_2 color as you swirl the flask gently and continuously during the titration. Continue the addition of the iodine, using progressively smaller volume increments, until the sample solution just turns a distinct blue. After reading the buret, titrate the other two sample solutions—being sure to add the starch indicator and to read your buret before and after each titration.

B. Analysis of an Unknown Containing Vitamin C

Given your experience with the standardization reaction, you should be able to devise an analogous procedure to determine the vitamin C content of your unknown sample. You will need to select a sample size, and you may need to carry out an initial treatment of the sample. In particular, if your instructor assigns you a fruit juice sample, it will be desirable to first filter the sample through cheese cloth, followed by rinsing of the filter with water.

It may be helpful in choosing the sample sizes to calculate an iodine solution parameter called the titer—the number of mg of ascorbic acid that reacts with 1 mL of iodine solution. This number is easily found from the I_2 concentration and the mass relationship in the reaction. It is desirable to have the volume of I_2 for each titration be at least 15 mL. Using a small initial sample will give you an indication of how much to scale up for your final titrations.

Report your results in percent vitamin C, if a solid sample was used. For liquid samples, report mg of vitamin C per 100 mL. In each case, calculate the size sample required to give the RDA of vitamin C. Finally, calculate the mean and the standard deviation of the concentrations (in mg of vitamin C per 100 mL).

DISPOSAL OF REACTION PRODUCTS.   All reaction products may be diluted with water and poured down the drain.

Experiment 43

Data and Calculations: Analysis for Vitamin C

A. Standardization of Iodine Solution

Sample	I	II	III
Mass of Ascorbic Acid Sample, g	_____	_____	_____
Moles of Ascorbic Acid (MM = 176 g)	_____	_____	_____
Initial buret reading, mL	_____	_____	_____
Final buret reading, mL	_____	_____	_____
Volume of I_2 added, mL	_____	_____	_____
Moles of I_2 consumed	_____	_____	_____
Molarity of I_2, moles/L	_____	_____	_____
Titer of I_2, mg Asc/mL I_2	_____	_____	_____

B. Analysis of an Unknown Sample Containing Vitamin C

	I	II	III
Mass or volume of unknown, g	_____	_____	_____
Initial buret reading, mL	_____	_____	_____
Final buret reading, mL	_____	_____	_____
Moles of iodine added	_____	_____	_____
Moles of vitamin C in sample	_____	_____	_____
Mass of vitamin C in sample, mg	_____	_____	_____
Percent vitamin C in solid sample	_____	_____	_____
Concentration of vitamin C in liquid sample, mg/100 mL	_____	_____	_____

Mean concentration _____ mg/100 mL   Standard deviation _____ mg/100 mL

	I	II	III
Amount of sample that will furnish RDA, mL or mg	_____	_____	_____

Name _____ Section _____

Experiment 43

Advance Study Assignment: Analysis for Vitamin C

1. Write a balanced equation for the reaction between I_2 and ascorbic acid. Identify the oxidizing agent and the reducing agent.

2. A solution of I_2 was standardized with ascorbic acid (Asc). Using a 0.1137-g sample of pure ascorbic acid, 26.75 mL of I_2 were required to reach the starch end point.

    a. What is the molarity of the iodine solution?

    _____ M

    b. What is the titer of the iodine solution?

    _____ mg Asc/mL I_2

3. A sample of fresh grapefruit juice was filtered and titrated with the above I_2 solution. A 100-mL sample of the juice took 10.26 mL of the iodine solution to reach the starch end point.

    a. What is the concentration of vitamin C in the juice in mg vitamin C/100 mL of juice?

    _____ mg/100 mL

    b. What quantity of juice will provide the RDA amount of vitamin C?

    _____ mL

Appendix I

Vapor Pressure and Density of Liquid Water

Temperature (°C)	Vapor Pressure (mm Hg)	Liquid Density (g/mL)	Temperature (°C)	Vapor Pressure (mm Hg)	Liquid Density (g/mL)
0.0	4.59	0.99979	26.0	25.2	0.99674
0.5	4.75	0.99982	26.5	26.0	0.99661
1.0	4.93	0.99985	27.0	26.8	0.99647
2.0	5.30	0.99989	27.5	27.6	0.99633
3.0	5.69	0.99992	28.0	28.4	0.99619
4.0	6.10	0.99993	28.5	29.2	0.99605
5.0	6.54	0.99992	29.0	30.1	0.99590
6.0	7.02	0.99989	29.5	31.0	0.99576
7.0	7.52	0.99986	30.0	31.9	0.99561
8.0	8.05	0.99980	32.0	35.7	0.99499
9.0	8.61	0.99974	34.0	39.9	0.99433
10.0	9.21	0.99965	36.0	44.6	0.99364
12.0	10.5	0.99945	38.0	49.8	0.99292
14.0	12.0	0.99920	40.0	55.4	0.99218
16.0	13.6	0.99890	45.0	72.0	0.99017
18.0	15.5	0.99855	50.0	92.6	0.98800
18.5	16.0	0.99846	55.0	118.2	0.98566
19.0	16.5	0.99836	60.0	149.6	0.98316
19.5	17.0	0.99826	65.0	187.8	0.98052
20.0	17.5	0.99816	70.0	234.0	0.97773
20.5	18.1	0.99806	75.0	289.5	0.97481
21.0	18.7	0.99795	80.0	355.6	0.97177
21.5	19.2	0.99784	85.0	434.0	0.96859
22.0	19.8	0.99773	90.0	526.4	0.96530
22.5	20.5	0.99761	92.0	567.7	0.96394
23.0	21.1	0.99750	94.0	611.6	0.96257
23.5	21.7	0.99738	96.0	658.3	0.96118
24.0	22.4	0.99725	98.0	708.0	0.95978
24.5	23.1	0.99713	99.0	734.0	0.95906
25.0	23.8	0.99700	99.5	747.2	0.95871
25.5	24.5	0.99687	100.0	760.7	0.95835

Wagner and Pruss, *J Phys Chem Ref Data*, 31-2: 387-535, 2002, in NIST Chemistry WebBook, NIST Standard Reference Database Number 69, Eds. P.J. Linstrom and W.G. Mallard, National Institute of Standards and Technology, Gaithersburg MD, 20899, http://webbook.nist.gov, (retrieved June 28, 2010).

Appendix II

Summary of Solubility Properties of Ions and Solids

Cation (color in aqueous solution)	Cl⁻, Br⁻ I⁻, SCN⁻	SO_4^{2-}	CrO_4^{2-}	PO_4^{3-}	$C_2O_4^{2-}$*	CO_3^{2-}
Na^+, K^+, NH_4^+	S	S	S	S	S	S
Ba^{2+}	S	I	A	A⁻	A	A⁻
Ca^{2+}	S	S⁻	S	A⁻	A	A⁻
Mg^{2+}	S	S	S	A⁻	A⁻	A⁻
Fe^{3+} (yellow)	S*	S	A⁻	A	S	D, A⁻
Cr^{3+} (blue-gray)	S	S	A⁻	A	S	A⁻
Al^{3+}	S	S	A⁻, C	A, C	A⁻, C	D, A⁻, C
Ni^{2+} (green)	S	S	S	A⁻, C	A, C	A⁻, C
Co^{2+} (pink)	S	S	A⁻	A⁻	A⁻	A⁻
Zn^{2+}	S	S	A⁻, C	A⁻, C	A⁻, C	A⁻, C
Mn^{2+} (pale pink)	S	S	S	A⁻	A⁻	A⁻
Cu^{2+} (blue)	S*	S	A⁻, C	A⁻, C	A, C	A⁻, C
Cd^{2+}	S	S	A⁻, C	A⁻, C	A, C	A⁻, C
Bi^{3+}	A	A⁻	A	A	A	A⁻
Hg^{2+}	S*	S	A	A⁻	A	A⁻
Sn^{2+}, Sn^{4+}	A, C	A, C	A, C	A, C	A, C	A, C
Sb^{3+}	A, C	A, C	A, C	A, C	A⁻, C	A, C
Ag^+	C*	S⁻	A, C	A, C	A, C	A⁻, C
Pb^{2+}	C, HW	C	C	A, C	A, C	A⁻, C
Hg_2^{2+}	O⁺	A	A	A	O	A

Key: S, soluble in water; no precipitate on mixing cation, 0.1 M, with anion, 1 M

S⁻, slightly soluble; tends to precipitate on mixing cation, 0.1 M, with anion, 1 M

HW, soluble in hot water

A⁻, soluble in 1 M $HC_2H_3O_2$

A, soluble in acid (6 M HCl or other nonprecipitating, nonoxidizing acid)

A⁺, soluble in 12 M HCl

O, soluble in hot 6 M HNO_3

O⁺, soluble in aqua regia

C, soluble in solution containing a good complexing ligand

D, unstable, decomposes to a product with solubility as indicated

I, insoluble in any common solvent

*Oxalates form many complex ions; oxides behave like hydroxides, but may be slow to dissolve; FeI_3 is unstable, decomposes to FeI_2 and I_2; CuI_2 is unstable, decomposes to CuI and I_2; AgCl and AgSCN dissolve in 6 M NH_3, but AgBr and AgI do not; most silver salts can be complexed with [additional] SCN⁻, $S_2O_3^{2-}$, or CN⁻ [USE IN BASE ONLY] (in order of increasing complexing strength); HgI_2 is insoluble, but dissolves in excess I⁻; Cr^{3+} and Co^{2+} can, under some conditions, form complexes with OH⁻ and NH_3, respectively.

Cation (color in aqueous solution)	SO_3^{2-}	S^{2-}	O^{2-}, * OH^-	NO_3^-, ClO_3^- $C_2H_3O_2^-$, NO_2^-	Complexing agents
Na^+, K^+, N_4^+	S	S	S	S	—
Ba^{2+}	A	S	S^-	S	—
Ca^{2+}	A^-	D, A^-	S^-	S	—
Mg^{2+}	S	D, A^-	A^-	S	—
Fe^{3+} (yellow)	D, S	D, A^-	A^-	S	—
Cr^{3+} (blue-gray)	S	D, A^-	A^-	S	(OH^-)
Al^{3+}	A^-, C	D, A^-, C	A^-, C	S	OH^-
Ni^{2+} (green)	A^-	O	A^-, C	S	NH_3
Co^{2+} (pink)	A^-	O	A^-	S	*
Zn^{2+}	S	A^-, C	A^-, C	S	OH^-, NH_3
Mn^{2+} (pale pink)	S	A^-	A^-	S	—
Cu^{2+} (blue)	A^-, C	O	A^-, C	S	NH_3
Cd^{2+}	A^-, C	A	A^-, C	S	NH_3
Bi^{3+}	A	A^+, O	A^-	A^-	Cl^-
Hg^{2+}	D, O	O^+	A^-	S	—
Sn^{2+}, Sn^{4+}	A, C	A, C	A, C	A, C	OH^-, Cl^-
Sb^{3+}	A, C	A, C	A, C	A, C	OH^-, Cl^-
Ag^+	A, C	O	A^-, C	S	NH_3, $S_2O_3^{2-}$, CN^-
Pb^{2+}	A, C	O	A^-, C	S	OH^-
Hg_2^{2+}	D, O	D, O^+	D, O	S	—

Key: S, soluble in water; no precipitate on mixing cation, 0.1 M, with anion, 1 M

   S^-, slightly soluble; tends to precipitate on mixing cation, 0.1 M, with anion, 1 M

   HW, soluble in hot water

   A^-, soluble in 1 M $HC_2H_3O_2$

   A, soluble in acid (6 M HCl or other nonprecipitating, nonoxidizing acid)

   A^+, soluble in 12 M HCl

   O, soluble in hot 6 M HNO_3

   O^+, soluble in aqua regia

   C, soluble in solution containing a good complexing ligand

   D, unstable, decomposes to a product with solubility as indicated

   I, insoluble in any common solvent

*Oxalates form many complex ions; oxides behave like hydroxides, but may be slow to dissolve; FeI_3 is unstable, decomposes to FeI_2 and I_2; CuI_2 is unstable, decomposes to CuI and I_2; AgCl and AgSCN dissolve in 6 M NH_3, but AgBr and AgI do not; most silver salts can be complexed with [additional] SCN^-, $S_2O_3^{2-}$, or CN^- [USE IN BASE ONLY] (in order of increasing complexing strength); HgI_2 is insoluble, but dissolves in excess I^-; Cr^{3+} and Co^{2+} can, under some conditions, form complexes with OH^- and NH_3, respectively.

Appendix IIA

Some Properties of the Cations in Groups I, II, and III

In the experiments on qualitative analysis we used the chemical and physical properties of the cations in Groups I, II, and III to separate the ions from each other. Here we have summarized most of the properties that were employed, and have included some others that help distinguish the cations we studied.

Lead ion, Pb^{2+}

Most lead compounds are insoluble in water; the only exceptions are $Pb(NO_3)_2$ and $Pb(C_2H_3O_2)_2$. Lead chloride is slightly soluble in water, much more soluble in hot water and in solutions where the Cl^- concentration is high. Lead ion forms stable complexes with OH^-, $C_2H_3O_2^-$, and, to a lesser degree, with Cl^-. $PbSO_4$, an acid-insoluble sulfate, is white, and will dissolve in 6 M NaOH or in a concentrated solution of acetate ions via the formation of the $[Pb(OH)_4]^{2-}(aq)$ complex ion or $Pb(C_2H_3O_2)_2(aq)$, a species which does not appreciably dissociate in solution. $PbCrO_4$ is a bright yellow insoluble solid, often used as a confirmatory test for lead. It will dissolve in 6 M NaOH.

Silver ion, Ag^+

Most silver compounds are insoluble, the nitrate being just about the only exception. Silver acetate is slightly soluble. Silver chloride is white and insoluble in water, but soluble in 6 M NH_3, due to the formation of $Ag(NH_3)_2^+$ complex ion. These properties of AgCl are usually used in the confirmatory test for silver ion. They may also be used to test for the presence of Cl^- ion. In NaOH solution, Ag^+ precipitates as brown Ag_2O; it is soluble in nitric acid and in strong ammonia solutions.

Mercurous ion, Hg_2^{2+}

No salts of Hg(I) ion are soluble in water. Mercury(I) nitrate solutions always contain fairly high concentrations of HNO_3. Calomel, Hg_2Cl_2, precipitates up on addition of HCl to any solution of Hg(I) ion. It is white, and essentially insoluble in all common solvents. If treated with 6 M NH_3, it turns black, due to a reaction to form Hg metal (black) and white, insoluble, $HgNH_2Cl$. This is the definitive test for the Hg(I) ion.

Bismuth ion, Bi^{3+}

There are no water-soluble bismuth salts. The usual solution is highly acidic and contains $Bi(NO_3)_3$ or $BiCl_3$. Bismuth ion forms complexes with Cl^-, but not with OH^- or NH_3. $Bi(OH)_3$ is white and insoluble in water but soluble in strong acids. If you add a few drops of acidic $BiCl_3$ solution to water, a cloudy white precipitate of BiOCl forms. This is the usual test for Bi^{3+} ion. Another confirmatory test is to add 0.1 M $SnCl_2$ solution to $Bi(OH)_3$ in a strongly basic solution. Bi(III) will be reduced to black Bi metal.

Tin ion, Sn^{2+} or Sn^{4+}

No common tin compounds are water soluble. The usual 0.1 M tin solution contains $SnCl_2$ in 0.1 M HCl. Tin exists in either the +2, stannous, or +4, stannic, state. The Sn(II) species, particularly in basic solution, is a good reducing agent, and can reduce Bi(III) and Hg(II) to the metals. Tin in either oxidation state forms many complex ions, and in solution is usually present as a complex ion. In basic solution Sn(II) exists as $Sn(OH)_4^{2-}$. In HCl the complex ion has the formula $SnCl_4^{2-}$. Tin compounds are white, for the most part. SnS is brown, and its formation at pH 0.5 as in Experiment 37 is good evidence for the presence of tin. The usual confirmatory test is to add $SnCl_2$ in HCl to a solution of $HgCl_2$. The tin ion reduces the mercury(II) to form a precipitate of white Hg_2Cl_2, which may darken to gray as the reduction proceeds all the way to metallic mercury.

Antimony ion, Sb^{3+}

There are no common antimony salts that are soluble in water. Antimony salts are not easily dissolved. The usual solution of Sb(III) contains $SbCl_3$ in 3 M HCl, in which the antimony exists as the $SbCl_6^{3-}$ complex ion. Antimony also forms a stable complex ion with hydroxide ion, $Sb(OH)_6^{3-}$. If a few drops of $SbCl_3$ in acidic solution are added to water, a white precipitate of SbOCl forms, very similar to that observed with $BiCl_3$ in solution. The confirmatory test for antimony is the formation of the characteristic red-orange sulfide on precipitation of Sb_2S_3 by addition of thioacetamide to antimony solution at pH 0.5. If the cation is Bi(III), the precipitate will be black Bi_2S_3.

Copper ion, Cu^{2+}

Copper(II) is colored in most of its compounds and solutions. The hydrated ion, $Cu(H_2O)_4^{2+}$ is blue, but in the presence of other complexing agents the Cu(II) ion may be green or dark blue. The sulfate, chloride, nitrate, and acetate are water soluble. If 6 M NH_3 is added to a solution of Cu(II) ion, light blue $Cu(OH)_2$ initially precipitates, but in excess reagent the hydroxide readily dissolves because of formation of the very characteristic dark blue copper(II) ammonia complex ion, $Cu(NH_3)_4^{2+}$. This ion serves as an excellent confirmatory test for the presence of copper. Copper ion is readily reduced to the metal by the more active metals, such as zinc.

Nickel ion, Ni^{2+}

Nickel is another of the colored aqueous cations. The chloride, nitrate, sulfate, and acetate are water soluble, and form green solutions containing the $Ni(H_2O)_6^{2+}$ complex ion. Nickel forms other complex ions, but the one most commonly observed is the blue $Ni(NH_3)_6^{2+}$ ion, formed on addition of 6 M NH_3 in excess to Ni(II) solutions. The color is much less intense than that of the copper ammonia complex ion. The usual test for the presence of nickel is the formation of a rose-red precipitate of nickel dimethylglyoxime on addition of that reagent to a slightly basic solution containing Ni^{2+}.

Iron(III) ion, Fe^{3+}

Iron has two common oxidation states, +2 and +3. Iron(II), called ferrous ion, is less commonly observed, since it is readily oxidized in air to the +3 state. We will limit our work to iron(III), or ferric, salts. Iron(III) salts are often colored, usually yellow in aqueous solution. The color is due to hydrolysis, since the $Fe(H_2O)_6^{3+}$ is itself colorless. The nitrate, chloride, and sulfate are water soluble, but tend to hydrolyze to form basic salts that may require a slightly acidic solution if they are to dissolve completely. Iron(III) hydroxide is very insoluble, and is the color of rust. Iron forms several complex ions, but the one most encountered in qualitative analysis is the $FeSCN^{2+}$ ion, which is dark red and often used in the confirmatory test for Fe(III) ion. Iron does not form complex ions with NH_3 or OH^-.

Chromium ion, Cr^{3+}

Chromium(III), as its name implies, is colored; it may be red-violet, green, or purple in solution. The chromium +3 ion usually exists as a complex which, unlike many complex ions, is sometimes slow to undergo ligand exchange. Chromium exists in two common oxidation states, +3 and +6. In the latter state it is found as the yellow chromate, CrO_4^{2-}, anion, or, if the solution is acidic, as the orange $Cr_2O_7^{2-}$ anion. The chloride, nitrate, sulfate, and acetate of Cr^{3+} are water soluble, but the dissolution process may be more rapid in acidic solution.

If excess 6 M NaOH is added to a solution of Cr(III), the initial hydroxide precipitate may dissolve and one may obtain a dark green solution due to the formation of the $Cr(OH)_4^-$ complex ion. With 6 M NH_3 an insoluble grey hydroxide is initially formed, which may slowly dissolve to yield a pink or violet complex ion. Boiling a solution of either complex causes insoluble $Cr(OH)_3$ to re-form. In most qual schemes chromium is oxidized to the +6 state, where in acidic solution it forms a characteristic, but fleeting, deep blue solution on addition of H_2O_2. Where chromium(III) is the only cation in a sample, the formation of the green hydroxide complex is definitive. If this test is inconclusive, oxidize Cr(III) as in Experiment 38 and use the alternate confirmatory test.

Aluminum ion, Al^{3+}

Aluminum salts are typically white. The nitrate, chloride, sulfate, and acetate are water soluble, but do show a tendency to form basic salts if no acid is added. Aluminum forms a very stable hydroxide complex ion, $Al(OH)_4^-$, so it is difficult to precipitate the otherwise insoluble $Al(OH)_3$ by addition of NaOH to Al(III) solutions. That hydroxide will come down on addition of 6 M NH_3 to a solution containing Al^{3+} buffered with NH_4^+ ion. The precipitate is characteristically light, translucent, and gelatinous. A good confirmatory test for aluminum is to precipitate $Al(OH)_3$ from a solution of the sample. Dissolve the solid in very dilute acetic acid. Add a few drops of catechol violet reagent. If aluminum is present, a blue solution will form.

Appendix III

Table of Atomic Masses (Based on Carbon-12)

	Symbol	Atomic No.	Atomic Mass		Symbol	Atomic No.	Atomic Mass
Actinium	Ac	89	[227]*	Iridium	Ir	77	192.2
Aluminum	Al	13	26.9815	Iron	Fe	26	55.847
Americium	Am	95	[243]	Krypton	Kr	36	83.80
Antimony	Sb	51	121.75	Lanthanum	La	57	138.91
Argon	Ar	18	39.948	Lawrencium	Lw	103	[257]
Arsenic	As	33	74.9216	Lead	Pb	82	207.19
Astatine	At	85	[210]	Lithium	Li	3	6.939
Barium	Ba	56	137.34	Lutetium	Lu	71	174.97
Berkelium	Bk	97	[247]	Magnesium	Mg	12	24.312
Beryllium	Be	4	9.0122	Manganese	Mn	25	54.9380
Bismuth	Bi	83	208.980	Mendelevium	Md	101	[256]
Boron	B	5	10.811	Mercury	Hg	80	200.59
Bromine	Br	35	79.909	Molybdenum	Mo	42	95.94
Cadmium	Cd	48	112.40	Neodymium	Nd	60	144.24
Calcium	Ca	20	40.08	Neon	Ne	10	20.183
Californium	Cf	98	[249]	Neptunium	Np	93	[237]
Carbon	C	6	12.01115	Nickel	Ni	28	58.71
Cerium	Ce	58	140.12	Niobium	Nb	41	92.906
Cesium	Cs	55	132.905	Nitrogen	N	7	14.0067
Chlorine	Cl	17	35.453	Nobelium	No	102	[253]
Chromium	Cr	24	51.996	Osmium	Os	76	190.2
Cobalt	Co	27	58.9332	Oxygen	O	8	15.9994
Copper	Cu	29	63.546	Palladium	Pd	46	106.4
Curium	Cm	96	[247]	Phosphorus	P	15	30.9738
Dysprosium	Dy	66	162.50	Platinum	Pt	78	195.09
Einsteinium	Es	99	[254]	Plutonium	Pu	94	[242]
Erbium	Er	68	167.26	Polonium	Po	84	[210]
Europium	Eu	63	151.96	Potassium	K	19	39.102
Fermium	Fm	100	[253]	Praseodymium	Pr	59	140.907
Fluorine	F	9	18.9984	Promethium	Pm	61	[145]
Francium	Fr	87	[223]	Protactinium	Pa	91	[231]
Gadolinium	Gd	64	157.25	Radium	Ra	88	[226]
Gallium	Ga	31	69.72	Radon	Rn	86	[222]
Germanium	Ge	32	72.59	Rhenium	Re	75	186.2
Gold	Au	79	196.967	Rhodium	Rh	45	102.905
Hafnium	Hf	72	178.49	Rubidium	Rb	37	85.47
Helium	He	2	4.0026	Ruthenium	Ru	44	101.07
Holmium	Ho	67	164.930	Rutherfordium	Rf	104	[261]
Hydrogen	H	1	1.00797	Scandium	Sm	62	150.35
Indium	In	49	114.82	Scandium	Sc	21	44.956
Iodine	I	53	126.9044	Selenium	Se	34	78.96

*A value given in brackets denotes the mass number of the longest-lived or best-known isotope.

(continued on next page)

	Symbol	Atomic No.	Atomic Mass		Symbol	Atomic No.	Atomic Mass
Silicon	Si	14	28.086	Thulium	Tm	69	168.934
Silver	Ag	47	107.870	Tin	Sn	50	118.69
Sodium	Na	11	22.9898	Titanium	Ti	22	47.90
Strontium	Sr	38	87.62	Tungsten	W	74	183.85
Sulfur	S	16	32.064	Uranium	U	92	238.03
Tantalum	Ta	73	180.948	Vanadium	V	23	50.942
Technetium	Tc	43	[99]	Xenon	Xe	54	131.30
Tellurium	Te	52	127.60	Ytterbium	Yb	70	173.04
Terbium	Tb	65	158.924	Yttrium	Y	39	88.905
Thallium	Tl	81	204.37	Zinc	Zn	30	65.37
Thorium	Th	90	232.038	Zirconium	Zr	40	91.22

*A value given in brackets denotes the mass number of the longest-lived or best-known isotope.

Appendix IV

Making Measurements—
Laboratory Techniques

For centuries people who made measurements set up systems of units to describe length, area, volume, and mass. These units often had rather romantic names that were created for a specific purpose. There was a myriad of such names, and among them we might mention:

acre	barrel	bolt	bushel
carat	chain	cord	cubit
dram	ell	em	fathom
furlong	gill	grain	hand
league	noggin	pace	perch
quire	rod	stone	ream

These days many of these units are still used, often in connection with only one application, such as horse racing (furlong), a gemstone's weight (carat), or printing (em).

Scientists early on realized that it would be advantageous to have one system of measurement, to be used by all, and several were suggested, and used by your authors when they were young students. These were mostly based on the metric approach, with different sizes related one to another by factors of 10. Thus we have the meter, equal to 100 centimeters, or 1000 millimeters. In 1960, a group of scientists set up the International System of Units, or SI, a coherent system of seven base units, which can be used to describe most measurements one is likely to make, and which can be used together in many natural laws. The five of these units we will use are the:

meter   kilogram   second   Kelvin   mole

Although in principle one might expect to use these units directly in reporting data, some of them turn out to have awkward magnitudes in many situations. So we use SI units when it is convenient, but for the first two we tend to work more with related metric units, or with unrelated, but conveniently sized, units. We will discuss our approach in the following sections. (Perhaps not surprisingly, none of the units in our first list are SI, or metrically related to SI.)

Mass

One of the most important measurements you will be making involves determining the mass of a sample. The unit we will almost always use is the gram, rather than the kilogram, since a kilogram is just too large for bench-scale chemistry, about two pounds. The gram is not a base unit in SI, but is easily converted to the kilogram when necessary.

Measurement of mass is almost always done by determining by one means or another the force exerted on a sample by gravity. This force is proportional to the mass of the sample. The device that is used is called a balance, because the mass of the sample is established by balancing it against a standard mass or force. In Figure IV.1 are shown several balances that were used in the past century. The balance on the left was used until about 1950. Using it required a set of weights that were put on one pan until the two pans came to the same height; with care one could weigh a sample to 0.0001 grams, but it would take several minutes to make one weighing. There were a lot of drawbacks to such a balance, but they were used for over a century in the form shown or more primitive ones. After World War II, instrument design developed rapidly. The balance in the middle is a mechanical one, but the weights are internal, and added to or taken from a pan inside the

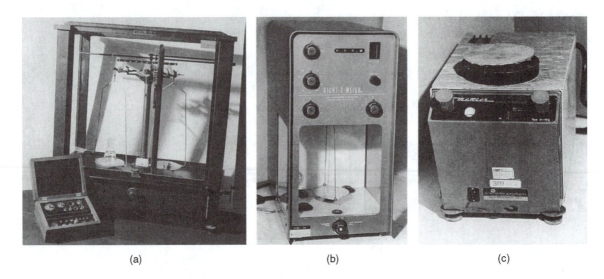

(a)   (b)   (c)

Figure IV.1 (a) An old mechanical analytical balance circa 1940—good to 0.0001 g, maybe. (b) A modern mechanical analytical balance circa 1960—maximum load 120 g, good to 0.0001 g. (c) A modern top-loading semi-automatic balance circa 1970—good to about 1 mg. *(Sherman Schultz)*

balance. Final readings are taken from a set of dials that add or take off weights, and an optically projected scale that furnishes masses in the milligram range. This type of balance is still used in some schools, and if you have one in your laboratory, your instructor will show you how it operates. The balance on the right is called a top-loading balance; mass is read directly from dials and an optical scale; this kind of balance gives rapid results but is limited in its precision to milligrams at best.

In recent years electronic balances have been developed that furnish the mass directly with a digital read-out (see Fig. IV.2). They do not have any weights, but depend on balancing the sample mass by a magnetic force that can be accurately related to mass. These balances can be accurate to 0.0001 g, and may arrive at the mass of a sample very quickly. In a very real sense, they are the ultimate weighing machines. In using such a balance, you first depress the control bar. This will zero the balance, and it will read 0.0000 g. Place your sample on the balance pan, close the balance door if it has one, and read the mass when the balance gives a steady reading. If your sample is to be in a container, you can find its mass by weighing the empty container, then depressing the control bar to re-zero, or tare, the balance. Then when you add your sample, its mass will appear directly as the digital readout. There are many brands of these balances, so if your lab has one, your instructor will describe the details of the operation of your balance before you use it.

Here are some guidelines for successful balance operation, which apply to the weighing of a sample on any balance:

1. Be certain the balance has been "zeroed" (reads 0 grams) before you place anything on the balance pan.
2. Never weigh chemicals directly on a balance pan. Always use a suitable container or weighing paper. In several experiments you will weigh a sample tube and measure the mass of sample removed, by difference.
3. Be certain that air currents are not disturbing the balance pan. In the case of an analytical balance, always shut the balance case doors when making measurements you are recording.
4. Never put hot, or even warm, objects on the balance pan. The temperature difference will change the density of the air surrounding the balance, and thus give inaccurate measurements.
5. Record your measurement, to the proper number of significant figures, directly on your Data page or in your notebook.
6. After finishing your measurements, be certain that the balance registers zero again and close the balance doors. Brush out any chemicals you may have spilled.
7. Be gentle with your balance. It is a sensitive, rather delicate instrument, and, like a person, responds best when treated properly.

(a)

(b)

Figure IV.2 (a) A modern electronic automatic analytical balance, circa 1985. Its maximum load is 120 g, and it is reliable to 0.0001 g. The beaker weighs 27.5056 g. (b) An electronic top-loading balance with a maximum load of 3.6 kg and reliable to 0.01 g. *(Mary Lou Wolsey)*

Volume

Another very common measurement we make in the laboratory is that of volume. The base SI unit of volume is the cubic meter (m^3). Again, this is much much larger than you ordinarily encounter. A large bathtub might contain a single cubic meter of water, roughly 250 gallons. The units of volume we ordinarily use are the liter, L, and the milliliter, mL. A liter equals 0.001 m^3, and a mL equals 1×10^{-6} m^3, so both are derived from the SI unit.

Another name for the milliliter, mL, is the cubic centimeter, cm^3, sometimes called the cc, pronounced "see see", a term often used in hospitals.

$$1 \text{ mL} = 1 \text{ cc} = 1 \text{ cm}^3$$

If we are asked to add 150 mL of water to a beaker during an experiment in which we are carrying out a chemical reaction, we don't need high accuracy, and might use a beaker or flask on which there are some volume markings (as shown in Fig. IV.3), which are reliable to about 5%. Somewhat more precise volumes can be obtained with a graduated cylinder, where one can get within about 1% of a needed volume.

A. Pipets

Still more precise volume measurements are often required; these may be obtained with pipets, which are available in many different volumes, from 1 mL up to about 100 mL. In several experiments in this manual pipets are used. Pipets are calibrated to deliver the specified volume when the meniscus at the liquid level is coincident with the horizontal line etched around the upper pipet stem. The following steps make for proper use of a pipet:

1. The pipet must be clean. When it drains, there should be no drops left on the pipet wall. It does not need to be dry.
2. Always use a pipet bulb, not your mouth, when pulling liquid into a pipet. You shouldn't run the risk of getting the liquid or its vapor into your mouth.
3. Place the pipet bulb on the upper end of the pipet and squeeze the air out. Immerse the tip of the pipet into the liquid and draw up enough liquid to get some into the main body of the pipet. Remove the bulb and place your index finger over the top of the pipet stem. Hold the pipet in the horizontal position and swirl

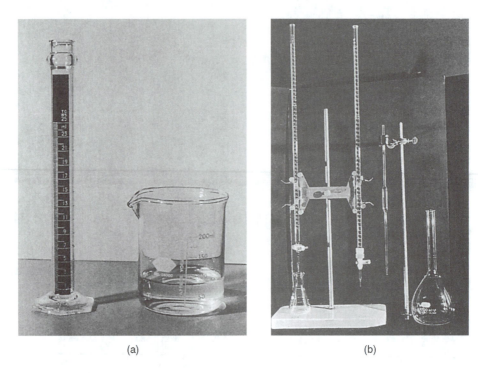

(a)   (b)

Figure IV.3 (a) A 25-mL graduated cylinder, good to 0.2 mL, and a 250-mL beaker with graduations, good to 20 mL. (b) A pair of 50-mL burets, a 10-mL pipet, and a 1000-mL volumetric flask. The burets and pipet are good to 0.01 mL, and the flask is good to about 0.1 mL. *(Sherman Schultz)*

the liquid around inside, rinsing the upper stem and body. Drain the liquid into a beaker; repeat the rinsing twice with small volumes of the liquid. This ensures that the contents of the pipet will be the liquid you wish to work with, and not water or a previously-used reagent. Finally, using the bulb, fill the pipet, drawing liquid a centimeter or two above the calibration mark. Put your index finger over the top of the stem and, carefully, with the tip of the pipet against the side, let liquid flow out into a beaker until the bottom of the meniscus is just at the level of the calibration mark. Wipe off the lower end of the pipet with a tissue, if a droplet is hanging there, and then place the tip inside the receiving container, with it touching the wall. Lift your index finger to release the liquid and let the liquid flow into the container. Hold the tip against the wall for about 10 seconds (longer for a thick liquid) after it appears that the liquid has stopped flowing, to ensure the correct amount of liquid drains out. Almost all glass pipets are designed "to deliver" and will be marked with the code "TD". These will accurately deliver the indicated amount of an *aqueous* liquid when a small amount remains in the tip—that last bit should *not* be blown out with a pipet bulb.

It takes skill and practice to make good use of a pipet. It is now possible to obtain autopipets and repipets, which more readily deliver precise volumes of reagents. These speed up the pipetting procedure enormously, and you may have them in your laboratory.

B. Burets

A buret is a long calibrated tube fitted with a stopcock to control release of liquid reagents. (See Fig. IV.3b) Burets are used in a procedure called titration, and are frequently used in pairs. This procedure allows one to add precisely measured volumes of reagents to a so-called "end point," at which point the amount of one reagent is equivalent chemically to another, which may also be a liquid in a buret, or a weighed solid. We have several titration experiments in this manual.

As with a pipet, working with a buret requires skill and practice. The following procedure should be helpful:

1. Check the buret for cleanliness by filling it with distilled water and allowing it to drain out. A clean buret will leave an unbroken film of water on the interior walls, with no beading or droplets. If necessary, clean your buret with soapy water and a long buret brush. Rinse with tap water, and finally with distilled water.

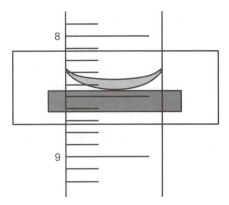

Figure IV.4 Reading a buret. The volume reading is 8.46 mL.

2. Rinse the buret three times with a few milliliters of the reagent to be used. Tip the buret to wash the walls with the reagent, and drain through the tip into a beaker. With the stopcock closed, fill the buret with reagent to a level a little above the top graduation, making sure to fill the tip completely. Open the stopcock carefully and let the liquid level go just below the top zero line. Read the level to the nearest 0.01 mL. To do this, place a white card with a sharp black rectangle on it (a piece of black tape is ideal) behind the buret, so that the reflection of the bottom of the meniscus is just above the upper black line, and at eye level (see Fig. IV.4). Then make the volume reading.

Add the reagent until you get to the end point of the titration, which is usually established by a color change of a chemical called an indicator. To hit the end point within one drop, you need to add reagent slowly when the titration is nearly complete, more rapidly at the beginning. Swirl the container to ensure the reactants are well-mixed. Often you get a clue from the indicator that you are near the end point. If you go past it, you can use the first titration as a guide for the second or third trial. Read the liquid level as before, to 0.01 mL.

C. Volumetric Flasks

A volumetric flask is one that has been especially calibrated to hold a specified volume: 10.00 mL, 25.00 mL, etc. These flasks are used when accurate dilutions are required in analytical experiments. Place the solute or solution in the previously cleaned, but not necessarily dry, flask. Add water to bring the volume up to the mark on the flask. After the liquid volume reaches the lower neck of the flask, add the water with a wash bottle. The last volume increments should be added dropwise until the bottom of the meniscus is coincident with the mark on the flask. Stopper the flask and mix thoroughly by inverting at least 13 times.

D. Pycnometers

The most precise volume measurements are made with a pycnometer, such as the one used in the first experiment in this manual. A pycnometer is simply a container with a very well-defined volume, such as a small flask fitted with a ground glass stopper. With such a device, and a liquid with an accurately known density, one can determine the volume of the pycnometer to 0.001 mL, which is much better than you can do with a pipet or buret. Knowing that volume, you can measure the density of an unknown liquid with an accuracy equal to that of the density of the calibrating liquid, usually within 0.01%.

Temperature

The SI unit of temperature is the Kelvin, which is equal in size to the Centigrade degree, °C. At 0°C, the Kelvin temperature is 273.15 K. The degree sign, °, is not used when reporting Kelvin temperatures.

Measurement of temperature is always done with a thermometer. Your lab drawer probably contains a standard laboratory thermometer suitable for temperature readings between about −10°C and 100°C, with 1° graduations. We will use this thermometer in experiments where high accuracy is not needed.

The typical glass thermometer contains a bulb connected to a very fine capillary tube. It contains a liquid, usually alcohol or mercury, which fills the bulb and part of the capillary. As the temperature goes up, the liquid expands and the level rises in the capillary. To make the graduations, one would in principle find the level at 0°C and at 100°C, and divide that interval into 100 equal parts. With mercury, the scale fits the one obtained from ideal gas behavior quite well. With other liquids, you must calibrate the thermometer at several places if you are to obtain reliable temperatures. Your standard thermometer can be read to about 0.2°C if you are careful, but the actual error is likely to be greater than that.

In some experiments it is desirable to be able to get more precise temperature values than are possible with the standard thermometer. You may be furnished with a digital thermometer that has a wide range and reads directly to 0.1°C. The sensing tip on that kind of thermometer contains a thermistor, similar to a transistor, which has electrical resistance that varies markedly with temperature. In recent years these thermometers have come into wide usage.

Glass thermometers are fragile and relatively expensive, so be careful when using them. If you slip a split rubber stopper around the thermometer and support it with a clamp you will minimize breakage, and lab fees. Should you ever break a mercury thermometer, contact your lab supervisor. Liquid mercury has a low vapor pressure, but that vapor is very toxic, so it is important to clean up mercury spills completely, which is not easy. With limited-range thermometers, do not heat them above their maximum readable temperature, since that can ultimately break the thermometer.

When making temperature readings, allow enough time for the level to become steady before noting the final temperature. This will probably take less than a minute, so if you check the reading over a period of time you should be able to get a reliable value.

To obtain temperatures with an accuracy better than 0.1°C is not a simple matter. Some mercury thermometers read to 0.01°C, but they have large bulbs and very fine capillaries and are very expensive, about $250 each. Still higher precision can be obtained using thermistors. By making careful resistance measurements, one can get temperatures accurate to about 0.001°C.

Pressure

In studying the behavior of gases one must be able to measure the gas pressure. In the chemistry lab this is done by comparing the pressure with that exerted by the atmosphere, using a device called a manometer. A manometer is a glass U-tube partially filled with a liquid, usually water or mercury.

The atmospheric pressure is measured with a mercury barometer (or, increasingly, its electronic equivalent). This consists of a straight glass tube about 80 cm long, which is initially completely filled with mercury. The tube is then inverted while the open end is immersed in a pool of mercury. The mercury level in the tube falls. The observed height of the Hg column above the pool, as read from a scale behind the tube, is equal to the pressure exerted by the atmosphere, and is called the barometric pressure. One can also measure the barometric pressure with an aneroid barometer, which consists of a spirally-wound, flexible-walled flat metal tube containing a fixed amount of air. As the external pressure changes, the tube flexes, which turns a needle that records the pressure.

Barometric pressure is often reported in mm Hg, but it may be given in other units, such as atmospheres, by comparing the value in mm Hg to 760 mm Hg, the standard atmospheric pressure. In SI the pressure is given in either Pascals or bars. One standard atmosphere equals 1.01325×10^5 Pascals, or 1.0325 bars. In this manual we will report gas pressures in mm Hg or mm H_2O, since these are most easily measured directly. We can convert pressures in mm Hg to mm H_2O, or vice versa, by using the conversion factor 1 mm Hg = 13.57 mm H_2O.

Having found the barometric pressure, it is a simple matter to measure the gas pressure in a flask such as that shown in Figure IV.5. In Experiment 8, the pressure inside the flask is equal to the barometric pressure, in mm H_2O, plus the pressure exerted by a water column of length $h_2 - h_1$, the difference between the water levels in the right and left arms of the U-tube. If you know those heights in mm,

$$P_{gas} = (P_{bar} + h_2 - h_1) \text{ mm } H_2O$$

When reading the H_2O levels in a manometer, take the height at the bottom of the water meniscus in the two arms, which you can read to 1 mm or better.

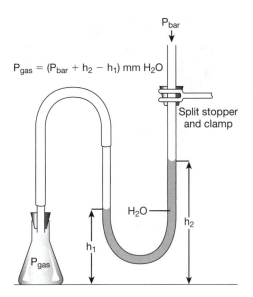

$$P_{gas} = (P_{bar} + h_2 - h_1) \text{ mm } H_2O$$

P_bar

Split stopper
and clamp

H₂O

h₂

h₁

P_gas

Figure IV.5   Measuring the pressure of a gas.

Light Absorption

In many chemical experiments the interaction of light with a sample can be used to furnish very useful information. In this laboratory course we will often determine the concentration of a species in a solution by measuring the amount of light at a given wavelength (color) that is absorbed by the sample. In an instrument called a spectrophotometer one can do this very easily. The spectrophotometer contains a device called a monochromator (usually incorporating a diffraction grating), which allows only one wavelength of light to pass through a sample. The absorbance, A, of the sample at that wavelength can be read directly from a meter on the instrument. By a famous equation called Beer's Law, the absorbance is proportional to the concentration, c, of the sample, usually expressed as a molarity:

$$A = K \times c$$

where K is a constant that depends on the sample, the container, and the wavelength of the light. Most samples obey Beer's Law. To find the concentration of a species in a sample, one measures the absorbance at an appropriate wavelength of a solution containing a known concentration of the species. From that absorbance, or several absorbances obtained at known concentrations, we can make a calibrating graph of concentration vs. absorbance and then use that graph to analyze unknown solutions containing that species. Your instructor will probably furnish you with such a graph in some of the experiments you perform.

A commonly used spectrophotometer is a Spectronic 20 (see Fig. IV.6). This instrument has two knobs on the front face. On the left is a zero transmittance adjustment knob and on the right a 100% transmittance adjustment knob. On the top surface at the left is a covered sample chamber, and at the right there is a wavelength selection knob. The absorbance is read from the upper meter or display register.

To use the spectrophotometer, turn it on about 15 minutes before you need to operate it, to give the components time to warm up. Set the wavelength to one where the sample absorbs a moderate amount of the incident light. Since the sample is probably colored, this wavelength will most likely be in the visible region of the spectrum. With the sample compartment closed, turn the zero adjustment knob until the needle or display register indicates 0% transmission. Open the sample chamber and insert a sample tube that is about 3/4 full of a "blank" reference solution (usually pure water). Shut the sample chamber and turn the 100% adjustment knob until the needle reads 100% transmission (this will be zero on the absorbance scale). Repeat the adjustment steps until no further changes are necessary. Rinse the sample tube several times with the solution being studied, then fill the sample tube about 3/4 full, insert it in the sample chamber, and close the chamber. Read the absorbance from the meter or register. Then use the calibration graph to obtain the concentration of the species in the solution.

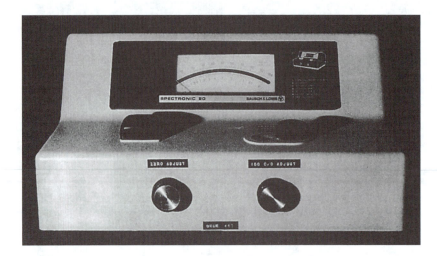

Figure IV.6 A Spectronic 20 spectrophotometer. The absorbance of a solution can be reliably read to within about 1%. *(Sherman Schultz)*

pH Measurements

The pH of a solution is used to describe the concentration of the H^+ ion in that solution. To determine pH one can use acid-base indicators, which change color as the pH changes. Each indicator has a characteristic pH at which the color change occurs, so can be used to indicate whether the pH is higher, or lower, than the characteristic value. By employing several indicators one can fix the pH of a solution within about one half of a pH unit.

More accurate pH measurements can be made with a pH meter. The pH meter is a very high resistance voltmeter which measures the difference in voltage between a reference electrode (usually a so-called calomel electrode) and a glass indicating electrode whose potential is a function of the H^+ ion concentration in the solution in which it is immersed. The meter indicates the pH directly. The two electrodes may be separate, but more commonly are together in one combination electrode. The electrode must be kept wet, in an appropriate storage solution, when not in use.

When you are asked to find the pH of a solution, you may first need to calibrate the pH meter, although it is likely that will have been done for you by your lab supervisor prior to the laboratory session. To calibrate the meter you will need a buffer solution with a well-defined pH as a reference. Put about 25 mL of the buffer in a small beaker. Remove the electrode from the storage solution, rinse it with a stream of distilled water from your wash bottle, and then blot the electrode with a clean tissue. Place the electrode in the buffer. (The electrode is fragile, and expensive, so treat it gently.) Allow the system to equilibrate for a minute or two, or until the pH reading becomes steady. Then use the pH adjust knob to bring the displayed pH to that of the reference buffer solution, or press the button used to indicate to the meter that the value has stabilized.

Remove the electrode from the buffer, rinse it with distilled water as before, blot with a tissue, and place the electrode in the solution being studied. Allow time for equilibration, and record the pH. Rinse the electrode and return it to the storage solution. Used properly, the average pH meter can reliably indicate pH to within one tenth of a pH unit. Many will indicate to the nearest hundredth of a unit, but are not reliable to that digit unless exquisite care is employed and a temperature correction is applied.

Separation of Precipitates from Solution

A. Decantation

Decantation is used to remove a liquid from a precipitate by pouring it off. Allow the solution to sit for a period of time until the precipitate has settled to the bottom of its container. (Some precipitates will not settle out without centrifugation—see below.) Position your stirring rod across the container with one end protruding beyond the lip. With the index finger of one hand holding the rod in place, pour the liquid slowly down the stirring rod into a receiving vessel. Try not to disturb the precipitate as the last bits of liquid are poured off.

B. Filtration—The Büchner Funnel

Filtration is used when you wish to recover the precipitate in pure form. This is first required in Experiment 3. You will use a Büchner funnel, which contains a piece of filter paper of appropriate size covering the holes at the base of the funnel. The Büchner funnel is connected by a rubber stopper or adaptor to the top of a side-arm filter flask. The side arm of the flask is connected via a series of rubber tubes and a safety trap to a water aspirator, a central vacuum system accessible via a jet on your lab bench, or other vacuum source (see Fig. 3.2) When the vacuum source is turned on, the reduced pressure created in the system by the water rushing by the side opening in the aspirator or by another vacuum source will draw air through the funnel.

When you are ready to filter, moisten the filter paper with water, then turn on the vacuum source. With the water faucet or valve to the vacuum source in the fully open position, pour the slurry of solution and precipitate down a stirring rod, as described under "Decantation," above. Wash out any remaining precipitate with a slow stream of liquid from your wash bottle. The procedure may call for washing the precipitate on the Büchner funnel with an appropriate volatile liquid before passing air through the filter cake for several minutes to complete drying. Turn off the water or vacuum source, disconnect the vacuum tube from the filter flask, and remove the funnel.

C. Centrifugation

Centrifugation is used to aid in the separation of a precipitate from a solution in a test tube. Spinning the test tube at several hundred revolutions per minute effectively increases the force of gravity experienced by the tube to many times normal earth gravity (see Fig. IV.7).

To use the centrifuge, be certain that the test tube containing the precipitate is not overly full—the liquid should be at least 3 cm below the top of the tube. Check to see that the test tube has no cracks, as these will cause the tube to break during centrifugation. Place the test tube containing your precipitate and solution in one of the centrifuge tubes. Place a blank tube containing the same amount of water as you have in your sample tube in the centrifuge tube that is opposite your sample tube. Turn on the centrifuge, allow it to spin for about a minute, and then turn it off. Keep your hands away from the spinning centrifuge top. After the spinning top has come to rest, remove the tube. Decant the supernatant liquid from the precipitate, as described above under "Decantation."

Qualitative Analysis—A Few Suggestions

In the course of your laboratory program you may do several experiments involving qualitative analysis. In those experiments we will use small test tubes and small beakers as sample containers. Separations of precipitates from solutions will (or can) be accomplished by centrifugation.

Figure IV.7   A lab centrifuge. *(Sherman Schultz)*

In many procedures you will be told to add 1 mL, or 0.5 mL, to a mixture. This is best accomplished by first finding out what 1 mL looks like in the test tube being used. So, measure out 1 mL in a small graduated cylinder, and pour that into a test tube, noting the height reached by the liquid in the tube. From then on, use that height to tell you the volume to add when you need 1 mL. Half that height indicates 0.5 mL.

Many of the steps in qualitative analysis involve heating a mixture in a boiling-water bath. To make the bath, fill a 250-mL beaker about 2/3 full of water. Support the beaker on a wire screen on an iron ring and heat the water to the boiling point with your Bunsen burner, adjusting the flame so as to keep the water at the simmering point. Or, place the beaker on an electronic hotplate, adjusting the heat setting to accomplish the same result. Use this bath to heat the test tube when that is called for. You can remove the tube with your test tube clamp.

In separating a solid from a solution, we first centrifuge the mixture. The supernatant liquid is decanted into a test tube or discarded, depending on the procedure. The solid must then be washed free of any remaining liquid. To do this, add the indicated wash liquid and stir with a glass stirring rod. Centrifuge again and discard the wash. Keep a set of stirring rods in a 400-mL beaker filled with distilled water. Return a rod to the beaker when you are done using it, and it will be ready when you next need it.

Pay attention to what you are doing, and don't just follow the directions as though you were making a cake. Try to keep in mind what happens to the various cations in each step, so that you won't end up pouring the material you want down the drain and keeping the trash. When you need to put a fraction aside while you are working on another part of the mixture, make sure you know where you put it by labeling the test tube or rack with the number of the step in which you will return to analyze that fraction.

Appendix V

Mathematical Considerations— Significant Figures and Making Graphs

In the laboratory you will be carrying out experiments of various sorts. Some of them will involve almost no mathematics. In others, you will need to make calculations based on your experimental results. The calculations are not difficult, but it is important you make them properly. In particular, you should not report a result that implies an accuracy that is greater, or smaller, than is consistent with the accuracy of your experimental data. In making such calculations we resort to the use of significant figures as a guide to proper reporting of our results. In the first part of this appendix we will discuss significant figures and how to use them.

In several experiments we will carry out a series of measurements on the same system under different conditions. For example, in Experiment 8 we measure the pressure of a sample of gas at different temperatures. In such experiments it is helpful to draw a graph representing our data, since it may reveal some properties of the sample that are not at all obvious if we just have a table of data. The second part of this appendix will discuss how to construct and interpret a graph.

Significant Figures

In the laboratory we make many kinds of measurements. The precision of the measurement depends on the device we use to make it. Using an analytical balance, we can measure mass to ± 0.0001 g, so if we have a sample weighing 2.4965 g, we have *five* meaningful figures in our result. These figures are called, reasonably enough, significant figures. If we measure the volume of a sample using a graduated cylinder, the volume we obtain depends on the cylinder we use. If we have a volume of about 6 mL and measure it with a 100-mL cylinder, we would have difficulty distinguishing between a volume of 6 mL and 7 mL, and would be unable to say more than that the volume was about 6 mL. That volume reading has only *one* significant figure. With a 10-mL cylinder, we could measure the volume more precisely, report a volume of 6.4 mL, and have confidence that both figures in our result were meaningful, and that it had *two* significant figures. Many measurements, perhaps most, can be interpreted as we have here.

Sometimes it is not clear how many significant figures there are in a number. Say, for example, we are told to add 1 mL of liquid from a pipet to a test tube. A pipet is a precisely made device, and can deliver 1.00 mL of liquid if used properly. In such a case, it would be sensible to say that there are really three significant figures in that volume, rather than one. It would probably have been better to be told to add 1.00 mL with a pipet, but often that is just not done, even by the authors of this manual.

Some numbers are exact and have no inherent error. There will be an integral number, like 12, or 19, students in your lab group. There is no way there will be 14.5 students, unless there is something strange going on. Conversion factors, used to convert one set of units to another, often contain exact numbers: 1 meter = 100 cm. Both numbers are exact, with no uncertainty at all. With volumes, 1 liter = 1000 mL. Again, both numbers are exact. If we convert from one system to another, then usually only one of the numbers in the conversion factor is exact: 1 mole = 6.022×10^{23} molecules. Here the 1 is exact, but the second number is not, and has four significant figures.

If you are in doubt as to the number of significant figures in a given number, there is a method for finding out that usually works. We write the number in exponential notation, expressing the number as the product of a number between 1 and 10, times a power of ten.

$$26.042 = 2.6042 \times 10^1 \quad 0.0091 = 9.1 \times 10^3 \quad 605.20 = 6.0520 \times 10^2$$

The first number in the product has the number of significant figures we seek. So, the first of the above numbers has 5 significant figures, the second has 2, and the third has 5 (a trailing zero is significant). There is one problem with this approach, and that is that sometimes we are given a number with no decimal point, like 2400. We cannot be sure how many significant figures are in that number, since, if the number were rough, there might be 2, but there could be 4 or even more. Here, you have to use some judgment, but in the absence of any guide, you would choose 4.

Students usually have little trouble deciding on the number of significant figures in a piece of data. The difficulty arises when the piece of data is used in a calculation, and they have to express the result using the proper number of significant figures. Say, for example, you are weighing a sample in a beaker and you find that:

$$\begin{aligned}
\text{Mass of sample plus beaker} &= 25.4329 \text{ g} \\
\text{Mass of beaker} &= 24.6263 \text{ g} \\
\text{Mass of sample} &= 0.8066 \text{ g}
\end{aligned}$$

Even though the two measured masses contain 6 significant figures, the mass of the sample contains only 4. In another case, where we mix some components to make a solution, weighing each of the components separately on different balances, we get:

$$\begin{aligned}
\text{Mass of beaker} &= 25.5329 \text{ g} \\
\text{Mass of water} &= 14.0 \text{ g} \\
\text{Mass of salt} &= 6.42 \text{ g} \\
\text{Total mass} &= ?
\end{aligned}$$

Here we have several pieces of data, each with a different degree of precision. The first mass is reliable to 0.0001 g, the second to 0.1 g, and the third to 0.01 g. If we take the sum, we get 45.9529 g, but we certainly can't report that value, since it could be off, not by 0.0001 g, but by 0.1 g, the possible error in the mass of the water. We can't improve the quality of a result by a calculation, so, using good sense, we can only report the mass to ± 0.1 g. Since the measured mass is closer to 46.0 g than to 45.9 g, we round up to 46.0 as the reported mass, and have a mass with 3 significant figures. Generalizing, *in adding or subtracting numbers, round off the result so that it has the same last decimal place as in the least precise measured quantity.* Round up the last digit if the number that follows is greater than 5, and don't round up if it is not.

When *multiplying* or *dividing* measured quantities, the rule is quite simple: *The number of significant figures in the result is equal to the number of significant figures in the quantity with the smallest number of significant figures.*

In the first experiment we measure the density of a liquid using a pycnometer, which is simply a flask with a well-defined volume. We find that volume by weighing the empty pycnometer and then weighing the pycnometer when it is full of water. We might obtain the following data:

$$\begin{aligned}
\text{Mass of pycnometer plus water} &\quad 60.8867 \text{ g} \\
\text{Mass of empty pycnometer} &\quad 31.9342 \text{ g} \\
\text{Mass of water} &\quad 28.9525 \text{ g}
\end{aligned}$$

We are asked to find the volume of the pycnometer, given that the density of water is 0.9973 g/mL:

$$\text{density} = \frac{\text{mass}}{\text{volume}} \quad \text{so} \quad \text{volume} = \frac{\text{mass}}{\text{density}} = \frac{28.9525 \text{ g}}{0.9973 \text{ g/mL}} = 29.030883 \text{ mL}$$

The mass contains 6 significant figures, the density has 4, and the results from a calculator have 8 or more. Since the density has the smallest number of significant figures, four, we report the volume to 4 significant figures, as 29.03 mL.

In most cases, the reasoning that is required to correctly report a calculated value of a property is no more complicated than that we used here. Remember, you can't improve the accuracy in a result by means of a calculation. If you note the quality of your data, you should be able to report your result with the same quality. As with most activities, a little practice helps, and so does a little common sense.

Graphing Techniques

In many chemical experiments we find that one of our measured quantities is dependent on another. If we change one, we change the other. In such an experiment we find data under several sets of conditions, with each data point associated with two measured quantities. In such cases, we can represent these points on a two-dimensional graph, which shows in a continuous way how one quantity depends on another.

Interpreting a Graph

The first graph in the lab manual appears in Experiment 3, and is a graph of this kind. It is reproduced in Figure V.1. The graph describes how the solubility of each of two compounds, KNO_3 and $CuSO_4 \cdot 5\,H_2O$, depends on temperature. The temperature, which was the controlled variable when the data for the graph was collected, is shown along the horizontal axis, called the abscissa, over the range from 0 to 100°C. The solubility, in grams per 100 grams water, was the measured quantity in the experiment that produced the data and is on the vertical axis, sometimes called the ordinate. The small x's along each of the two curved lines in the graph are data points, showing measured solubility at given temperatures.

The graph has some implied features that are not shown. We actually made the graph on a piece of graph paper on which there was a grid of parallel lines. On the vertical lines the temperature is constant. If you draw a vertical line upwards from the 50°C hashmark on the vertical axis, the temperature on that line is always 50°C. On the vertical line at 73°C, three-tenths of the distance between the 70 and 80 hashmarks, it is always 73°C, and so on. Along horizontal lines the solubility is constant. The horizontal line drawn through the 80 hashmark represents a solubility of 80 grams in 100 grams of water all along its length. When we obtained the data for the solubility of KNO_3, we found that, at 60°C, we could dissolve 105 g KNO_3 in 100 g of water. We entered the data point we show as A on the graph at 60°C and 105 g KNO_3. The rest of the data points for the two compounds were entered in the same way. We then drew a smooth curve through those points, and took out the grid.

Given a graph like that we have shown, you can extract the data by essentially working backwards. To find the solubility of KNO_3 at 60°C, draw the dashed line up from 60°C. The solubility is given by the point at which that line intersects the KNO_3 graph line. Draw a horizontal line from that intersection to the vertical axis and read off the solubility of KNO_3, which you can see is about 105 g/100 g H_2O. If you wanted to find

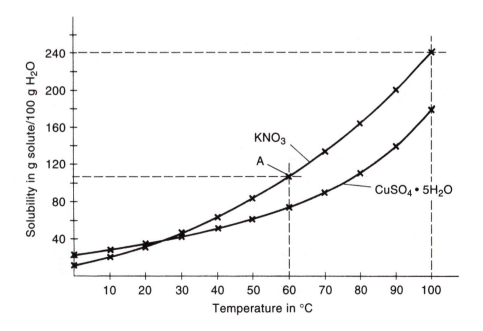

Figure V.1

the amount of water you would need to dissolve 21 g KNO_3 at 100°C, you would draw the vertical line at 100°C up to the KNO_3 line; then draw the horizontal line at the intersection with the KNO_3 curve, and you find that at 100°C, about 240 g KNO_3 will dissolve in 100 g H_2O. Then a simple conversion factor calculation would give you the amount of water needed to dissolve 21 g KNO_3. Clearly, there is a lot of information in Figure V.1 if you interpret it completely. Its meaning is almost, but not quite, obvious.

Making a Graph

Now let us construct a graph from some experimental data. Given the grid in Figure V.2, we measure the pressure of a sample of air as a function of temperature. One of our students obtained the following set of data:

Pressure in mm Hg	674	739	784	821
Temperature in °C	0.0	24.5	42.0	60.1

To make the graph we need to decide what goes where. Since we are controlling the temperature and measuring the pressure, we put the pressure on the vertical axis, and the temperature on the horizontal axis, labeling those axes. The temperature goes from 0°C to 60°C; we want the graph to fill the grid, not just be in one corner, so we select a temperature interval between grid lines that will fill the grid. Since we have about 30 vertical grid lines available, and need to cover a 60°C interval, we put the 10° hashmarks at intervals of five grid lines, with 2° between grid lines. Since there are about 20 horizontal grid lines, and we need to cover a pressure change of about 150 mm Hg, we make the lowest pressure 650 mm Hg and the interval between grid lines equal to 10 mm Hg. We put in the hashmarks at 50-mm intervals. Having laid out the graph, we insert each of the data points, using x's or dots. The points fall on a nearly straight line, so we draw such a line, minimizing as best we can the sum of the distances from the data points to the line. If we wish, we can find the equation for that line, which, since it is straight, is of the form $y = mx + b$ where m is the slope of the line and b is the value of y when x equals zero. Plugging two points on the line, $(x_1, y_1) = (0, 674)$ and $(x_2, y_2) = (60, 830)$, into the point-point formula for a line, $y - y_1 = [(y_2 - y_1) / (x_2 - x_1)] (x - x_1)$, and simplifying, our

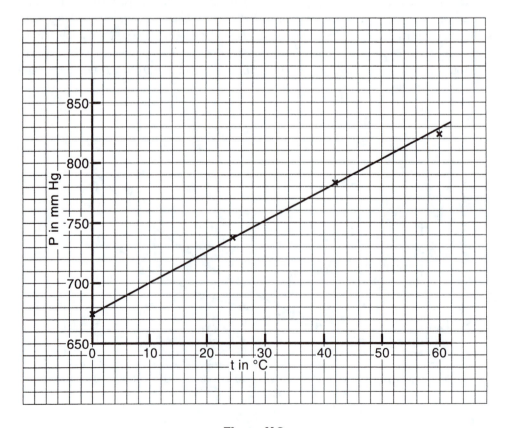

Figure V.2

student found that the equation was y = 2.6x + 674. This equation can be used to find another useful piece of information about the properties of gases, but we need not go into that here.

The key to successful construction of graphs is to properly select the variables and assign them to the two axes. Label those axes. Then, given the changes in the variables, select grid intervals that will cause your data to fill the area of the grid. Put in the hashmarks for each variable at appropriate intervals and label them with the values of the variable. Take the data points and enter them on the grid. If the line is theoretically straight, draw a straight line through the data points in such a way as to minimize the sum of the distances from the points to the line.

Many natural laws are not as simple as the one relating gas pressure to temperature. However, by a few rather easy tricks, involving mathematical manipulation of variables, one can obtain data points that will lie on a straight line. For example, in Experiment 15, we find that the vapor pressure of a liquid depends on temperature according to the following equation:

$$\ln VP = \frac{-\Delta H_{vap}}{RT} + C$$

where ln VP is the logarithm of the vapor pressure to the base e, T is the Kelvin temperature, and ΔH_{vap} and R are constants and equal to the heat of vaporization and the gas constant, respectively. The equation looks formidable, and it is. However, if we let x equal ln VP and y = 1/T, then the equation takes the form y = mx + b, a straight line with a slope m, equal to $-\Delta H_{vap}/R$. By using those variables, we can find the slope of the straight line and determine ΔH_{vap} for the liquid. It is very possible that the equation above was initially found by an imaginative scientist who measured the vapor pressure as a function of temperature and tried plotting the data against different functions of the variables, seeking a straight line dependence. From such efforts many great discoveries have been made.

Appendix VI

Suggested Locker Equipment

2 beakers, 30 or 50 mL
2 beakers, 100 mL
2 beakers, 250 mL
2 beakers, 400 mL
1 beaker, 600 mL
2 Erlenmeyer flasks, 25 or 50 mL
2 Erlenmeyer flasks, 125 mL
2 Erlenmeyer flasks, 250 mL
1 graduated cylinder, 10 mL
1 graduated cylinder, 25 or 50 mL
1 funnel, long or short stem
1 thermometer, $-10°C$ to $150°C$
2 watch glasses, 75, 90, or 100 mm
1 crucible and cover, size #0
1 evaporating dish, small

2 medicine droppers
2 regular test tubes, 18×150 mm
8 small test tubes, 13×100 mm
4 micro test tubes, 10×75 mm
1 test tube brush
1 file
1 spatula
1 test tube clamp, Stoddard or equivalent
1 test tube rack
1 pair of lab tongs
1 sponge
1 towel
1 plastic wash bottle
1 chemical casserole, small

Appendix VII

Introduction to Excel

O ver the past decade a type of software called a spreadsheet has come into general use for processing scientific data obtained in the laboratory, or, indeed, data involved in business activities. Essentially, a spreadsheet contains many cells in which data can be entered and manipulated. It is particularly useful when there are several, or many, experimental values for a property, such as the volume of a gas obtained at different temperatures under constant pressure. In this introduction we will offer a short primer on how to use the most widely-used spreadsheet package, called Excel, in some of its simpler applications. You will find it very helpful, but it will require some practice. To get started, you will need to have a computer on which the software has been installed. There are many versions of this program, but we try here to provide a version-independent set of instructions. These instructions are specific to Excel, but should be readily adapted to other spreadsheet packages with only minor changes.

Basic Layout

Click on the Excel icon on your computer. If there is none, your instructor will tell you how to call it up. You should see a screen on which there is a grid containing many rectangular boxes, called cells. Each cell has an associated label, with a letter indicating its column, and a number indicating its row. If you click on cell B5, it will be highlighted. If you put the cursor in that cell, type a number or a word, and click the enter or return key on the keyboard, that data will appear in the cell. You can change the width of the cell by clicking and holding on the grid line in question, up where it intersects with the rectangle containing a letter.

Setting Up Columns of Data

Let's proceed by entering the heading "Temperature (C)" in cell A1. Change the cell width, if necessary, to make the title fit. In cell A3, type in the number 20 and press enter or return. The highlighted cell will automatically become A4. Enter the formula "= A3 + 20" in that cell, press return or enter, and 40 should appear in A4. Click cell A4 *once* with the mouse, and move the cursor to the lower right-hand corner of that cell. You will find that the cursor will change from a white plus symbol to a solid black plus symbol or a square. While the cursor is a black plus or a square, click and drag the cursor down to select all the cells you want to include with that formula; in this case, let's go down to cell A12. When you release the mouse button, the numbers 60, 80, 100, and so on, up to 200, will appear in the column. This method is very useful if you have a column of data that are related by a formula.

Now enter the heading "Temperature (K)" in cell B1, adjusting the width of that column as necessary to make it fit. In cell B3, enter the formula "= A3 + 273", press enter, and 293 will appear in cell B3. Using the same approach we used for column A, copy this formula down into cells B4 through B12 to produce the Kelvin temperatures in column B for each of the Celsius temperatures in column A.

In cell C1, type the heading "Volume (L)". Let's use the Ideal Gas Law, $PV = nRT$, to find the volume in liters of a mole of gas at one atmosphere pressure at the Celsius temperatures in column A. By the Ideal Gas Law, $V = nRT/P$; in our example, $P = 1$ atm, $R = 0.0821$ L·atm/mole·K, and T is the Kelvin temperature. So in column C, V equals 1*0.0821*T/1, or just 0.0821*T. (The mathematical symbols used for operations in Excel do not correspond exactly to those you learned in school. + and − do mean add and subtract, but * means multiply, / means divide, and ^ means raise to a power. We write 2^3 as 2^3 in Excel. A number like $3.45*10^5$ must be written as 3.45E5.)

To find the values of Volume in column C, we first enter the gas constant into a cell and label it: type "R =" into cell A14, and "0.0821" in cell B14. We should specify the units for R, so in cell C14, enter "L atm / mol K".

In cell C3, write the formula "= B14*B3", and press enter. In cell C3 the number 24.055 should appear (possibly with more or fewer significant figures shown, as Excel knows nothing of significant figures and will simply display what it has sufficient room to display). Highlight cell C3 by clicking on it once, position the cursor over the lower right-hand corner of the cell, and when the black plus symbol appears, click and drag down to cell C12 and release. The volumes in liters of one mole of an ideal gas at the various temperatures should appear in column C.

You may wonder why we used B14 instead of just B14 in the formula in cell C3. Try replacing C10 with B14 in the formula in cell C3, and see what happens! At first, everything will look fine, but if you drag down the cursor to copy this cell, the program will increment not only B3 to B4, but also B14 to B15, and zeroes will appear in column C, since B15 is zero. In many formulas, you will find that you wish to have a constant retain its value when the formula is copied. You can accomplish this by preceding the letter, or number, or both, in the cell references that refer to the cell containing the constant with $ signs.

Some Comments on Excel

In Excel, the program will assign the contents to a cell to one of three categories: a number, text, or a formula. Numbers and text only cause trouble when one is mistaken for the other. To avoid this, when entering numerical values you intend to use in subsequent calculations or a plot, be certain to enter *only* the number—do not, for example, append it with units within the same cell. (Specifying units is an excellent habit, but to do so in a spreadsheet use a column heading or an adjacent cell, as demonstrated above.). Formulas most often cause trouble when Excel does not realize you are entering one. You must get into the habit of beginning every formula you enter into Excel with = or +, lest Excel interpret your entry as text. (Some spreadsheets will assume a formula, and you may need to indicate text entries by beginning them with " or "). The = sign used to denote a formula must be the first character in a cell, with no space to its left. The equation for the formula may be complex, but it must include the Excel symbol(s) for any operation to be carried out. All references to particular cells must be to cells with known numerical contents. Parentheses should be carefully used to make clear to Excel the desired order of the mathematical operations you wish to have done.

If you mis-type or mis-click, and this is inevitable, you may find nothing happens when you hit enter, or you may get one of several error messages, which may or may not be helpful. If you get into trouble, the delete or esc button may clear the screen to its form before the error occurred. If that fails, using the "Undo" option (on the Edit menu in older versions, or indicated by a looping arrow icon in newer versions) at the top of the screen will often put you back on track. Do not be reluctant to seek help, either from your lab supervisor or a fellow student more familiar with Excel than you are. Working with Excel can be frustrating, but sooner or later you will able to use it effectively, and learning to use it will pay huge dividends.

Using Excel to Make Graphs

In some cases it is helpful to create graphs that show the relationships between experimentally measured variables.

Given two columns of data for two variables, like temperature and volume, obtained under different conditions, we can use Excel to construct a graph (Excel calls it a "chart") that shows how the variables are related.

Before you proceed, click on View on the top line of the screen, and select Normal from the list.

Next, choose Chart from the Insert menu on the top edge of the screen. If you are using one of the newest versions of Excel, you will need to select the type of chart you want at this point—in science, the default selection you should make is Scatter, and specifically the flavor that displays only the data points. If you are using a version of Excel dating from before 2007 (your authors prefer these, frankly), choosing Chart from the Insert menu will bring up a screen showing several possible graph types, from which you will want to select XY Scatter and then click on "Next >".

Now you need to specify the data you wish to use in the chart. In an older version of Excel, click on the "Series" tab; in a newer version, choose "Select Data" from the top of the screen. Click on Add. In the data entry regions that appear, labelled Name, X Values, and Y Values, delete the contents of any boxes that are not already empty by selecting the contents and hitting backspace or delete.

Now, to specify the y values, put the cursor in the Y Values Box, and click and drag the mouse down over the column of data containing the dependent variable, in our case the Volume, in column C (specifically, cells C3 through C12). When you release the mouse button, that data range will appear in the Y Values box.

To specify the x values, put the cursor in the X Values Box, and click and drag the mouse down over the column of data containing the independent variable, in our case the Kelvin temperature, in column B (specifically, cells B3 to B12). When you release the mouse button, that cell range appear in the X Values box. A small graph showing the relation between the variables will appear on the screen. The scale will probably be incorrect, but we will fix that soon. For now, hit enter or return on the keyboard to go on to the next step.

If you are using a new version of Excel, click on Cancel, as you are done specifying the data you wish to use, and then click on the first of the "Chart Layouts" that appear at the top of your screen. Click on Title. Type "One Mole of an Ideal Gas at One Atmosphere Pressure" as the title of the graph [do not hit enter or return yet]. Now click once on X Axis title and type in the name of the independent-variable, "Temperature (K)". Click once on y-axis title and type the name of the dependent-variable, "Volume (L)".

In a newer version of Excel, click on the word Layout at the top of the screen. Click on Gridlines. Add or remove them as seems reasonable.

If you are using an older version of Excel, you will now want to click on "Next >", for Step 4, and decide where to place the graph, either on a separate page or as an object on the spreadsheet (usually, you'll want the latter) before clicking on "Finish" to bring down the final graph, labeled and titled, but probably not scaled as you would like.

To fix the scale, *right*-click (press down on the side of the mouse *opposite* that you normally click) on the numbers of the axis you wish to change, and select Format Axis. With an older version of Excel, click on Scale. Proceed by disabling automatic axis limits and entering appropriate fixed limit values (300 to 500 for x, 24 to 40 for y, in our example here). This is easier to do if you can see the chart and the data as you do so—you can move the data entry window out of the way by clicking and holding on the window name ("Format Axis") and dragging the window to a more convenient location. The graph will update as you make changes, which makes it easy to fine-tune your adjustments.

In a chart displaying a single series, such as our example, the legend serves no useful function and can be removed by clicking on it once and pressing the backspace or delete key on the keyboard. This frees up more room for the plot. Were the legend useful, it could be moved onto the plot region by clicking and dragging it there.

To change the physical dimensions and position of the graph you have created, click once on the edge of its frame, to select it. Then position the mouse over any of the "handles" that appear on the frame (a single black square in older versions of Excel, a collection of groups of tiny dots in newer versions) and click and drag to change the dimension to which that handle corresponds. To move the entire plot, click on an empty region of the chart, just inside one of the upper handles, and click and drag. Especially with these operations, it is easy to have things go awry. . . if they do, press the Esc key and then, if needed, with an older version try Edit>Undo from the menu at the top of the screen, or with a newer version click on the undo icon, a looping arrow just to the right of the disk image at the very top left edge of the screen.

If you wish to have Excel calculate the optimal line through your data, select your chart (see above) and then get to the Trendline menu. (Chart>Add Trendline. . . with older versions, Layout>Trendline with newer ones.) You can specify that Excel should display the equation of this line (as well as customize several other aspects of it) from the Options tab on the Trendline menu.

Appendix VIII

Statistical Treatment of Laboratory Data

Whenever there are multiple determinations of a piece of data (called "replicate data"), as is the case in several experiments in this laboratory manual, it is customary for scientists to express a central value and some measure of the deviation of the replicate data points from the central value—thus indicating something about the reliability of the experimental results. In chemistry, the kinds of experiments that are treated statistically in this fashion generally involve the analysis of a substance to determine the proportions of its chemical constituents, often by either titration or the weighing of precipitates. This branch of chemical science is referred to as quantitative analysis.

It is necessary to have some means of expressing the central value—the best estimate of the real value that one can obtain from the replicate data. You may have heard terms which are frequently used to express the distribution of exam scores—the (arithmetic) mean, median, and mode, along with the range. Let's consider a set of laboratory data, such as the percent iron by mass in an ore sample:

Trial number	$\%_{mass}$ Fe
1	15.66
2	15.79
3	15.22
4	15.66

An inspection of this data set shows that the **median** value (that result about which all others are equally distributed) is 15.66%, the **mode** (the value having the largest number of replications) is 15.66%, and the **mean** (the arithmetic mean, or "average") is 15.58%. The mean is calculated by adding up all of the percentage values (62.33%) and dividing by the number of trials (4). The **range** (difference between high and low values) is 0.57%.

You may have used the terms precision and accuracy in a variety of settings. In science, **precision** is a measure of the reproducibility of a measurement—the agreement among several determinations of the same value. On the other hand, **accuracy** properly refers to the closeness of a measurement to the accepted (or true) value. It appears that the above data set has decent precision, but nothing can be said about its accuracy in the absence of additional information. Your instructor will have access to an accepted value. If we assume that the known value for this particular iron ore is 15.88%, it is apparent that the accuracy is poor—all of the measured values fall below the true value, suggesting a **systematic** discrepancy. Your instructor may use a combination of your accuracy and your precision in grading some experiments.

One widely used expression of the precision of a data set is the **standard deviation**, which, in a manner of speaking, averages the deviations of each individual measurement from the mean value. The standard deviation is calculated by the following formula:

$$\sigma = \sqrt{\frac{\Sigma (x_i - \bar{x})^2}{N-1}}$$

In which σ is the standard deviation, N is the number of trials, x_i is the value of an individual measurement, $\bar{x}$ is the mean value, and Σ is a mathematical symbol denoting a summation, the addition of the members of a series of terms.

The table below illustrates this calculation.

Trial	$x_i - \bar{x}$	$(x_i - \bar{x})^2$
1	0.08	0.0064
2	0.21	0.0441
3	−0.36	0.1296
4	0.08	0.0064
		$\Sigma = 0.1865$

Dividing 0.1865 by (4 − 1 =) 3, and then taking the square root, yields 0.25 as the standard deviation, σ, for this data set. A low value of σ is an indication of good precision. While the standard deviation can always be calculated as we just did, you will most likely find it more convenient to do this on your personal calculator (which may have an internal program which will evaluate this quantity) or with a spreadsheet, such as Excel—your instructor or another student may be able to provide some assistance.

In considering the original data set, you might wonder if the low value of 15.22% should have been included. If you were the experimenter who gathered the data, this low value could be thrown out if a known mistake was made, such as spilling some of a solution, or if you know a weighing error was definitely made. However, if such a personal error was not known to have been made, then the value probably should be included. There is a statistical test which can be used as an unbiased guide for rejection of a potential outlier value. This test, called the Q test, is based on the assumption that a significantly large data set would approach a normal distribution. If the assumption is made that our data set is large enough that it resembles a normal distribution, then a $\mathbf{Q_{rej}}$ **(rejection quotient)** can be calculated from the following equation

$$Q_{rej} = \frac{|x_q - x_n|}{r}$$

where x_q is the questionable value, x_n is the nearest neighbor value, and r is the range, inclusive of the questionable value. Note that the absolute value of the difference is used. The calculated rejection quotient, Q_{rej} is then compared with $\mathbf{Q_c}$, **the critical quotient**, listed in the table below.

N	2	3	4	5	6
Q_c	—	0.94	0.76	0.64	0.56

The value of N corresponds to the number of trials and Q_c, the critical quotient, represents a threshold at the 90% confidence level that the value of n_q, the questionable value, can be discarded. Thus, if $Q_{rej} > Q_c$, the questionable value can be discarded with a 90% degree of confidence, provided the data resembles a normal distribution. Note that no value of Q_c is given for 2 trials. Can you figure out why? HINT: compare the difference between n_q and n_n with the range.

Let's go back to the data set of iron percentages. We have identified trial 3, with 15.22% Fe, as the questionable one. As we calculate $Q_{rej} = |15.22 − 15.66| /0.57 = 0.44 / 0.57 = 0.77$, we find that the Q_c value for four trials is 0.76. Therefore $Q_{rej} > Q_c$ (although barely), and we can discard the 15.22% Fe value with a 90% degree of confidence, subject to the validity of the assumption that our dataset follows a normal distribution.

With the 15.55% value excluded, we can calculate a new mean value of 15.70% Fe with a new standard deviation of 0.075%. Note the significantly lower standard deviation. You should do these calculations yourself to verify these data and to gain some practice.

The operational introduction to the statistical analysis of experimental data provided here is intentionally brief. You may wish to do additional reading to gain further insight. Go to your library and consult textbooks used for Analytical Chemistry courses—these books usually have a section on statistics in the first few chapters or in an appendix at the end. Other sources include any introductory text on statistics. Alternatively, you can find a number of websites which give an overview of statistics. You may find that your physics and/or biology courses will incorporate a treatment of statistics in the laboratory portion of the course.